Implementing Identity Management on AWS

A real-world guide to solving customer and workforce IAM challenges in your AWS cloud environments

Jon Lehtinen

BIRMINGHAM—MUMBAI

Implementing Identity Management on AWS

Group Product Manager: Wilson Dsouza

Publishing Product Manager: Yogesh Deokar

Senior Editor: Athikho Sapuni Rishana

Content Development Editor: Sayali Pingale

Technical Editor: Sarvesh Jaywant

Copy Editor: Safis Editing

Project Coordinator: Neil Dmello

Proofreader: Safis Editing

Indexer: Tejal Daruwale Soni

Production Designer: Alishon Mendonca

First published: August 2021

Production reference: 1120821

Published by Packt Publishing Ltd.

Livery Place

35 Livery Street

Birmingham

B3 2PB, UK.

978-1-80056-228-8

www.packt.com

A big thank you to everyone who encouraged me to take this on, had to deal with me while I was doing it, and kept me going when major life events made me think that I had finally stretched myself too far. To my family, Aleeta, Calvin, Annabelle, and Syd, and to the Hombres, Hutch, Sean, Anthony, and Pat, I am dead serious when I say I could not have done this without you.

– Jon Lehtinen

Foreword

"Is it okay if I share your table?"

I first met Jon Lehtinen in Napa, California as we were both attending our first-ever Cloud Identity Summit. This simple meeting turned out to be one of the few key moments in my life when an act of pure serendipity caused my career to ricochet in a direction I could not have foreseen. An exchange of pleasantries turned into a couple of hours of conversation, an exchange of business cards, and me urging Jon to apply for one of the open positions on our IAM Operations team. Soon after, Jon packed up his family in Arizona and moved out to Virginia to join our team at GE's Information Security Technology Center.

Jon's guiding star has always been in designing his identity services to better serve the communities of developers and practitioners who consumed them. Early on, my role at GE was SSO service leader in charge of an engineering team tasked with aiding the business in integrating their applications with our IAM services. An important lesson I learned from Jon early on is that it's not enough to say that "the business is our customer." Jon has always known that the real customers were those that were building products around our services, administrators within those businesses, and the end users trying to get their work done. When I moved into a principal architect role, Jon was promoted into my old role and he quickly transformed it into one of servant leadership to support those communities. He established open office hours, built on the foundations of an online forum, and took all of the input from those communication streams to build a self-service portal where developers were expertly guided in the implementation of our IAM services in their applications in minutes and hours instead of days and weeks. Jon was truly the voice of the identity practitioners and their champion on the engineering side of the house.

Jon also wrote extensive developer and end user documentation on how the services worked, how to request them, and best practices for integrating them in different application scenarios. He continued to maintain that focus on the practitioner after leaving GE and moving on to roles at Thomson Reuters and at Okta, where today he is responsible for maximizing the value of their identity products through the internal utilization of those same services. It comes as no surprise to me that when Jon decided to make his foray into writing a book, he would choose to once again shine his light of expertise into a field that can be difficult to traverse, even for the most experienced users.

The **Amazon Web Services** (**AWS**) team deserves high praise for building such an extensive identity and access management service at the heart of AWS security. It gives developers and administrators the ability to control access by creating users and groups, designing centralized policies based on both RBAC and ABAC models, assigning those policies to specific populations, integrating existing users via federation, setting up additional controls such as multi-factor authentication, and numerous other features. Because AWS recognizes that identity services are the foundation of any security service, it offers the IAM services for free.

These services allow organizations to overcome the many security hurdles to cloud adoption but, even with Amazon's extensive documentation, the sheer number and granular nature of IAM services available can make that adoption difficult to navigate. The same wide array of authentication and authorization controls that make it possible to deploy highly secure environments using AWS can also frustrate administrators and security professionals alike.

In this book, Jon does what he has always done best: take the mystery out of a complicated service and provide practical guidance on and examples of its utilization. In the pages that follow, Jon provides a complete overview of IAM, shares details of each of the services, demonstrates the benefits of various combinations, and teaches you best practices on how to protect your AWS accounts. He will also guide you through the latest services and coach you through the configuration of important controls such as MFA.

Jon has once again provided me with knowledge and expert guidance on a service I still periodically struggle with. It's always rewarding, and comforting, to be shown the way by a professional who has been there before and has worked out the best avenue to success. I wish you well on your own journey, both through this book and on your successful path to securing your cloud workloads with AWS identity services.

Steve "Hutch" Hutchinson
VP for Security Architecture, MUFG
Board Member, IDPro

Contributors

About the author

Jon Lehtinen has 16 years of enterprise identity and access management experience and specializes in both the strategy and execution of IAM transformation in global-scale organizations such as Thomson Reuters, General Electric, and Apollo Education Group. In addition to his work in the enterprise space, he has held positions on Ping Identity's Customer Advisory Board and as an advisor to identity verification start-up EvidentID. He currently owns the workforce and customer identity implementations at Okta.

Jon is dedicated to the growth and maturity of IAM as a profession and serves on the Board of Directors for IDPro.org. He is also a member of the Kantara Initiative, ISC2, OpenID Foundation, and Women in Identity. Jon has presented his work at several conferences, including RSA, Identiverse, and KuppingerCole's European Identity and Cloud Conference.

Currently, he owns Okta's workforce and customer IAM implementations as their Director of Okta on Okta.

About the reviewers

Surendra Singh Khatana is a seasoned **Identity and Access Management (IAM)** professional with more than a decade of experience in building security solutions using public clouds, container technologies, and IAM tools.

He has worked in the past with TCS and Accenture as a security specialist and currently works as a digital identity senior consultant at Nixu, Finland. At Nixu, he helps customers embrace digitalization securely.

Punit Kumar (an open source lover), is a Certified Cloud Solution Architect – Professional (AWS) with 9 years of experience in the IT industry and an understanding of both public and private clouds.

He has spent a major part of his time on AWS in the areas of infrastructure design and architecting, DevOps automation, infrastructure as code, security frameworks, operations and cloud governance, and private cloud to public cloud transformations.

Punit has worked and helped various industry start-ups in their cloud adoption journey, right from inception. He is a qualified AWS Well-Architecture reviewer and cost-optimizer. Punit is an ambivert and likes socializing with his large circle of friends. All his decisions in life are based on deep thinking and analysis and, of course, his friends' support.

The completion of this book is the very first milestone in my journey as a technical reviewer.

I am very grateful to my colleagues and mentors who really helped me through my learning journey, encouraging me to start work.

Also, I would like to acknowledge my gratitude to family members – my parents and my wife, Kajal Behl, for their support and trust, which really encouraged me during this work.

Maurício Harley is doing a master's in computer science with cyber Security at the University of York, UK. He holds a BSc in electrical engineering from Universidade Federal do Ceará, Brazil and a telematics degree from Faculdade Integrada do Ceará, Brazil.

On his blog (itHarley), he writes about security, the cloud, and networks. He also writes about offensive security for PenTest Magazine and Hakin9. He is also the founder of an OWASP chapter.

He has more than 25 years of experience in IT. He holds industry certifications such as (ISC)2 CISSP, CCIE Routing and Switching, CCIE Service Provider, AWS Certified Solutions Architect Associate, and AWS Certified Security Specialty.

He has given talks at Latin American conferences such as RootDay and OWASP LATAM@Home. He works as a senior consultant at Amazon Web Services in France.

His research is split between cloud security, threat intelligence, and malware analysis.

To my wife, Paula, for the never-ending support and belief in all my career endeavors.

To my parents, Bartolomeu and Leuzete, who went out of their way to build a healthy home and to give us all education resources.

To my siblings, Raquel and Robson, for the friendship, partnership, and love.

Table of Contents

3

IAM User Management

4

Access Management, Policies, and Permissions

5

Introducing Amazon Cognito

6

Introduction to AWS Organizations and AWS Single Sign-On

7
Other AWS Identity Services

Section 2: Implementing IAM on AWS for Administrative Use Cases

8
An Ounce of Prevention – Planning Your Administrative Model

9
Bringing Your Admins into the AWS Administrative Backplane

10
Administrative Single Sign-On to the AWS Backplane

Section 3: Implementing IAM on AWS for Application Use Cases

11
Bringing Your Users into AWS

12

AWS-Hosted Application Single Sign-On Using an Existing Identity Provider

Other Books You May Enjoy

Index

Preface

Amazon Web Services (**AWS**) is the largest cloud platform in the world. It was also the first modern cloud services provider and the first to achieve broad enterprise penetration. Whereas being the successful first mover in a market has its advantages, it can also limit a service's flexibility. Compared to its biggest peers, which logically extend enterprise identity architectures (in part because they came to market years later), AWS' IAM capabilities can appear slightly alien. Like an archaeologist examining a dig site, we can see artifacts that suggest the service had a history of differing access mechanisms and strategies over the years. Given the success of the service, perhaps it was deemed too great a risk to the growing user base to make sweeping, foundational changes to align more with the familiar IAM patterns found in other organizations.

As AWS predates many enterprise identity best practices and reference architectures, bridging the paradigms of modern enterprise IAM and AWS' custom approach to IAM is often a difficult leap. Fortunately, with the advent of services such as AWS Organizations, AWS SSO, and Amazon Cognito, the service has never been more approachable. In this book, we will begin by examining the core services and components of identity on AWS in a manner designed to take the rough edges off its more eccentric components. Once we have built up our foundational knowledge, we will then apply what we have learned by solving familiar enterprise use cases.

Who this book is for

Identity on AWS may be well-trodden ground, but that doesn't make it any more inviting for the uninitiated. AWS' own documentation, while comprehensive, can be abstruse to those who are not approaching the platform or their own use case from a developer's perspective. The experience of many enterprise IAM practitioners is being thrown into the deep end to figure out how to solve a business identity problem on AWS through an implementation. With so many services appearing to offer such similar capabilities for similar use cases, getting up to speed, especially under a deadline, can be daunting.

This book was written to introduce IAM practitioners to identity on AWS with a use case-driven perspective. By using the language, patterns, and perspective of an enterprise identity practitioner, I aim to make this topic much more approachable.

What this book covers

Chapter 1, An Introduction to IAM and AWS IAM Concepts, introduces essential IAM concepts and AWS IAM as a suite of capabilities that provides user management and access control to AWS resources.

Chapter 2, An Introduction to the AWS CLI, introduces the AWS CLI, which is the primary programmatic method to interact with AWS resources.

Chapter 3, IAM User Management, addresses best practices around AWS user account security, life cycle management, governance, and authentication/password policies.

Chapter 4, Access Management, Policies, and Permissions, provides an overview of the authorization framework of AWS.

Chapter 5, Introducing Amazon Cognito, introduces Cognito and explores what it can do as an application identity service.

Chapter 6, Introduction to AWS Organizations and AWS Single Sign-On, explores the tools for applying organizational policies and managing access to multiple AWS accounts.

Chapter 7, Other AWS Identity Services, provides an overview of a few other identity and identity-adjacent services that, while important, did not get their own chapter.

Chapter 8, An Ounce of Prevention – Planning Your Administrative Model, provides guidance on designing an administrative and authorization policy model that addresses an organization's use cases.

Chapter 9, Bringing Your Admins into the AWS Administrative Backplane, walks through methods for bringing existing administrative accounts into AWS.

Chapter 10, Administrative Single Sign-On to the AWS Backplane, walks through federated authentication into the AWS console for administrative accounts and methods for applying fine-grained access control.

Chapter 11, Bringing Your Users into AWS, examines the distinction between administrative and standard user accounts, explores solution architectures for bringing user accounts into AWS, and demonstrates how to extend an on-premises AD forest into AWS using a trust.

Chapter 12, AWS-Hosted Application Single Sign-On Using an Existing Identity Provider, addresses configuring an application with Cognito user pools against a federated provider and using identity pools to authorize users to interact with AWS services on behalf of the application.

To get the most out of this book

To get the most out of this book, we expect familiarity with some basic IAM concepts and tools. Strictly speaking, there are no significant prerequisites to follow along with most of the exercises as they leverage existing AWS services and tools. As we look at some enterprise use cases in the later chapters, you will benefit from either setting up or using an existing non-production enterprise-grade identity provider and Active Directory domain controller.

Software/hardware covered in the book	OS requirements
AWS CLI	Windows, macOS, or Linux
AWS Account	
SAML2 and an OIDC-compliant identity provider, such as Okta, Azure AD, or PingOne	
Active Directory domain controller	Windows Server 2012 or later

Most cloud-based identity providers offer developer accounts and/or basic functionality free of charge that are more than sufficient for the examples in this book. Some chapters include references to guides for configuring a test environment for low or no cost in AWS, such as building a domain controller using Amazon EC2.

It would be impossible to cover every use case or scenario of AWS in a single book. However, we believe those who read it will not only be ready to contribute their identity expertise to most enterprise use cases but will be armed with the foundational knowledge and experience needed to solve increasingly complex ones as well.

Download the example code files

You can download the example code files for this book from GitHub at `https://github.com/PacktPublishing/Implementing-Identity-Management-on-AWS`. In case there's an update to the code, it will be updated on the existing GitHub repository.

We also have other code bundles from our rich catalog of books and videos available at `https://github.com/PacktPublishing/`. Check them out!

Download the color images

We also provide a PDF file that has color images of the screenshots/diagrams used in this book. You can download it here: `http://www.packtpub.com/sites/default/files/downloads/9781800562288_ColorImages.pdf`.

Conventions used

There are a number of text conventions used throughout this book.

`Code in text`: Indicates code words in text, database table names, folder names, filenames, file extensions, pathnames, dummy URLs, user input, and Twitter handles. Here is an example: "We can use resource tags and the `ec2:ResourceTag` variable to enforce this."

A block of code is set as follows:

```
{
    "Version": "2012-10-17",
    "Statement":
```

Any command-line input or output is written as follows:

```
$ aws iam delete-virtual-mfa-device --serial-number
arn:aws:iam::451339973440:mfa/rbis3
```

Bold: Indicates a new term, an important word, or words that you see onscreen. For example, words in menus or dialog boxes appear in the text like this. Here is an example: "Let's start with the **IAM_NonProd** AWS account. We tick the box next to that account and hit the **Assign users** button."

> **Tips or important notes**
> Appear like this.

Get in touch

Feedback from our readers is always welcome.

General feedback: If you have questions about any aspect of this book, mention the book title in the subject of your message and email us at `customercare@packtpub.com`.

Errata: Although we have taken every care to ensure the accuracy of our content, mistakes do happen. If you have found a mistake in this book, we would be grateful if you would report this to us. Please visit www.packtpub.com/support/errata, selecting your book, clicking on the Errata Submission Form link, and entering the details.

Piracy: If you come across any illegal copies of our works in any form on the Internet, we would be grateful if you would provide us with the location address or website name. Please contact us at copyright@packt.com with a link to the material.

If you are interested in becoming an author: If there is a topic that you have expertise in and you are interested in either writing or contributing to a book, please visit authors.packtpub.com.

Share Your Thoughts

Once you've read *Implementing Identity Management on AWS*, we'd love to hear your thoughts! Scan the QR code below to go straight to the Amazon review page for this book and share your feedback.

https://packt.link/r/1800562284

Your review is important to us and the tech community and will help us make sure we're delivering excellent quality content.

Section 1:
IAM and AWS – Critical Concepts, Definitions, and Tools

Identity is the most granular unit of security. To ensure the confidentiality, integrity, and availability of a system, that system's infrastructure, applications, APIs, and endpoints must all be identifiable, authenticated, and authorized in order to perform its functions. The AWS platform operates under a rigid identity-centric model. Bridging that model with your own organization's identity implementation can be daunting. At the end of this section, you will understand the industry-standard and AWS-specific IAM terminology that will be referenced throughout this book. You will also learn about best-practice access management patterns and the tools available to implement said patterns within AWS.

This part of the book comprises the following chapters:

- *Chapter 1, An Introduction to IAM and AWS IAM Concepts*
- *Chapter 2, An Introduction to the AWS CLI*
- *Chapter 3, IAM User Management*
- *Chapter 4, Access Management, Policies, and Permissions*
- *Chapter 5, Introducing Amazon Cognito*
- *Chapter 6, Introduction to AWS Organizations and AWS Single Sign-On*
- *Chapter 7, Other AWS Identity Services*

1
An Introduction to IAM and AWS IAM Concepts

Identity is the perimeter of security, and every transaction, capability, administrative event, and infrastructure component of cloud providers such as **Amazon Web Services (AWS)** ultimately depends upon identity services to govern all its capabilities. If that scope wasn't large enough already, tying AWS' native capabilities to an existing enterprise, customer, administrative, or infrastructure identity deployment can seem so complex as to make it difficult for cloud identity administrators to know how or where to start. This book will help you overcome the paralysis caused by the capabilities of the platform by approaching the implementation of AWS **IAM (IAM)** in a use case driven fashion, informed by real experiences working in large enterprise AWS environments.

By the end of this chapter, you will be familiar with the foundational concepts of IAM and see how they are applied within an organization. You will learn the purpose of the AWS IAM service, its components, and how they all work together to secure access to AWS resources. Finally, you'll use the AWS Management Console to create and manage AWS IAM resources, including IAM users, groups, and policies.

This chapter will cover the following topics:

- Understanding IAM
- Exploring AWS IAM
- Putting it all together

Technical requirements

To get the most out of this chapter, you will need the following:

- A web browser
- An AWS account

Understanding IAM

Identity is the most granular unit of security. The users, services, and systems that interact with infrastructure, applications, APIs, and endpoints must all be identified, authenticated, and authorized in order to perform their functions. The AWS platform operates under a rigid identity-centric model. Bridging that model with your own organization's identity implementation can be daunting.

Identity practitioners can (and do) argue about the minutiae and nuances of the terminology used within **IAM**. However, for our purposes, we can afford to use a broad definition of IAM in AWS:

> *"Identity & Access Management is the discipline of managing the life cycle*
> *of digital accounts that correspond to and are under the control of a person*
> *and ensuring that only the correct resources are accessed by the correct*
> *actor under the correct context."*

For something purported as a simple definition, that sure is a mouthful. However, if we break the statement down into its constituent components and consider a typical use case, it affords us an opportunity to see how many technical disciplines you may already be familiar with that relate to IAM:

> *"Managing the life cycle of digital accounts that correspond to and are under the control of a person..."*

In layman's terms, we have these digital accounts that can be used to access computer systems. These accounts either directly or indirectly map to a person. This means that the account is either a digital representation of that person or the person owns and controls those accounts. That person can demonstrate proof of control of those accounts and is accountable for actions taken with those accounts. And those accounts have a life cycle, meaning under certain conditions they are created, under other conditions they may change, and at some point, they may eventually cease to be.

This is called **identity management**. Identity management is responsible for the following:

- Keeping accounts up to date
- Keeping downstream consumers of those accounts synchronized with the authoritative sources that define the account
- Provisioning and deprovisioning accounts entirely from various data stores

In short, it's a collection of processes responsible for managing account life cycle events in accordance with business, legal, or technical controls. These controls trigger life cycle events for accounts, such as account creation, modification, and disablement. What those life cycle events are will vary depending upon the event, type of account, business, and requirements of the system using those identities.

Now, let's look at the rest of the definition:

> *"...and ensuring that only the correct resources are accessed by the correct actor under the correct context."*

Those accounts, having been created, can be used to execute specific activities. What they can do is determined by rules and policies. In order to do anything, the account must first provide proof that whoever or whatever is using it to perform an activity is actually allowed to do so. That proof comes through a shared secret that validates the authenticity of the actor behind the account. This second part of our IAM definition addresses something called **access management**. Access management addresses the authentication of the account (proving you are who you say you are) and the authorization of that account (proving that you are allowed to do what you are trying to do with that account) to access resources or to perform certain tasks in accordance with established policies.

IAM applied to real-world use cases

To understand this better, and to provide a flimsy pretext to introduce some additional concepts that are not so easily derived from that definition of IAM, let's imagine what happens when someone joins a new company. To help visualize all the actors, systems, and life cycle events in play, take a look at the diagram in *Figure 1.1*.

In this example, Bob has applied for a sales role at a large identity services organization called Redbeard Identity, which has a reasonably mature internal IAM program, application portfolio, and cloud platform capabilities. Bob's identity experience actually began long before he got to the point where the hiring manager was prepared to make an offer, because in order to apply for the position, he had to create a profile inside of Redbeard Identity's candidate management system.

> **Important note**
>
> The Redbeard Identity organization will be the organization referenced for several use cases and scenarios throughout this book. Whereas real organizations typically have a fixed enterprise architecture, we will adjust the architecture, capabilities, services, user accounts, and other characteristics of the Redbeard Identity organization from chapter to chapter in whatever ways we need to best demonstrate the material of that chapter. Please don't be confused if our example organization's characteristics are not entirely consistent throughout the book.

This marks the first identity life cycle event in Bob's onboarding journey: **user account creation**. Bob, as a user of the candidate management system, is providing self-issued, unverified information about himself such as his name, contact information, and details about his work history. As there is neither external proof nor an outside source of control validating the information he enters into this system, his candidate account is considered a low-assurance record. As long as Bob remains merely a candidate for the sales role, that low level of assurance is sufficient for the purpose that the candidate record system account serves:

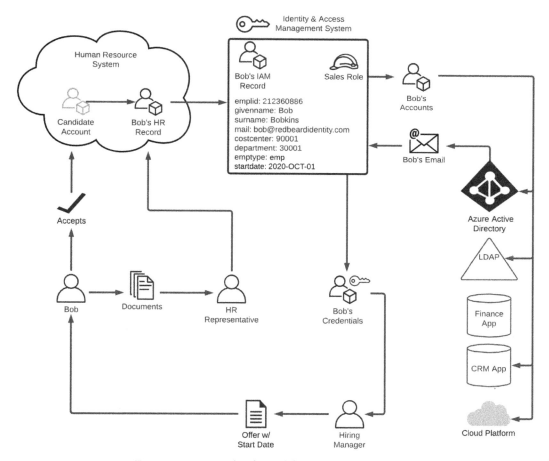

Figure 1.1 – Example of IAM life cycle events and flows

Bob knows his craft well, is an impressive salesman, and aces his interviews. After the details are agreed upon, the hiring manager sends Bob the offer letter confirming the details of his role, along with instructions for accepting the offer. Bob accepts by signing into the candidate portal and accepting the job offer. Now that Bob is more than just a candidate, the authenticity of the details that Bob provided when populating his candidate account must now be verified. To ensure that he is who he says he is, the HR representative will start a process called **identity verification**. This process is defined by the US Department of Commerce's National Institute of Standards and Technology as a process "*to ensure the applicant is who they claim to be to a stated level of certitude*" (*NIST Special Publication 800-63A, Digital Identity Guidelines, Section 4, NIST*).

The HR representative asks Bob to provide some identifying documents to facilitate his onboarding and help corroborate the information that he already entered as part of his candidate profile, such as a copy of his passport, a state-issued identification card, and his tax information. For the sake of argument, let's just say Bob hands the HR representative these documents in person to ensure that Bob himself has been compared against these artifacts. Thus, he sidesteps any concerns about him stealing valid credentials from someone else to use in his efforts to secure employment. The HR representative will finally validate these artifacts against their authoritative sources to ensure their authenticity, proving that Bob really is who he says he is. With the confidence that Bob is Bob and that the information Bob entered into the candidate management system is accurate, the HR system creates Bob's employee record and sets it to become active on Bob's start date.

As we said earlier, this organization has a reasonably mature IAM program. As part of a nightly process, the IAM system checks the HR system for any discrepancies in the data between the records stored there and its own corresponding identity records that it maintains in order to keep them in sync. When a change is made to an existing HR record that has a corresponding identity record, such as in the case of an employee changing departments, the *department* attribute on that employee's identity record also gets updated with the new *department* value. This is an example of **attribute and metadata synchronization** being used to ensure the consistency of identity data across data stores. In this case, the HR system is acting as an **authoritative source** for the IAM system, meaning that the records, attribute values, and other information from that system will overwrite any changes made directly against the records in downstream systems.

This organization uses business logic that tells the IAM system to create new identity records for new joiners one week from the start date listed on the new joiner's HR record. Once Bob's start date is less than a week away, that logic triggers the IAM system to **provision**, or create, his identity record. This will be the authoritative account record for all downstream systems, which in turn look to the IAM system as their own respective authoritative source. The IAM system will create Bob's identity record based upon an established pattern of attributes and characteristics, or a **schema**. It contains certain attribute types and values based upon the kind of account that Bob's identity record is. In *Figure 1.1*, we see a sample of (an admittedly spartan) schema for Bob's identity record. Let's pretend that we can actually take a look at the identity schema for Bob's record within the IAM system using *Table 1.1*:

Authoritative Source	Attribute Name	Attribute Value
HR System	`emplid`	212360886
HR System	`givenname`	Bob
HR System	`surname`	Bobkins
Azure Active Directory	`mail`	
HR System	`costcenter`	90001
HR System	`department:`	30001
HR System	`emptype`	emp

Table 1.1 – Sample schema record for Bob within the IAM system

This shows us the attribute names, their current values, and the authoritative sources for each of the attributes in this schema. You'll notice that for the most part, the HR system provides the bulk of the authoritative data for the attributes, with the exception of "mail," which is currently null (or without a value), and which also uses Azure **Active Directory (AD)** as its authoritative source.

You aren't constrained to a single authoritative source for your identity schema. In fact, you can have nearly infinite combinations of conditional clauses, secondary sources, and compound sources when defining your schema and the authoritative sources used to populate the schema's attributes. Beyond that complexity, you can also have several distinct schemas depending on the type of identity you are defining. We've only been examining Bob's identity journey as he gets onboarded at Redbeard Identity, and he is an employee as denoted by the `emptype` attribute. Contractors will likely have distinct schemas, as will bot process automation accounts, service accounts, business-to-business accounts, and customer accounts. But to keep things simple, we will stick with Bob the employee.

Bob works in sales, but it is doubtful that Redbeard Identity is a pure-sales organization given that they have enough technical wherewithal to run their own IAM infrastructure. Even if they were *that* operationally lean, there are regulations that demand evidence that some workers with certain job responsibilities cannot perform other, complementary responsibilities in order to reduce the risk of malfeasance. The go-to example for this is the protection control between accounts receivable and accounts payable in financial services organizations in order to prevent someone entitled to issue invoices from also approving their payment.

Separation of duties requires more than one person in order to complete a business task. Organizations implement separation of duties by applying technical controls that restrict or enable what a person can do based on business and regulatory requirements. Those rules, restrictions, and permissions are called **policies**, and a collection of policies that grants somebody the full range of access that they are entitled to depending upon their responsibilities is called a **role**. Aligning policies to roles, and roles to users through attributes or business logic is one part of access management. Providing evidence that those controls function as designed and comply with business and regulatory requirements is **identity governance and audit**. Identity governance and audit, access management, roles, and policy, all work to ensure that Bob will only be able to access the systems and resources that are appropriate for him to access, or in other words, that he is authorized to access.

This "all or nothing" approach to access is an example of **coarse-grained authorization**. Here, access is determined on a seemingly binary "yes/no" level based on the role that Bob was assigned provisioning him an account in the system. In Bob's case, he received the Sales role because, as we've said more than a few times now, he works in sales. However, there was no attribute labeled "role" that indicated which role he would be assigned. And there doesn't need to be. The logic that determines which entitlements get applied to an identity upon creation can vary wildly. In this scenario, Redbeard Identity's IAM system assigns roles based on the combination of the "costcenter" and "department" attributes. There could also be application-level roles and policies that provide **fine-grained authorization** to certain application-specific functions.

Now that Bob's identity has been provisioned and the IAM system has determined what role aligns to that identity, the next step is for the IAM system to begin provisioning Bob's accounts in the various downstream systems that he is entitled to access. Users with the Sales role get certain **birthright entitlements**, which are accounts and access that everyone gets just by being active employees within Redbeard Identity with that basic Sales role. *Figure 1.2* shows the provisioning process from the IAM system into these account stores in greater detail, with information about the schema for each of the accounts that Bob will be getting:

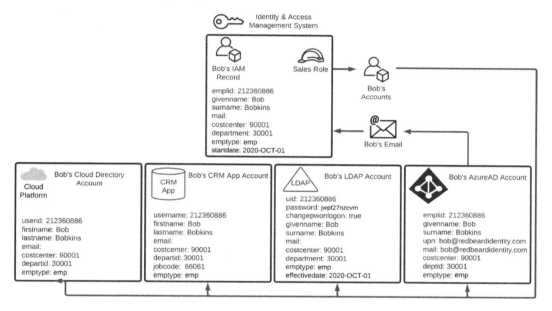

Figure 1.2 – Different account schemas across different identity stores

The IAM system provisions the following:

- Bob's Azure AD account

- An LDAP account in the company's directory

- A user account in Redbeard Identity's customer relationship management system where Bob will be spending most of his workdays

- An account in the cloud directory used by Redbeard Identity's cloud-hosted applications

Each one of these account stores is an example of an **identity store**. This is the place where an application or system can store its own instance of Bob's account with all the application-specific attributes added on. Just like how the HR system was the authoritative source for the IAM system, the IAM system is the authoritative source for these accounts and for many of the attributes within these identity stores. Now that Bob has an Azure Active Directory Account, he can get a mailbox and email address. If you recall from *Table 1.2*, Bob's main identity record did not have a value under the *mail* attribute when it was first provisioned. It is only now that the IAM system will detect Bob's email address when checking Azure AD for any new account updates. Upon detecting the discrepancy between the null value for the mail attribute in the identity record it has for Bob and the email attribute it reads on Bob's Azure AD account, the IAM system imports that update into Bob's IAM record with the new information obtained from that authoritative source. But the updates don't stop there! Look at *Figure 1.3*:

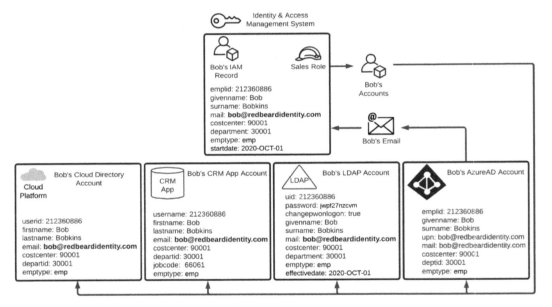

Figure 1.3 – Account attribute synchronization across authoritative sources

Remember that the IAM record itself is an authoritative source for several of the attributes on the downstream accounts that were provisioned as part of Bob's Sales role. In this instance, Bob's LDAP, CRM, and Cloud Directory account will each get Bob's new email address value written to the attribute that each has mapped to correspond to the IAM record's mail attribute value. Now that Bob has all of his accounts provisioned and synchronized with their authoritative sources, Bob is poised to be productive on his first day on the job.

That is to say Bob could be productive, assuming he knew how to identify himself as the owner of the account in each of these systems. This takes us to the last life cycle event depicted in *Figure 1.1*, which is the issuance of Bob's **credentials**. Credentials are the evidence used to attest that the person accessing a resource is who they say they are.

When talking about user accounts, credentials most often take the form of a unique identifier (such as a username) and a shared secret. This shared secret is between the person attempting to access a resource and the system that is trying to validate the identity of the person attempting to access a resource (such as a password). Bob's username plus his account password are his credentials to access these Redbeard Identity systems. Let's take a look at how that credential was created and delivered to Bob, as well as how the downstream applications can also verify Bob's identity despite not necessarily needing to maintain a set of their own for Bob to use.

Within Redbeard Identity, the mechanics of creating Bob's credentials are fairly straightforward. As part of the initial account creation process, the IAM system generates a random password to use as the password value on Bob's account. As you can see in *Figure 1.3*, though the password was generated by the IAM system, the IAM system is not acting as the authoritative source for the password, nor is it even storing the password attribute in its main identity record for Bob. Looking more closely at the schemas on those downstream accounts, we see that the only system that stores Bob's password value (or some form of hash of this value) is the LDAP directory.

In addition to being the only place where that value is stored, there is another unique attribute on that LDAP account called `changepwonlogon`, which is currently set as `true`. When the `changepwonlogon` value is set to `true`, it will force the person who entered the username and password to enter a new value for the password. When `changepwonlogon` is `false`, the person who correctly enters the account's username and password will simply be permitted to access the system or resource they were attempting to access when challenged for their credentials.

Providing the credentials is how a user can **authenticate** themselves, or how they prove that they are who they say they are. As Bob can't receive that initial password directly from Redbeard Identity's systems since he does not have access to Redbeard Identity's network yet, the IAM system instead issues the first password for Bob's account to Bob's hiring manager.

So why isn't the password written into all of the other identity stores where Bob has an account? In the specific situation we are examining using Bob's onboarding into the Redbeard Identity organization, they are maintaining a single authoritative identity store for all of their application authentication. This means that Bob will use a single, centrally managed username and password to access the applications and systems he needs to use to perform his job. This is as opposed to a system where he would be required to memorize a unique username and password stored and managed by each individual application. This is **single sign-on (SSO)**.

Applications maintain application-specific user records for each user that they use for their own purposes (such as authorization). However, the application delegates authentication to a central identity store using a directory services protocol such as LDAPS or Kerberos, or in the case of many modern web apps, a federated web-based protocol such as **SAML** or **OpenID Connect**. Using SSO reduces the number of credentials and the locations where those credentials are stored. This reduces the attack surface that a malicious actor can try to exploit to steal a credential. Using SSO also helps keep Redbeard Identity workers happy since they only have one password to manage.

Bob's first day at Redbeard Identity arrives. He shows up at the office for new hire orientation, receives his laptop, and his hiring manager shares the initial password for his account with him so he can sign into his account. After his credentials are validated, the `changepwonlogon` attribute triggers the life cycle event responsible for ensuring that the initial password gets changed. Bob enters a new password.

Once that is accepted and written to his LDAP account, the `changepwonlogon` value flips to `false`, and Bob becomes the sole owner of his account, which is essential for **non-repudiation**. From now on, any actions logged under `emplid` can be tied solely to him since he is the only one who can access resources and applications by authenticating using those credentials. And with that, Bob's identity onboarding experience is complete:

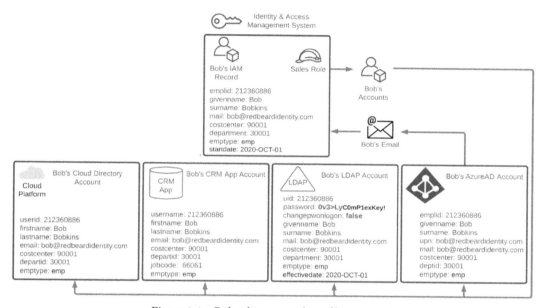

Figure 1.4 – Bob takes ownership of his account

Now that Bob has his account, he needs to sign into the applications he will use to perform the majority of his job duties. As we mentioned earlier, the Redbeard Identity organization maintains its users' passwords exclusively in its LDAP directory. Though Bob has accounts in the user stores of other systems and applications, those applications have delegated their user authentication to that LDAP directory. Applications can perform lookups and password validations directly against the LDAP using LDAPS, but that model has constraints that limit its usefulness as a modern authentication pattern.

Modern applications should rather use **identity federation** for user authentication, which is a model where the application looks to an external identity authority to receive trusted identity information. The CRM application that Bob will be spending most of his time in uses identity federation to authenticate its users. The process for the CRM app receiving an authentication token for Bob's identity from the identity provider is shown in *Figure 1.5:*

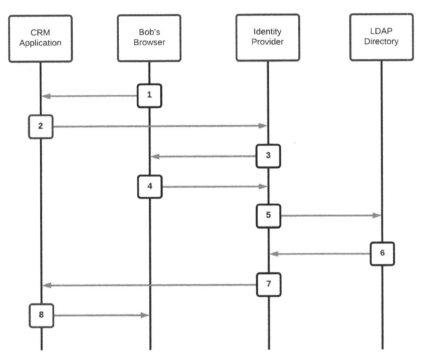

Figure 1.5 – Federated authentication transaction using an identity provider

Let's break down the steps:

1. From a browser, Bob goes to the CRM application.

2. Since Bob doesn't have a session cookie, the CRM application redirects the browser to the Identity Provider that it uses for user authentication.

3. The Identity Provider redirects Bob's browser to a logon form to collect Bob's username and password.

4. Bob's username and password are posted to the Identity Provider.

5. The Identity Provider performs the password validation on Bob's account against the authoritative source it uses for authentication – in this case, the LDAP directory where the Redbeard Identity organization stores its user credentials.

6. The LDAP directory responds that the credentials are valid and may optionally send along some additional attributes that the CRM application may need to reference at authentication time.

7. The Identity Provider creates a signed authentication token using its private signing certificate and posts that to the CRM application. The CRM application is assured that Bob has been authenticated by the external Identity Provider due to the unique cryptographic signature on the authentication token.

8. The CRM application looks at the subject of the authentication token, which in this instance is the `emplid` from the LDAP directory, and matches it to its local account under that same `username` value. The CRM application examines its local user record for Bob for his `jobcode` value to determine what application role he can assume. Job code `66061` corresponds to a sales representative role. The CRM app establishes an application session for Bob under that authorization context, and Bob is now logged in.

> **Important note**
>
> It is important to remember that the example we just walked through was meant to highlight IAM concepts, not necessarily IAM architecture or engineering best practices. Organizations' IAM and security maturity can vary greatly as they balance the risk equation of facilitating their core business against the monetary and opportunity costs of identifying and remediating potential security threats.

The Redbeard Identity scenario has provided us with an example of IAM principles in action and shows how the various components of the IAM system combine to form a platform that facilitates business outcomes and secure organizational resources. Now that we have an idea of what an IAM is, let's begin our exploration of it within AWS.

Exploring AWS IAM

"Wait," you may be saying, "I thought this book was supposed to be about AWS IAM? What's the deal with the overwrought organizational identity scenario I just spent the last several pages reading? When do we get to the AWS stuff?" **AWS** is a cloud provider that is ultimately governed by identity. AWS environments are owned by Amazon accounts and organizations, and each of the resources created within those environments has life cycle events governing its creation, modification, and eventual termination.

Additionally, the scope and scale of what someone or something can do with those resources are governed by identity, access management policies, and delegated authorization models. Where organizations and technologists encounter difficulty is in understanding the how, what, and why of AWS in the context of identity in light of its rich, and seemingly overlapping, infrastructure-as-a-service and platform-as-a-service components.

Taking a moment to recontextualize your organization's use of AWS through the lens of identity, and especially in the context of the business, security, and governance challenges you may have already solved in on-premises infrastructure in ways similar to the Redbeard Identity scenario, will aid us as we demystify this seemingly complex topic.

According to Amazon (*What is IAM?*, at `https://docs.aws.amazon.com/IAM/latest/UserGuide/introduction.html`),

> *"AWS IAM (IAM) is a web service that helps you securely control access to AWS resources. You use IAM to control who is authenticated (signed in) and authorized (has permissions) to use resources."*

This reads very similarly to our initial, high-level definition of IAM that we outlined in *Understanding IAM*. AWS IAM creates and manages the accounts used to sign in to the AWS Management Console and handles credential management and strong authentication capabilities for the accounts it manages.

Access management and authorization for users, services, and even resources, including fine-grained authorization to AWS resources, are managed through access policies that are defined, governed, and validated against AWS IAM. Governance, compliance, and audit are also reported through AWS IAM and presented through other AWS services. AWS IAM and its supporting identity security services offer a complex and feature-rich IAM capability for administrating and controlling who has access to what and under what context that access is authorized.

IAM for AWS and IAM on AWS

AWS IAM is not the only tool that is capable of providing IAM services inside of the AWS cloud. As we saw in the Redbeard Identity scenario, a comprehensive IAM solution at an organizational level is composed of several different systems and services. These fulfill the business and security use cases required for that business. There is not a monolithic "Redbeard Identity IAM Service." Rather, it is a mix and match of various provisioning, governance, authentication, and directory services.

AWS IAM is the service to govern access to AWS services, but there are several other services that can be mixed and matched in pursuit of solutioning business IAM challenges. When speaking about AWS IAM, we are referring to IAM *for* AWS, specifically in the context of AWS as infrastructure-as-a-service. When we use other AWS services to solve IAM challenges, we are applying IAM *on* AWS and using those services in the context of AWS as a platform-as-a-service. We will address what some of those other services are and their use cases in later chapters.

> **Tip**
>
> AWS IAM provides identity services for AWS as an infrastructure-as-a-service platform. Other AWS identity services provide identity capabilities for AWS as a platform-as-a-service.

The AWS IAM dashboard

With so much capability, it can be difficult to see how all the pieces fit together, let alone figure out how to mash all of those pieces together in order to develop solutions to identity challenges in the cloud without first familiarizing yourself with the tool directly. Let's start by taking a quick look at the tool as it appears from inside of an AWS environment:

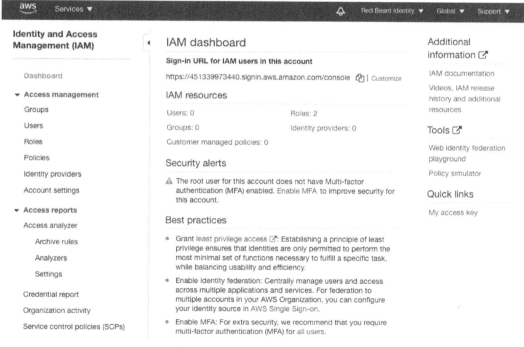

Figure 1.6 – AWS IAM dashboard

The center panel offers a single-pane-of-glass scoreboard for the counts of critical identity objects that are live in the environment. Already, you may recognize some of the IAM terms and concepts that we touched upon in the Redbeard Identity organization example earlier in this chapter, particularly roles, users, and policies. In the context of AWS, each of these terms has a specific definition, which we will discuss in more detail momentarily.

Links to the individual administrative panes for **Groups**, **Users**, **Roles**, **Policies**, **Identity providers**, and **Account settings** are on the left side of the AWS IAM dashboard. Each one of those links will allow you to individually view and administrate the components.

Further down, we have some reports and analytics tools designed to facilitate policy administration and aid in the creation and audit of policy structures. They govern access to AWS resources in the environment. The Credential Report details the status and age of the credentials for every AWS IAM user managed by the environment. Finally, the **Organization activity** and **Service control policies (SCPs)** sections are special administrative sections. They are only activated when an AWS account is part of something called an AWS organization. This is a construct that allows large organizations to govern multiple AWS accounts in line with a single, centralized policy.

Principals, users, roles, and groups – getting to know the building blocks of AWS IAM

In case you couldn't tell, we are already experiencing some namespace collision on the terms we used earlier to describe Bob's onboarding and authentication journeys at Redbeard Identity and those used within AWS. For example, we could get away with interchangeably referring to "Bob's identity record" and "Bob's identity" in the example. The definitions used when referring to the components that compose and interact with AWS IAM have very specific definitions. You will need to understand that taxonomy to ensure you understand how AWS IAM, and AWS as a platform at large, operates. The following definitions are taken from the AWS IAM User Guide (`https://docs.aws.amazon.com/IAM/latest/UserGuide/intro-structure.html#intro-structure-terms`):

- **Principals**: A person or application that uses the AWS account root user, an IAM user, a federated user, or an IAM role to sign in and make requests to AWS.

- **Entities**: The IAM resource objects that AWS uses for authentication. These include IAM users, federated users, and assumed roles.

- **Identities**: The IAM resource objects that are used to identify and group. You can attach a policy to an IAM identity. These include users, groups, and roles.

- **Resources**: The user, group, role, policy, and identity provider objects that are stored in IAM. This can be an action in the AWS Management Console or an operation in the AWS CLI or AWS API.

To ensure topical clarity moving forward, we will be using the AWS definitions of these terms unless a distinct definition is specifically referenced in context. Speaking of topical clarity, those definitions are far from clear. A principal authenticates with an entity, but is that entity considered an AWS identity? An identity can be an AWS IAM role, user, or group, and an entity can be an AWS IAM user or role, but can both of them have an access policy attached to them? Both entities and identities are resources, and both entities and identities can be roles or users, but are roles and users themselves resources? Perhaps it will make things easier to approach these definitions with an old-fashioned Venn diagram. Take a look at *Figure 1.7*:

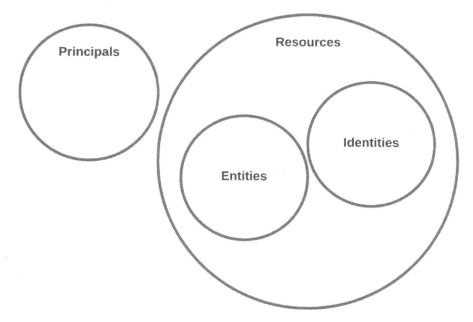

Figure 1.7 – The relationship amongst AWS IAM principals, entities, identities, and resources

Let's start with the circle that's all by itself, **Principals**. The best way to think about **Principals** is in terms of the subject of a sentence, or the *who* or *what* that performs the action in the sentence's predicate. In the case of AWS, individual users act as principals when they sign into their AWS IAM user accounts. However, principals don't have to be tied to a flesh-and-bone person.

Consider service accounts, bot process automation accounts, or even programmatic access when calling APIs – each one of those use cases acts as the principal when either signing into a corresponding account or assuming a delegated role that permits that service to access a resource. What's more, other AWS services, such as S3 or EC2, may assume a service-level role to act as principal and authenticate and get access to manipulate resources.

On the topic of resources, resource is a very broad term in AWS. In this particular AWS IAM context, it refers specifically to all of the things stored and managed in AWS IAM that appeared on the at-a-glance dashboard on the landing screen of AWS IAM we saw in *Figure 1.6*, such as users, groups, and roles. If you create it in AWS IAM, it is an AWS IAM resource. Similarly, S3 bucket objects are resources within the S3 service. In the broader context of AWS as a platform, a resource is an object created and managed within an AWS service in an AWS environment.

That is also the reason why both entities and identities are fully encapsulated within the resources circle in the diagram, as they both only exist in the context of AWS IAM. The entity is the AWS IAM resource that a principle uses during authentication, and as such provides information about the principal through policy objects attached to that IAM user, assumed role, or federated user. If entities can be described as AWS IAM resources used by a principal during authentication, then identities are the corresponding resources assumed by principals for authorization.

Since identities cover users, groups, and roles under the auspice of grouping and identification, this means that the foundation for authorization decisions within AWS IAM are actually the user, groups, and role identity resources. The two act in conjunction when a principal is engaged with the service, with the entity as the surrogate for the principal within AWS IAM correlating the principal to an authenticator or credentials, and the user identity's attached policy supplies the information for AWS IAM to determine what that principal is and isn't allowed to do:

Figure 1.8 – AWS IAM evaluates principal authentication and request context to permit or deny actions on a resource

But how does AWS IAM decide what the principal can and cannot do? AWS IAM evaluates each request in light of a request's context, which is to say a combination of characteristics that can be evaluated against a policy. Take a look at *Figure 1.8* to see a breakdown of what contributes to a request's context. AWS IAM considers who is attempting to take action, or the principal. Any time that principal interacts with the AWS Management Console, they are performing operations against resources.

The AWS IAM service has about 40 specific operations that a principal can perform against its resources. For the most part, they align with the familiar CRUD acronym, (create, read, update, delete), but the actions specify the AWS IAM resource targeted by that operation in the operation name, for example, create-user, update-group, get-role, delete-policy. Further details about a specific resource that will be the target of the operation narrow the scope of action further. Finally, there are the environmental details in which the request takes place, such as the time of day or originating IP address.

AWS IAM considers the full request context against the policy applicable to the principal's identity resource and decides whether the action is permitted or authorized, assuming the principal's entity has been sufficiently authenticated.

Authentication – proving you are who you say you are

In the Redbeard Identity scenario, we made several references to both "verifying the authenticity" of things, such as Bob's personal information, and "authenticating" that Bob really was the account holder entitled to access the CRM application by providing his password, a shared secret.

The first, while an authenticating activity, is identity verification. Identity verification ensures that the principal you are issuing credentials to really is who they say they are through the validation of that identifying information by an authoritative source. Conversely, proving possession of a shared secret or token to demonstrate ownership and control of an account in the context of gaining immediate access to a system with that account is authentication as we will refer to it from here on out.

As we briefly touched upon in the previous section, before any principal is permitted to take action on an AWS resource, they must first authenticate themselves through the AWS IAM service. The most common way to do this is with a username and password pair through the AWS Management Console.

We will discuss the differences between the root user and IAM users and best practices on securing and administrating your AWS administrative users in *Chapter 3*, *IAM User Management*. But not every principal is a human behind a keyboard. For other principals, such as applications requiring programmatic access where a username and password validation flow would not serve, there is also the option to authenticate via an access key ID and secret key ID. You have the option of granting either access type to new IAM users.

Authorization – what you are allowed to do and why you are allowed to do it

Truth be told, we've already discussed authorization at length throughout this chapter. AWS IAM's primary function is arguably making authorization decisions based upon a policy evaluation against a request's context. That said, we've mentioned "policy" several times without defining what it means both in the broader context of IAM and specifically as a component of AWS IAM.

Policies are rules that define a course of action. IAM policies are rules that determine whether a user or system can access or manipulate a resource based on their attributes, role, or security context. AWS IAM authorization policies are a variety of rules and evaluation logic that combine to determine whether a given request is authorized based upon the information present in its request context. We will be diving very deeply into the various policy types and the anatomy of AWS IAM's JSON-based policy structure in a future chapter, but the policies that may be evaluated based on a request's context include the following:

- Identity-based policies: These are inline policies that are attached to IAM identity resources, namely users, groups, and roles.

- Resource-based policies: These are inline policies that are attached to AWS resources, such as a policy on an S3 bucket that indicates what a specific principal can do with that specific bucket's contents.

- Permissions boundaries: A policy that sets limits on what a specific IAM user or role can do with a service or resource. This policy represents a "boundary" for the IAM user or role it is applied to, meaning that other policies outside of that boundary will not be respected.

- The organization's service control policies: A policy that is similar to permissions boundaries but applies to AWS accounts governed by that organization.

- **Access Control Lists (ACLs):** ACLs restrict the resources that principals from different AWS accounts can access within your AWS account. This policy is unique as it does not use AWS IAM's JSON-based policy structure.

- Session policies: Session policies create a hybrid policy that lasts only the duration of the principal's session based upon attributes programmatically passed during authentication time and an identity-based policy. This is an "advanced" policy according to the AWS IAM User Guide.

AWS evaluates all applicable policies based on the request context to determine how the request should be evaluated. Generally speaking, if the request context fails any evaluation criteria for any of the applicable policies, the entire request is rejected unless a policy includes an explicit "allow" statement.

The AWS IAM dashboard is the jumping-off point for applying identity to AWS services and provides administrators an at-a-glance view of the IAM objects that currently exist within their AWS account. Don't be intimidated by the flood of terminology, or the obtuse relationships between the various IAM objects and authorization policies. These things may be difficult to fully grasp right now as they are devoid of context. This will become clearer as we work through some examples of how IAM objects are governed by AWS IAM.

Putting it all together

Now that we've seen the AWS IAM dashboard, familiarized ourselves with the terminology used with the service, and examined the relationship between principals, entities, identities, roles, groups, and policies, let's create some AWS IAM resource objects using the AWS Management Console. In order to complete this exercise, you will need to sign up for an AWS account at `https://aws.amazon.com`.

Signing in with the root user

If you have signed up with a new account, the first and only option you have to sign in to the AWS Management Console is with the Root user. The Root user is the owner of the AWS account, and similar to a root user in a Linux system, it is a super administrator with full access to all the services and resources available. Just as one would when configuring a server, we should only use the Root user for as long as it takes to set up a different administrative account to use:

1. From the AWS IAM dashboard, expand **Access management** on the left and click on **Users**. From this screen, you can see every non-root user in your account, including important security information such as group membership information, access key age, password age, last activity, and whether or not that account has multifactor authentication enabled:

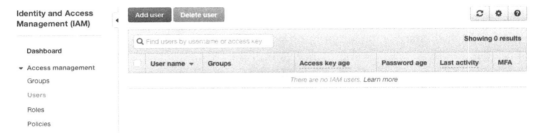

Figure 1.9 – AWS IAM user administration console

2. As this is a new environment, our user list is empty. We create a new user by clicking **Add user**:

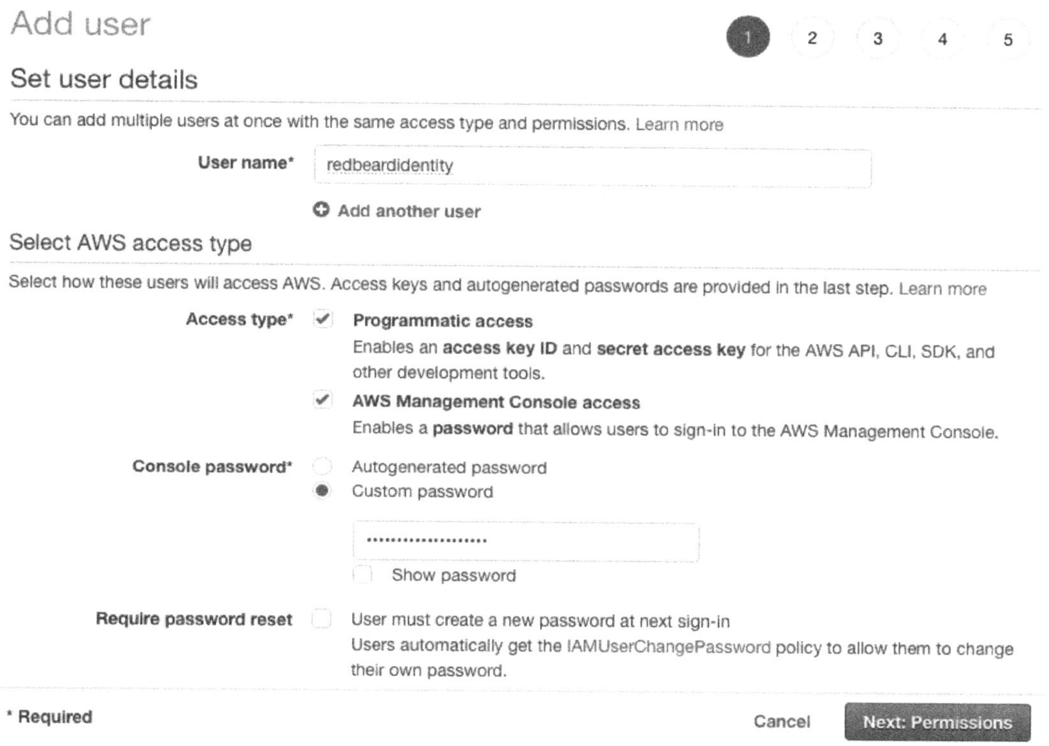

Add user

Set user details

You can add multiple users at once with the same access type and permissions. Learn more

User name*   redbeardidentity

⊕ **Add another user**

Select AWS access type

Select how these users will access AWS. Access keys and autogenerated passwords are provided in the last step. Learn more

Access type*   ✔   **Programmatic access**
Enables an **access key ID** and **secret access key** for the AWS API, CLI, SDK, and other development tools.

✔   **AWS Management Console access**
Enables a **password** that allows users to sign-in to the AWS Management Console.

Console password*   ○   Autogenerated password
●   Custom password

....

☐   Show password

Require password reset   ☐   User must create a new password at next sign-in
Users automatically get the IAMUserChangePassword policy to allow them to change their own password.

* Required   Cancel   **Next: Permissions**

Figure 1.10 – User configuration and access type

3. Let's name the new account `redbeardidentity` and give it both programmatic access and AWS Management Console access. This means the account will be issued two sets of credentials, a password for console access, and the access key ID and secret key ID for use with the AWS command-line interface:

4. Since we will be using this account, we can select the option to populate our own password and uncheck the box that requires a new password on first login. If we were provisioning an account for another administrator, we would leave the "password reset on first logon" requirement in place to ensure that the other administrator was the only person who knew their password. Click on the **Next: Permissions** button:

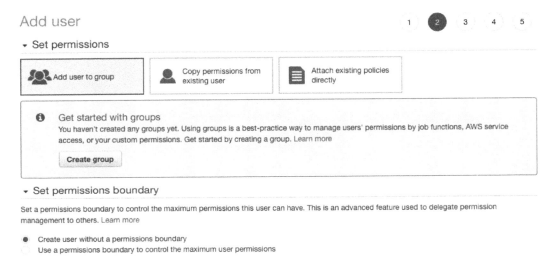

Figure 1.11 – Permissions options

On the next screen (*Figure 1.11*), we see several options for granting permissions for the new account. Let's examine the options available to us. If this were a shared account with several different administrators performing different job functions, we could set up a group for each one of those job functions and attach policies to the group. Then by adding the new user accounts to the appropriate group, those users inherit the policies from the group. Alternatively, we could just copy the permissions from an existing user. This is a non-starter for our use case as we are currently creating the very first non-root user account in the environment and have no other account from which to copy permissions. Finally, we can create and attach a policy directly to the user. Since the wizard is selling groups as a "best-practice way to manage users' permissions," we'll do that. This is also where we can optionally set a permissions boundary for this user. Since this user is an administrator, we don't need to set such a boundary:

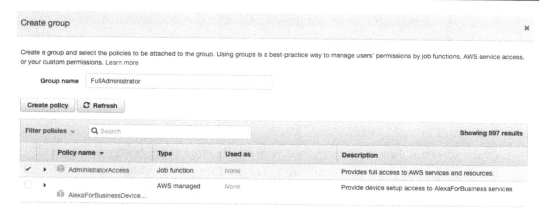

Figure 1.12 – Create group and attach policy

5. Clicking **Create group** takes us to the group creation screen where we can name the group and attach AWS-created policies to it. We also have the option to create our own custom policy for the group, but as the goal for this group is to grant full administrative privileges to the environment, and AWS already has a policy that grants those entitlements, we'll spare ourselves the administrative overhead.

6. We give our administrator's group a name that will help ourselves and others recognize its purpose and click **Create group**:

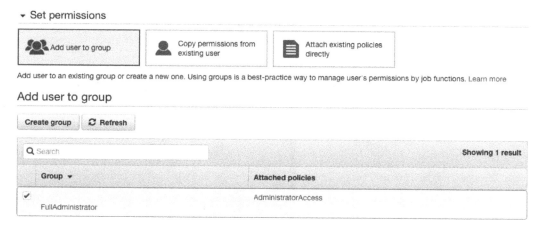

Figure 1.13 – Create group and attach policies

The group is created, and we are returned to the user creation screen. The form now shows the new user as a member of the `FullAdministrator` group. Click on the **Next: Tags** button. On the next screen, we can optionally create some tags to associate with this user. **Tags** are customizable attributes in the form of key-value pairs that you can define on nearly every resource object type in AWS, and you can use tags for reporting, searching, and perhaps most importantly, authorization policy.

7. Tags are powerful tools for governance, so we will define some `costcenter` and `jobcode` tags and populate them with values that we may be able to use to define some session policies later. As we type, the console opens new rows for other tags. Type something like what is shown in *Figure 1.14* and click on the **Next: Review** button:

Figure 1.14 – Attaching tags to the new IAM user

8. After that, we can review all of our selections and create the user. Simply click on the **Create user** button and the operation is finished:

Review

Review your choices. After you create the user, you can view and download the autogenerated password and access key.

User details

User name	redbeardidentity
AWS access type	Programmatic access and AWS Management Console access
Console password type	Custom
Require password reset	No
Permissions boundary	Permissions boundary is not set

Permissions summary

The user shown above will be added to the following groups.

Type	Name
Group	
	FullAdministrator

Tags

The new user will receive the following tags

Key	Value
costcenter	90001
jobcode	1701

Cancel   **Previous**   Create user

Figure 1.15 – Review and create the new user

The AWS IAM dashboard has been updated to reflect the new user and group creation, and the **Users** and **Groups** control panels now give us options to administrate the new IAM resource objects:

Figure 1.16 – Updated IAM dashboard

If we check the list of users, we see the new IAM user we've created, complete with an at-a-glance view of the group membership, the age of its credentials, its last activity, and whether it has multifactor authentication enabled:

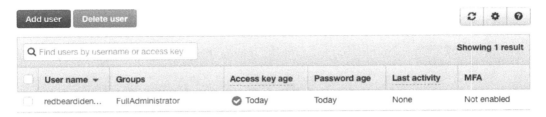

Figure 1.17 – Updated IAM user administration console

9. Now we can sign in using the non-root account. Note the **Sign-in URL for IAM users in this account** in *Figure 1.16*. It is an account-specific sign-in link for IAM users to use when signing into this particular AWS account so we will not need to memorize and enter the account number each time we sign in through `https://aws.amazon.com`:

Figure 1.18 – Root user on the left, IAM user on the right

Once signed in under the `redbeardidentity` IAM user account, and despite it having full administrator permissions to the AWS account just like the root account, we can see that it is an IAM user account based on the differences in the account information displayed in the menu bar.

Now that we've created our first AWS IAM user, let's recall once again why we bothered to do so in the first place. IAM is the discipline of managing the life cycle of digital accounts that correspond to and are under the control of a person and ensuring that only the correct resources are accessed by the correct actor at the correct time and under the correct context. Understanding how identity life cycle events and processes interact to achieve a specific business or technological outcome helps us understand how to achieve those same outcomes within the cloud. AWS IAM is the service that an AWS account uses to authenticate and authorize users and applications that use the account's services and handles the IAM use cases for AWS services.

AWS IAM controls IAM resource objects, including the entities that users, applications, or federated users use to authenticate themselves to the service. IAM users, roles, and groups are identity objects used to identity or group IAM resource objects for the application of authorization policy. AWS IAM assesses requests to take actions on AWS objects using the request context, which is a combination of details about the request, in conjunction with authorization policy objects that apply to the user, role, group, or resource that the principal is trying to manipulate. Each AWS account gets a root user, which is the superuser for the account. Best practice recommends that you use an appropriately scoped IAM user when accessing your AWS account, and not the root account.

Summary

Over the course of this chapter, we've learned the basics of IAM and seen it applied to a typical enterprise worker scenario. We also learned about the AWS IAM service, and how it applies IAM to AWS itself. Additionally, we learned about the types of objects managed by the AWS IAM service and got a high-level overview of how AWS IAM uses authentication, authorization, and request context to evaluate every request that comes into an AWS account. Finally, we took a tour of the IAM dashboard inside of the AWS Management Console and created our first non-root administrative user.

The next chapter will introduce us to an alternate method for interacting with our AWS account – the AWS CLI. We will learn how to install and configure it on Windows and macOS/Linux, including setting up different profiles for different use cases. We'll also examine the command syntax and get introduced to some tools that will help us discover command syntax and administrate objects. By the end of the chapter, we will have learned how to make programmatic calls to an AWS account.

Questions

1. What is IAM?

2. What is authentication?

3. What is authorization?

4. An IAM user, a role, a federated user, and an application are examples of what in an AWS service request?

5. What are some examples of AWS IAM objects?

6. How does the AWS IAM service decide whether a request to access a resource is permitted?

7. Name the AWS policy types that can impact an authorization decision.

2
An Introduction to the AWS CLI

The AWS CLI is the programmatic interface for administrating resources within an AWS account. You can use the AWS CLI to access any number of AWS accounts, operate under any number of IAM user objects under each of those accounts, and execute complex commands to perform all manner of functions on all of your AWS objects. Accessing an AWS account through the AWS CLI requires an IAM user object with programmatic access to its AWS account, which is a type of credential distinct from the user object's Management Console password. With this access, you will be able to do everything via the CLI that you can do in the Management Console.

In this chapter, you will learn all about the AWS CLI. This includes the installation and configuration of the utility, including configuring different user profiles. You will also learn about several methods to discover the syntax of AWS CLI commands, including the built-in `help` command, auto-prompt, and creating and importing templates. The AWS CLI reveals how many distinct AWS resources can contribute to what may be perceived as a single operation, such as creating an administrative account. You will see that what makes them work as we expect them to work is their relationships with each other. We can use scripts to efficiently execute vast numbers of AWS CLI commands to administrate an AWS account.

By the end of this chapter, we will have covered the following:

- Exploring the AWS CLI basics

- Using the AWS CLI

- Putting it all together – creating a functional IAM user with the AWS CLI

Technical requirements

To get the most out of this chapter, you will need the following:

- A workstation running either Windows 10, macOS, or Linux

- The AWS CLI v2 installer for your operating system, available at `https://aws.amazon.com/cli/`

- An AWS account

- A text editor or IDE to edit JSON/YAML files, such as Microsoft Visual Studio Code

Exploring the AWS CLI basics

In this section, we will get introduced to the AWS CLI utility. We will learn what it is used for, how to install it on various operating systems, and how to configure it for use with our AWS account.

What is the AWS CLI?

The **AWS CLI** is a command-line utility that enables programmatic access to the services in your AWS account. After you sign in to the AWS Management Console with an IAM user (or assume an IAM role), the actions you take within the console to look at the objects within each service, create new objects, or delete existing objects are all called via backend APIs invoked through your interactions with that GUI. The AWS CLI is another mechanism to administrate your AWS account, services, and resources just like you do in the Management Console. You authenticate yourself by presenting the credentials of an IAM user or with a session token and assumed role. You can list all of the services available to you based on your authorization, create new resources, modify existing resources, and delete resources. The AWS CLI commands manipulate the AWS platform in the exact same way as the GUI.

So then, why would someone prefer to use a command line and painstakingly type their commands when the Management Console offers useful wizards, tooltips, and a point-and-click interface? In a single word – speed. The AWS CLI facilitates programmatic and script-based resource administration that allows you to manipulate the environment at a much quicker pace than can be achieved using the GUI. Additionally, the programmatic nature of the CLI lends itself to infrastructure-as-code and DevOps processes that are not compatible with a GUI. Real-deal AWS administration is done using the command line.

While it usually doesn't make sense to dive straight into a new concept with demos and practical application, in order to explore the features and functionality of the AWS CLI, you need to connect it to an AWS account, and that in turn requires us to perform some basic configuration. So, let's start there.

Installing the AWS CLI

Since February 2020, there are two versions of the AWS CLI. Version 2 introduces several new features, including a self-contained installer, support for AWS Single Sign-On when using the CLI, and several quality-of-life improvements around documentation, tooltip documentation, and auto-completion suggestions. While version 1 remains supported, we will constrain our scope to the version 2 CLI with this book. Although there is considerable overlap between version 1 and version 2 commands, version 2 is not considered backward-compatible with version 1.

The following instructions represent the most concise method of installing and configuring the AWS CLI, but there are detailed instructions for niche use cases (such as specifying a specific point release to be downloaded) and additional operating systems available from AWS at `https://docs.aws.amazon.com/cli/latest/userguide/install-cliv2.html`.

macOS

To install the AWS CLI on macOS, use the following steps:

1. Open Terminal and navigate to a directory where you can save files. First, enter the following command to download the latest version of the command-line installer package:

```
$ curl ''https://awscli.amazonaws.com/AWSCLIV2.pkg'' -o ''AWSCLIV2.pkg''
```

2. Then, run the following command to install the AWS CLI for all users on the workstation:

```
$ sudo installer -pkg AWSCLIV2.pkg -target /
```

Here's what the output for *steps 1* and *2* looks like:

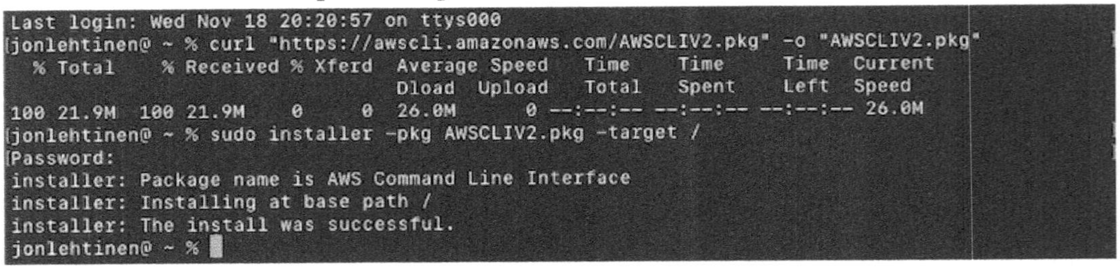

```
Last login: Wed Nov 18 20:20:57 on ttys000
[jonlehtinen@ ~ % curl "https://awscli.amazonaws.com/AWSCLIV2.pkg" -o "AWSCLIV2.pkg"
  % Total    % Received % Xferd  Average Speed   Time    Time     Time  Current
                                 Dload  Upload   Total   Spent    Left  Speed
100 21.9M  100 21.9M    0     0  26.0M      0 --:--:-- --:--:-- --:--:-- 26.0M
[jonlehtinen@ ~ % sudo installer -pkg AWSCLIV2.pkg -target /
[Password:
installer: Package name is AWS Command Line Interface
installer: Installing at base path /
installer: The install was successful.
jonlehtinen@ ~ %
```

Figure 2.1 – Terminal window output after downloading and installing the AWS CLI v2 package

3. To verify that the installation was successful and that AWS CLI commands are now included in your command $PATH, run the following command:

```
$ which aws
```

4. This will reveal the folder path where the symbolic links for the user-executable commands were added to your $PATH variable:

```
$ aws --version
```

This will show the AWS CLI version, the Python version, and additional host kernel information:

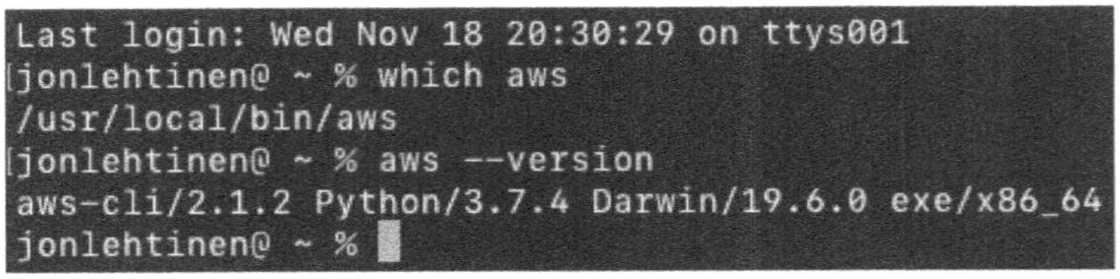

```
Last login: Wed Nov 18 20:30:29 on ttys001
[jonlehtinen@ ~ % which aws
/usr/local/bin/aws
[jonlehtinen@ ~ % aws --version
aws-cli/2.1.2 Python/3.7.4 Darwin/19.6.0 exe/x86_64
jonlehtinen@ ~ %
```

Figure 2.2 – Terminal window output after verifying the AWS CLI installation

Now that we have seen how it is done on macOS, let's move on to Windows.

Windows 10

Follow these steps to install the AWS CLI on Windows 10:

1. Open a browser and download the AWS CLI installer from https://awscli. amazonaws.com/AWSCLIV2.msi.

2. Open and run the installer and follow the instructions to complete the installation.

3. Once the installer has completed, open Command Prompt to verify the installation by entering the following command:

```
C:\> aws --version
```

This will show the AWS CLI version, the Python version, and additional host operating system information:

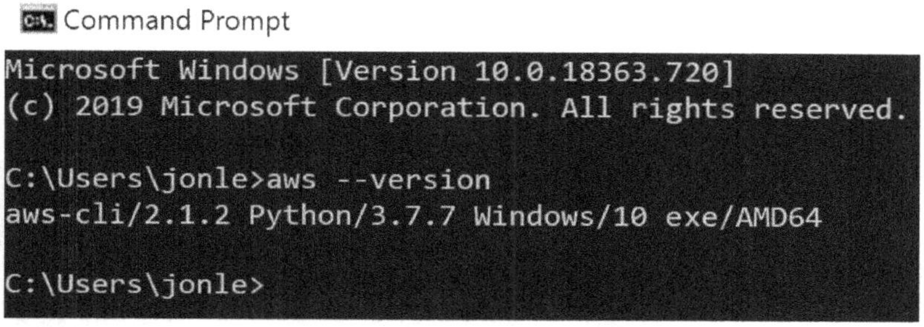

```
CMD Command Prompt

Microsoft Windows [Version 10.0.18363.720]
(c) 2019 Microsoft Corporation. All rights reserved.

C:\Users\jonle>aws --version
aws-cli/2.1.2 Python/3.7.7 Windows/10 exe/AMD64

C:\Users\jonle>
```

Figure 2.3 – The Command Prompt window output after verifying the AWS CLI installation

This concludes the configuration under Windows 10. Let's move on to Linux.

Linux

The installation process is very similar to the one we used for macOS, given macOS' Debian heritage. To install the AWS CLI on Linux, use the following steps:

1. Open Terminal and navigate to a directory where you can save files. First, enter the following command to download the latest version of the command-line installer package:

```
$ curl ''https://awscli.amazonaws.com/awscli-exe-
linux-x86_64.zip'' -o ''awscliv2.zip''
```

2. Then, run the following command to unzip the package. If your Linux install does not have unzip installed, install unzip using your distribution's package manager and try again:

```
$ unzip awscliv2.zip
```

3. Run the following to install the AWS CLI for all users:

```
$ sudo ./aws/install
```

4. To verify that the installation was successful and that AWS CLI commands are now included in your command $PATH, run the following command:

```
$ which aws
```

5. This will reveal the folder path where the symbolic links for the user-executable commands were added to your $PATH variable:

```
$ aws --version
```

This will show the AWS CLI version, the Python version, and additional host kernel information:

```
jonlehtinen@ ~ % aws configure
AWS Access Key ID [None]: AKIAWSFPVONAKWL37BGN
AWS Secret Access Key [None]: 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
Default region name [None]: us-east-1
Default output format [None]: json
jonlehtinen@ ~ %
```

Figure 2.4 – Output of installation on Ubuntu

The preceding should work for most versions of Linux, though it is possible you may run into issues with custom compilations. Additionally, it is discouraged to install from your distribution's package manager utility, such as yum or apt, as the versions maintained there may not be the most recent release from AWS.

AWS CLI configuration

As we have mentioned before, every request into an AWS account is evaluated by the AWS IAM service, which authenticates the principal making the request via the credentials presented with the call, and by examining the policy and request context to determine whether that request should be authorized to proceed. Requests from the AWS CLI are no different. While there are several advanced patterns for identity and authorization management that can be deployed with the AWS CLI, if you are operating in a new AWS environment or are simply looking to get started quickly, you need to define your basic configuration. Now that we have the AWS CLI installed, we will run the configuration command to define the default values for how the AWS CLI will present data to us, where it will send our requests, and the IAM user we will use to authenticate and authorize our calls into the service. This most basic configuration is done using the aws configure command.

The aws configure command will prompt for four settings:

- **AWS Access Key ID**: This the first half of the credentials issued to AWS IAM users that have been granted programmatic access. This value is shown in the **Security Credentials** section of the user object on the AWS IAM dashboard.

- **AWS Secret Access Key**: This is the second half of the credentials issued for programmatic access and is considered a secret. You can only see the secret value or download a .csv file with both the access key ID and the secret access key at the credential's creation time, so keep that in mind when creating your programmatic credentials for the IAM user you intend to use with the CLI.

- **Default region name**: This is the default region where requests will be sent. AWS has dozens of regions across the globe. While some services, such as AWS IAM, operate globally without a regional distinction, most services require you to specify the region where you want the request to take place. The default region will be the region where requests are sent unless a specific region is included within the syntax of the request. The best practice is to select the region closest to you, assuming you have no other region requirements for your use case.

- **Default output format**: This setting determines how the AWS CLI presents object information. Options include json, yaml, yaml-stream, text, and table. Whatever you select here is your preference, though json, yaml, and yaml-stream lend themselves to more advanced programmatic use cases as the format is machine-readable. Text and table options may improve human readability at the cost of programmatic utility.

It is best practice not to use the root account's credentials in your AWS CLI configuration as the loss of control of the root account's credentials could compromise the integrity of the entire account. It is better to configure your AWS CLI using the programmatic credentials of an IAM user whose access has been scoped appropriately for the work that will be performed from the CLI. In the previous chapter, we created a full administrator IAM user called redbeardidentity. We will be using that IAM user's programmatic credentials for our initial CLI configuration, and to explore the capabilities of the CLI throughout the chapter.

From the terminal, enter the following:

```
$ aws configure
```

You will be prompted to enter values for the four settings we listed earlier, and the current configuration values for each of those settings will be listed in the brackets next to those settings. Enter your desired value for each setting and hit *Enter*. Once all four are entered, you will be returned to Command Prompt:

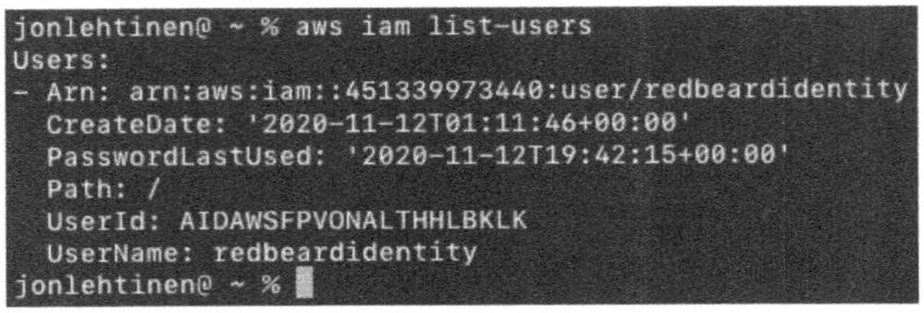

```
jonlehtinen@ ~ % aws iam list-users
Users:
- Arn: arn:aws:iam::451339973440:user/redbeardidentity
  CreateDate: '2020-11-12T01:11:46+00:00'
  PasswordLastUsed: '2020-11-12T19:42:15+00:00'
  Path: /
  UserId: AIDAWSFPVONALTHHLBKLK
  UserName: redbeardidentity
jonlehtinen@ ~ %
```

Figure 2.5 – Current and new values entered when performing basic AWS CLI configuration

If you need to modify any of these values, you can run aws configure again. It will now show the values you entered as the current settings. Hitting *Enter* on any value you do not want to change will leave that setting unchanged and entering a new value at the prompt will update the configuration. Let's say we would prefer to use us-east-2 as our default region; we just run the configuration again and update only that value:

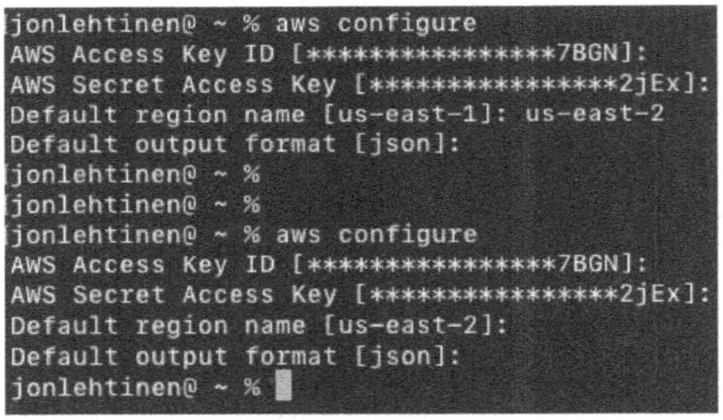

```
jonlehtinen@ ~ % aws configure
AWS Access Key ID [****************7BGN]:
AWS Secret Access Key [****************2jEx]:
Default region name [us-east-1]: us-east-2
Default output format [json]:
jonlehtinen@ ~ %
jonlehtinen@ ~ %
jonlehtinen@ ~ % aws configure
AWS Access Key ID [****************7BGN]:
AWS Secret Access Key [****************2jEx]:
Default region name [us-east-2]:
Default output format [json]:
jonlehtinen@ ~ %
```

Figure 2.6 – Changing our configuration to reflect us-east-2

The new values are committed and, assuming the authorization policy applied permits us to, we can now begin issuing commands to our AWS account under this IAM user context. These will be the default settings used with the CLI unless other values are specified.

> **Important Note**
>
> **Do not expose your Secret Key ID!** *Figure 2.5* commits a cardinal sin of IAM, AWS administration, and general information security; *never* reveal your secret key ID! The keys used in the preceding demonstration are dummy values for instructional purposes. In the event that your secret access key ever becomes exposed, disable it using the AWS IAM dashboard immediately and issue a replacement.

Testing out the CLI

Let's perform a basic test of the CLI to make sure our configuration works. Since the environment we are working in is brand new and given that we haven't yet gone through exactly *how* to use the AWS CLI yet, we are just going to run a command that identities the user object used to make the call into the AWS account. In Command Prompt, enter the following:

```
$ aws sts get-caller-identity
```

I immediately received the following response:

```
[jonlehtinen@TR-C02V71BSHTDD ~ % aws sts get-caller-identity
Account: '451339973440'
Arn: arn:aws:iam::451339973440:user/redbeardidentity
UserId: AIDAWSFPVONALTHHLBKLK
jonlehtinen@TR-C02V71BSHTDD ~ %
```

Figure 2.7 – Response from the aws sts get-caller-identity command

This test validates that our default AWS CLI configuration is valid. You may have noticed that we never defined an endpoint, nor an account, or any of the other things you typically do when making web service calls. As an identity-defined service, our access key ID and secret access key are sufficient identifiers for AWS to ensure that our requests are directed and evaluated against the appropriate AWS account and AWS IAM policies within that account.

Profiles

So, what happens if you need to use the AWS CLI with multiple AWS accounts? Or if you have different IAM users with different entitlements for different use cases? The AWS CLI manages your configuration settings by referencing a pair of files it manages in your operating system user's home directory, called `config` and `credentials`. You can differentiate between the configuration or credentials you want to use in your request from the AWS CLI by specifying something called a **profile**. You can define a profile for an AWS CLI configuration by adding the `--profile` operator with a label for the profile when running the `aws configure` command:

```
jonlehtinen@ ~ % aws configure --profile redbeardidentity
AWS Access Key ID [****************JXVU]:
AWS Secret Access Key [****************4OtZ]:
Default region name [us-east-1]:
Default output format [yaml]:
jonlehtinen@ ~ %
```

Figure 2.8 – Defining a redbeardidentity profile with specific credentials and region/output settings

We can see the profile name and profile-specific values for the `redbeardidentity` profile inside of the config and credentials files found in the hidden directory at `~/.aws/` on macOS and Linux, and `%USERPROFILE%\.aws\` on Windows 10. As the `redbeardidentity` IAM user has full administrator access to my AWS account, I am also going to configure a profile under a new IAM user I have created that is scoped only to administrate S3 objects and add that profile to my configuration, and another with similar access to EC2:

```
[jonlehtinen@ ~ % aws configure --profile rbi_s3
AWS Access Key ID [None]:
AWS Secret Access Key [None]:
Default region name [None]: us-east-1
Default output format [None]: yaml
jonlehtinen@ ~ %
```

Figure 2.9 – Defining a second profile

When I look at my config file, I can see the `redbeardidentity`, `rbi_s3`, and `rbi_ec2` profiles. The credentials file has corresponding entries for each of the profile names:

```
[jonlehtinen@ ~ % cat ./.aws/config
[default]
output = yaml
region = us-east-1

[profile personalaws]
output = json
region = us-east-1

[profile redbeardidentity]
region = us-east-1
output = yaml

[profile rbi_s3]
region = us-east-1
output = yaml

[profile rbi_ec2]
region = us-east-1
output = yaml
jonlehtinen@ ~ %
```

Figure 2.10 – Multiple profiles defined for use with the CLI

Speaking of the credentials file, as we have already emphasized the criticality of protecting the access key ID and secret access keys, it should go without saying that this file can be a high-risk target. When these AWS credentials are stored locally, you must take care to secure your workstation lest the integrity of your AWS account is put into jeopardy. This is also another reason why setting up restrictive access policies and IAM user objects with appropriately scope access is critical to AWS security, as limiting access will reduce the potential impact to your AWS account in the event that your credentials are leaked. Once again, for emphasis, do not use the AWS root account's programmatic credentials with the AWS CLI.

To get the most out of these profiles, you can either suffix each of your commands with the appropriate profile as you issue them or export an environmental variable within the terminal session that will change the default profile for all AWS CLI commands issued during that session. This is useful if you don't want to modify your true default AWS CLI configuration settings but want to do prolonged work under a specific IAM user.

For macOS and Linux, use the following command. This example uses the `rbi_ec2` profile we created earlier:

```
$ export AWS_PROFILE=rbi_ec2
```

Windows users can run the following:

```
C:\> setx AWS_PROFILE rbi_ec2
```

With that, we've completed the basic setup for accessing our AWS account using the CLI. We will be using the CLI for increasingly complex tasks throughout this book, so consider the configuration preferences you've already made, and don't be afraid to adjust your configuration to match your preferences and use case.

Using the AWS CLI

Now that we have our profiles set up, let's begin exploring how we can use the AWS CLI to do things inside of our AWS account. Unfortunately, the AWS CLI does not have a standard syntax of verbs similar to what you may be familiar with if you have ever worked with RESTful protocols; there is no command structure that is universal across every AWS service accessible from the CLI that can be recalled by applying a verb-like operator against a service. So, in a RESTful service, you may have a format such as the following:

Operator	Endpoint	Result
GET	/user-info/users	Get all users
POST	/user-info/users	Create new user

However, the AWS CLI has specific operations available on a per-service basis that will not align service to service. However, the syntax across all of the operations across the services does follow a basic syntax:

Base Call	Top-Level Command	Subcommand	Options or Parameters	More Options or Parameters (n+1)
aws	s3	ls		
aws	s3	cp	/Users/jonlehtinen/ Documents/ HeadshotQuarantine2020. png	s3://rbi-s3-bucket-1/

The emphasis here is on the *N+1 options or parameters* that can quickly pile up depending upon the specific commands you are issuing. The preceding two examples are relatively simple commands for the S3 service. The first lists all the buckets on the account that I am allowed to see under my current IAM user role or profile (and since I am using the redbeardidentity IAM user for these exercises, which is a member of the FullAdministrator IAM group, I will see them all). That command doesn't require any additional parameters past the ls subcommand since it is listing all of the buckets on the account. The second command is a bit different, in that it will copy a local file into an S3 bucket. The cp subcommand is followed by two additional parameters, the first indicating which file is to be copied, and the second specifying the bucket and path where that file is to be copied:

```
[jonlehtinen@ ~ % aws s3 ls
2020-11-22 13:14:39 rbi-s3-bucket-1
2020-11-22 13:17:57 redbeardidentity-bucket-1
[jonlehtinen@ ~ % aws s3 cp /Users/jonlehtinen/Documents/HeadshotQuarantine2020.png s3://rbi-s3-
bucket-1/
upload: Documents/HeadshotQuarantine2020.png to s3://rbi-s3-bucket-1/HeadshotQuarantine2020.png
jonlehtinen@ ~ % █
```

Figure 2.11 – Examples of the AWS CLI command syntax with differing amounts of parameters

Figure 2.11 shows us the results of those commands in action.

Discovering command syntax

Perhaps you are wondering how you can familiarize yourself with each of the commands, subcommands, and countless options and parameters required to enter a command as a typed string in Command Prompt. Unlike the AWS Management Console, there doesn't appear to be any tooltips or prompts to guide your input, so the odds of making a mistake and needing to go through the tedium of re-typing a long command seems high. Fortunately, there are a few options to help you discover the available commands from the CLI itself so you will not need to learn through trial and error, in addition to the detailed, per-service documentation available at https://docs.aws.amazon.com/cli/latest/reference/. Let's create a new user in the Redbeard Identity AWS account using the CLI and see how each of these options provides details on the commands and parameters required to do it successfully.

aws help

The most basic tool to aid you is the help command. By typing help after nearly anything, you get a contextual help article with details on the command or subcommand that precedes it. Since we are going to create a new IAM user, we can assume the first part of our command will be as follows:

```
$ aws iam
```

But after that, I am at a loss. Let's see how the help command reveals the additional information to us:

```
$ aws iam help
```

With that, a document listing the available commands opens up:

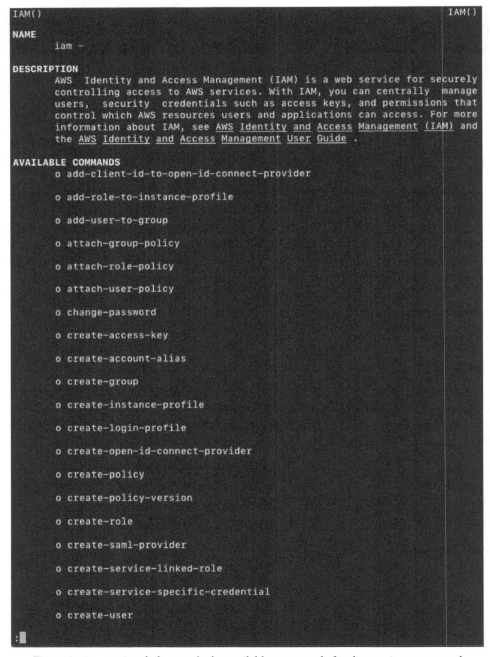

```
IAM()                                                                    IAM()

NAME
       iam -

DESCRIPTION
       AWS  Identity and Access Management (IAM) is a web service for securely
       controlling access to AWS services. With IAM, you can centrally  manage
       users,   security  credentials such as access keys, and permissions that
       control which AWS resources users and applications can access. For more
       information about IAM, see AWS Identity and Access Management (IAM) and
       the AWS Identity and Access Management User Guide .

AVAILABLE COMMANDS
       o add-client-id-to-open-id-connect-provider

       o add-role-to-instance-profile

       o add-user-to-group

       o attach-group-policy

       o attach-role-policy

       o attach-user-policy

       o change-password

       o create-access-key

       o create-account-alias

       o create-group

       o create-instance-profile

       o create-login-profile

       o create-open-id-connect-provider

       o create-policy

       o create-policy-version

       o create-role

       o create-saml-provider

       o create-service-linked-role

       o create-service-specific-credential

       o create-user

:
```

Figure 2.12 – aws iam help reveals the available commands for the aws iam command `create-user` seems like the ticket. Repeat this process again and again and again with each new subcommand and parameter until we have a properly formatted statement that will create a new IAM user and we are good to go.

Auto-prompt

Of course, capturing the command syntax up until we don't know what is next, entering help, reading the document, adding the next piece of the command, then repeating is awfully time-consuming. Fortunately, as of the AWS CLI version 2, there is a setting that will preemptively reveal the commands and variables as you type them called auto-prompt. There are several ways to enable auto-prompt. The first is by setting a parameter in the command line:

```
$ aws --cli-auto-prompt
```

This will enable the prompt for the duration of the command we are entering.

The second is by setting and exporting an environmental variable in our command-line terminal that enables auto-prompt for the duration of that session. This is done by entering the following commands:

```
$ AWS_CLI_AUTO_PROMPT=on
$ export AWS_CLI_AUTO_PROMPT
```

The third is by updating the config file where our profile settings are stored with an additional parameter to leave auto-prompt enabled under a specific profile. The config file is found in the hidden AWS directory within the user's home directory at ~/.aws/ on macOS and Linux, and %USERPROFILE%\.aws\ on Windows 10. We open the config file and add the following configuration key and value to the profile under which we want auto-prompt enabled:

```
cli_auto_prompt = on
```

When completed, our profile will look like this:

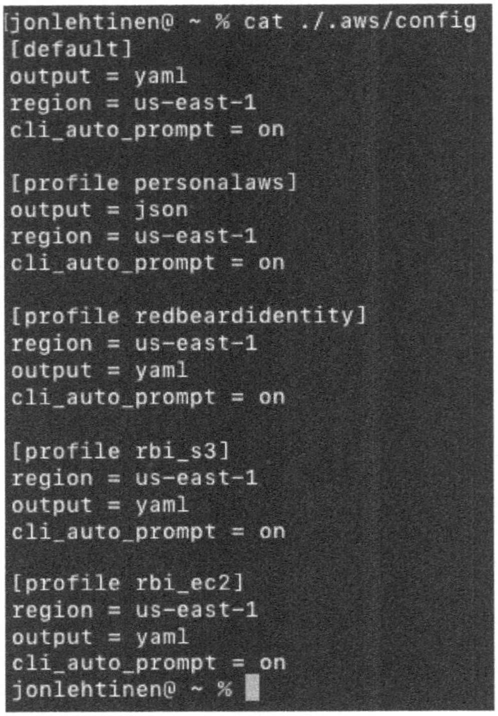

```
jonlehtinen@ ~ % cat ./.aws/config
[default]
output = yaml
region = us-east-1
cli_auto_prompt = on

[profile personalaws]
output = json
region = us-east-1
cli_auto_prompt = on

[profile redbeardidentity]
region = us-east-1
output = yaml
cli_auto_prompt = on

[profile rbi_s3]
region = us-east-1
output = yaml
cli_auto_prompt = on

[profile rbi_ec2]
region = us-east-1
output = yaml
cli_auto_prompt = on
jonlehtinen@ ~ %
```

Figure 2.13 – cli_auto_prompt enabled on every profile in the AWS CLI config file

Since there are various ways to enable auto-prompt, there is the chance that we could set ourselves up with a conflicting configuration between the CLI config parameter, the environmental variable, and the command parameter. While this will not cause an error, it is important to know that whichever method we choose for our use case, there is a hierarchy of precedence that will determine which configuration is ultimately respected. The command-line parameter takes utmost precedence – we can toggle auto-prompt on or off using that parameter regardless of any environment or config parameters. This is useful for spot-checking commands or disabling auto-prompt for the commands that you are well versed in using. An environment variable will override the config variable. The way we choose to enable auto-prompt is ultimately a matter of personal preference, but it is important to understand whether a workstation policy or an errant line in a script disables the functionality that we were expecting.

Now that we have set up auto-prompt, let's take a look at that new user creation command we were puzzling over earlier using auto-prompt as the guide. Since we have auto-prompt enabled as a parameter on the config file, we see our command line change once we hit *Enter* after entering any portion of an AWS CLI command:

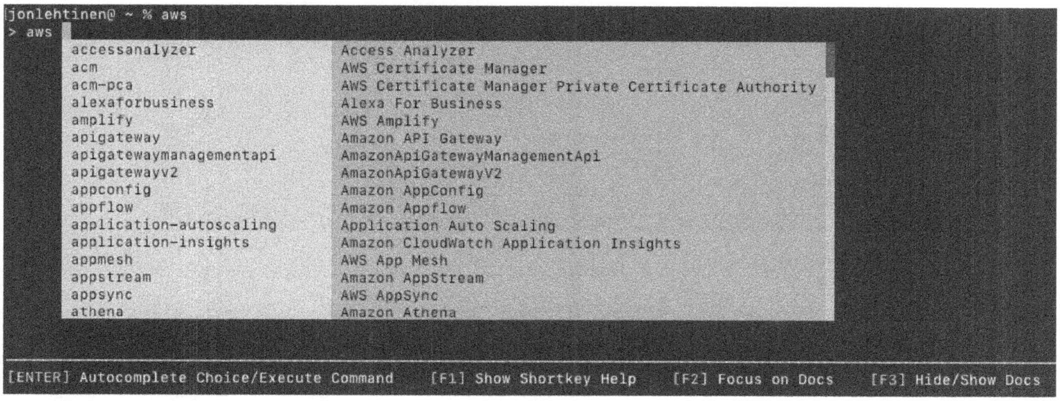

Figure 2.14 – The auto-prompt menu shows available commands and parameters

As we are trying to create a new IAM user object, let's enter `iam` and then `create-user` as the next commands and see what additional parameters appear:

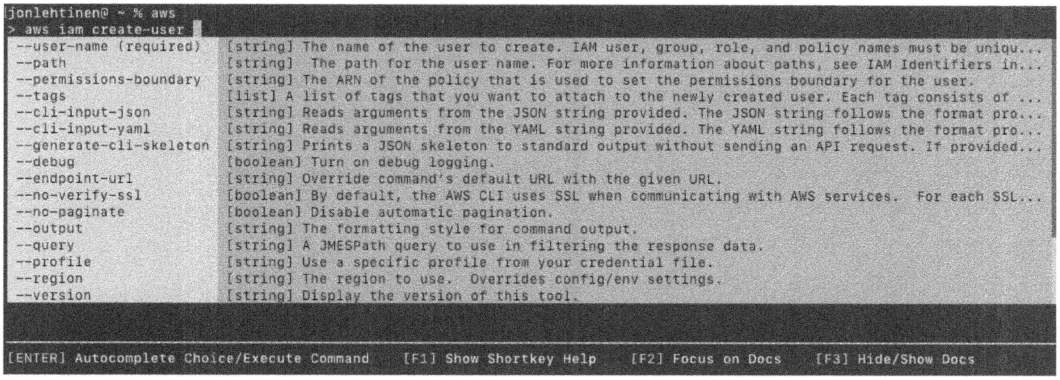

Figure 2.15 – Contextual parameters, including required parameters, indicated by auto-prompt

And with that, auto-prompt begins showing its value as we see a contextual list of parameters and explanations of what those parameters do. Most significantly, it reveals the required parameters, or the minimally required parameters, for the command to execute. Let's name this user something unique and move on.

We don't see any other parameters labeled as `required` on the list, so let's execute the command:

```
[jonlehtinen@ ~ % aws
[> aws iam create-user --user-name rbi_cliuser
User:
  Arn: arn:aws:iam::451339973440:user/rbi_cliuser
  CreateDate: '2020-11-25T00:07:11+00:00'
  Path: /
  UserId: AIDAWSFPVONAIQGOC47E2
  UserName: rbi_cliuser
jonlehtinen@ ~ % █
```

Figure 2.16 – The output from a successfully created IAM user object using the AWS CLI

And there is our user. We can even verify that the IAM user object was created by looking at the IAM dashboard within the AWS Management Console:

	User name ▾	Groups	Access key age	Password age	Last activity	MFA
	rbi_cliuser	None	None	None	None	Not enabled
	RBI_EC2	None	✓ 2 days	2 days	None	Not enabled
	RBI_S3	None	✓ 2 days	2 days	2 days	Not enabled
	redbeardidentity	FullAdministrator	✓ 2 days	12 days	Today	Not enabled

Showing 4 results — Find users by username or access key

Figure 2.17 – CLI-created user appears in the Management Console

> **Tip**
> `aws_cli_auto_prompt=on` versus `aws_cli_auto_prompt=on-partial`: Auto-prompt is useful, but as you grow in experience, it may become tedious to wade through the menus on every command you enter in the CLI. By setting the `aws_cli_auto_prompt` value to `on-partial`, you can set auto-prompt to only trigger on malformed commands. This balances the usefulness of its guidance with the intrusiveness of its prompts for more experienced users.

Looking at *Figure 2.17* there are a few differences with that user compared to the previous three that we created using the Management Console. We were never prompted to set an initial password for Management Console access, nor enable programmatic access for the AWS CLI, nor did we associate this user with an IAM role, group, or policy that governs its access. While the auto-prompt tool walked us through the minimum requirements to create an IAM user, the user we got at the end of that process isn't very useful. Was this a function of missing non-required parameters, or something else? We can examine this issue more easily by looking at the IAM user object prior to issuing the command to create it by using CLI-generated object templates to examine, edit, and then create an IAM user object when using the CLI.

But before we do that, let's delete this user using the CLI. As we don't know exactly how to do that, let's take a guess at the command and see how auto-prompt clues us in:

```
jonlehtinen@ ~ % aws iam
> aws iam delete-user --user-name rbi_cliuser
                                   rbi_cliuser
                                   RBI_EC2
                                   RBI_S3
                                   redbeardidentity
```

Figure 2.18 – Deleting the IAM user object under guidance from auto-prompt

Once we execute, the user is gone.

Generating an AWS CLI skeleton and loading a template

So far, we have created a new IAM user object using the CLI, but that user was not in a very usable state – we couldn't use it to sign in to the Management Console as it had no assigned password, nor could we use it to access our AWS account using the CLI as it had no programmatic access. While auto-prompt helped us figure out what was required to successfully execute the `aws iam create-user` command, that the user was unusable suggests there is more configuration required to perform the full suite of commands to create a fully usable IAM user via the AWS CLI than just that single one. Fortunately, we can examine existing objects to see how they may differ using the CLI and, using the `--generate-cli-skeleton` parameter, can even create a template that we can edit using a text editor or IDE and import via the CLI to simplify the creation of these objects:

1. First, let's look at an existing IAM user object that we know is good, such as our current non-root administrator `redbeardidentity`. From the AWS CLI, we can list the IAM users in our environment with the following command:

    ```
    $ aws iam list-users
    ```

This produces a list of all the IAM user objects in the account. As this is a new tenant, there are not very many users here for us to sort through, so we can find all the relevant details on the `redbeardidentity` account relatively easily:

```
> aws iam list-users
Users:
- Arn: arn:aws:iam::451339973440:user/rbi_cliuser
  CreateDate: '2020-11-25T00:07:11+00:00'
  Path: /
  UserId: AIDAWSFPVONAIQGOC47E2
  UserName: rbi_cliuser
- Arn: arn:aws:iam::451339973440:user/RBI_EC2
  CreateDate: '2020-11-22T18:23:15+00:00'
  Path: /
  UserId: AIDAWSFPVONACROBSEOSS
  UserName: RBI_EC2
- Arn: arn:aws:iam::451339973440:user/RBI_S3
  CreateDate: '2020-11-22T17:59:25+00:00'
  PasswordLastUsed: '2020-11-22T18:13:38+00:00'
  Path: /
  UserId: AIDAWSFPVONACP5HVGP3C
  UserName: RBI_S3
- Arn: arn:aws:iam::451339973440:user/redbeardidentity
  CreateDate: '2020-11-12T01:11:46+00:00'
  PasswordLastUsed: '2020-11-25T00:11:12+00:00'
  Path: /
  UserId: AIDAWSFPVONALTHHLBKLK
  UserName: redbeardidentity
```

Figure 2.19 – All IAM user objects in the AWS account

2. However, if we were working in an environment with thousands of users, this would be terribly inefficient. Instead, we can pull all the details on a specific user by using the `get-user` command:

```
$ aws iam get-user --user-name redbeardidentity
```

You'll notice that using that command provides more specifics on the user compared to the earlier list command:

```
jonlehtinen@ ~ % aws iam get-user
> aws iam get-user --user-name redbeardidentity
User:
  Arn: arn:aws:iam::451339973440:user/redbeardidentity
  CreateDate: '2020-11-12T01:11:46+00:00'
  PasswordLastUsed: '2020-11-25T00:11:12+00:00'
  Path: /
  Tags:
  - Key: costcenter
    Value: '90001'
  - Key: jobcode
    Value: '1701'
  UserId: AIDAWSFPVONALTHHLBKLK
  UserName: redbeardidentity
jonlehtinen@ ~ %
```

Figure 2.20 – Increased object detail using the get-user command versus the list-users command

3. Let's use the --generate-cli-skeleton parameter to create a full template for the object that we can manipulate using an IDE or text editor and import using the create-user command to ensure that we create the IAM user object with every attribute available to it. To do this, you attach the --generate-cli-skeleton parameter to the object for which you are running a create command, which in this case is an IAM user:

```
$ aws iam create-user --generate-cli-skeleton
```

This produces a JSON template for the object. Alternatively, you can specify YAML by further adding the following:

```
$ aws iam create-user --generate-cli-skeleton yaml-input
```

We see the skeleton templates in both JSON and YAML format in *Figure 2.21* as follows:

```
jonlehtinen@ ~ % aws iam create-user --generate-cli-sk
> aws iam create-user --generate-cli-skeleton
{
    "Path": "",
    "UserName": "",
    "PermissionsBoundary": "",
    "Tags": [
        {
            "Key": "",
            "Value": ""
        }
    ]
}
jonlehtinen@ ~ % aws iam create-user --generate-cli-sk
> aws iam create-user --generate-cli-skeleton yaml-input
Path: ''  #  The path for the user name.
UserName: '' # [REQUIRED] The name of the user to create.
PermissionsBoundary: '' # The ARN of the policy that is used to set the permissions boundary for the user.
Tags: # A list of tags that you want to attach to the newly created user.
- Key: ''  # [REQUIRED] The key name that can be used to look up or retrieve the associated value.
  Value: '' # [REQUIRED] The value associated with this tag.
jonlehtinen@ ~ %
```

Figure 2.21 – JSON and YAML templates for the IAM user object

4. With those templates, we can copy whichever format we are more comfortable with using into a text editor or IDE and create a file that we will direct the `create-user` command to look to for all the parameters it requires for the creation of the object. Since I prefer YAML, I will copy the YAML template into my preferred IDE, Microsoft Visual Studio Code, and start inserting values for a new user, which I will name `RBI_Admin`:

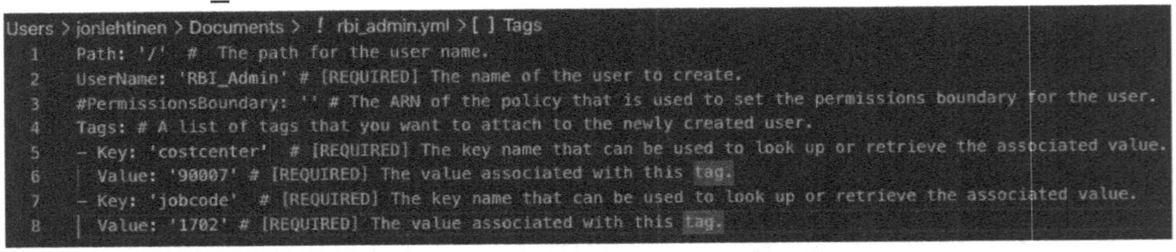

```
Users > jonlehtinen > Documents > ! rbi_admin.yml > [ ] Tags
  1   Path: '/'  #  The path for the user name.
  2   UserName: 'RBI_Admin' # [REQUIRED] The name of the user to create.
  3   #PermissionsBoundary: ''  # The ARN of the policy that is used to set the permissions boundary for the user.
  4   Tags: # A list of tags that you want to attach to the newly created user.
  5   - Key: 'costcenter'  # [REQUIRED] The key name that can be used to look up or retrieve the associated value.
  6     Value: '90007' # [REQUIRED] The value associated with this tag.
  7   - Key: 'jobcode'  # [REQUIRED] The key name that can be used to look up or retrieve the associated value.
  8     Value: '1702' # [REQUIRED] The value associated with this tag.
```

Figure 2.22 – Creating rbi_admin.yml using an IDE

5. For the values that we are uncertain about (such as `Path`), we can look into the details we got when we ran the `get-user` command on the known-good IAM user object, `redbeardidentity`. As we have not specified a permissions boundary in our AWS account, we will need to remove or comment out that optional parameter.

6. Finally, we can copy/paste additional key/value pairs for all the tags we want to attach to this user; we are not restricted to just what came out of the `--generate-cli-skeleton` command. Once we save this file, we can return to the AWS CLI and run the `create-user` command again, only this time telling it to refer to the file for the parameters, and the location of the file:

```
$ aws iam create-user --cli-input-yaml file://rbi_admin.
yml
```

The output tells us the operation was a success:

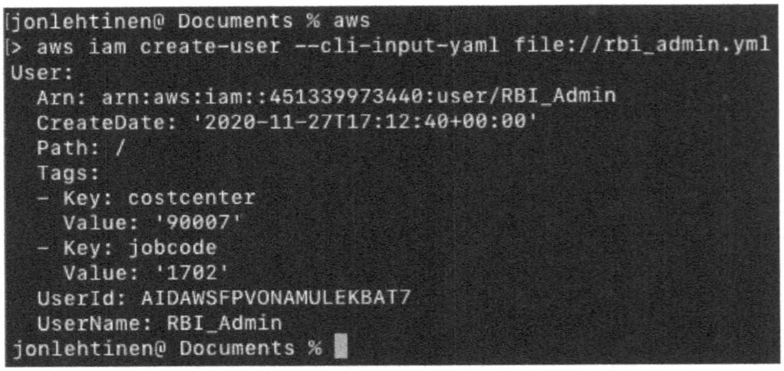

```
[jonlehtinen@ Documents % aws
[> aws iam create-user --cli-input-yaml file://rbi_admin.yml
User:
  Arn: arn:aws:iam::451339973440:user/RBI_Admin
  CreateDate: '2020-11-27T17:12:40+00:00'
  Path: /
  Tags:
  - Key: costcenter
    Value: '90007'
  - Key: jobcode
    Value: '1702'
  UserId: AIDAWSFPVONAMULEKBAT7
  UserName: RBI_Admin
jonlehtinen@ Documents %
```

Figure 2.23 – New IAM user object created from a template file

That said, there is nothing else there that gives us insight into what additional required parameters may have been missed when we initially created our user using the `create-user` command that would explain why our new IAM user object has a password for Management Console access, nor programmatic access, nor authorization policy. Rather than draw this out unnecessarily, I will spoil how this will end: the IAM user object, its credentials, and the authorization policy that governs its access are all distinct objects within AWS. What links them all together is a binding relationship through attributes that we can manipulate using the AWS CLI.

Though we may not have fully figured out how to create a functional administrative user object quite yet, don't let that diminish the value of CLI skeleton templates for your future use cases. When using the templates with the CLI, there are a few important things to remember:

- Required attributes are marked as `REQUIRED` and must be populated.

- Comment out or delete any non-required attribute that you are not using in the template or the process will fail.

- The default output when using the `--generate-cli-skeleton` command is JSON. To specify a YAML template, add the `yaml-input` parameter at the end of the command.

- When importing a template, you must align the `import` command to the format of the file using either `--cli-input-json` or `--cli-input-yaml`, followed by `file://<file path>`.

- The `file://` portion of the command starts relative to your current working directory in the command line, so if you are running your AWS CLI in a command line and your current working directory is `/Users/yourname/Documents/`, you must either move the template to that location or adjust the path after `file://` to be correct relative to your current working directory.

Now that we have learned how to navigate the AWS CLI to the degree that we can discover commands, create and import templates, and create objects, let's add the missing pieces we need to make this IAM user a full administrator with programmatic and Management Console access.

Putting it all together – creating a functional IAM user with the AWS CLI

Now that we have created `RBI_Admin` as a new IAM user object, let's use the AWS CLI to assign it credentials for both the Management Console and the AWS CLI, and give full administrator access to our AWS account. As I mentioned earlier, identity objects used for authorization decisions (groups, permission boundaries, user policies, and so on), those used for authentication (credentials), and those used for identification (user objects) are all fully independent IAM objects within AWS IAM. What makes them work as we expect them to work is their relationships with each other. This relationship is most readily seen through attributes on one of those objects referencing another. We will be using the AWS CLI to establish those relationships. Before we begin, let's take a moment to map out what it is we want to achieve, as this may help us understand how and why certain AWS CLI commands are invoked:

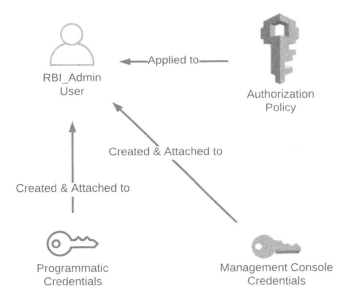

Figure 2.24 – A map of the objects and actions needed to make the IAM user a full administrator

Looking at *Figure 2.24*, we seem to have three overarching objectives:

- Apply an authorization policy that grants the `RBI_Admin` user object full administrator access.

- Create and attach Management Console credentials to the `RBI_Admin` user object.

- Create and attach programmatic credentials for the AWS CLI access to the `RBI_Admin` user object.

Now that we have an idea of what we are aiming to do, let's take a look at the commands at our disposal to execute those tasks. Rather than hunt and peck through the AWS CLI for the relevant commands, we can instead peruse all of the available commands available to use at `https://docs.aws.amazon.com/cli/latest/reference/iam/index.html` to see which ones look to get us close to the mark.

Attaching an administrator policy

First, let's make `RBI_Admin` an administrator. There are several ways this can be accomplished using AWS IAM policies, but groups are easy to conceptualize and administrate, so let's accomplish this task by adding `RBI_Admin` to a group that will grant it that access:

1. First, we list the groups available in our AWS account to see what may already be present:

    ```
    $ aws iam list-groups
    ```

    Fortunately, we had previously created a group for use with non-root administrators on the first account we created (and are presently using it to issue these commands from the CLI):

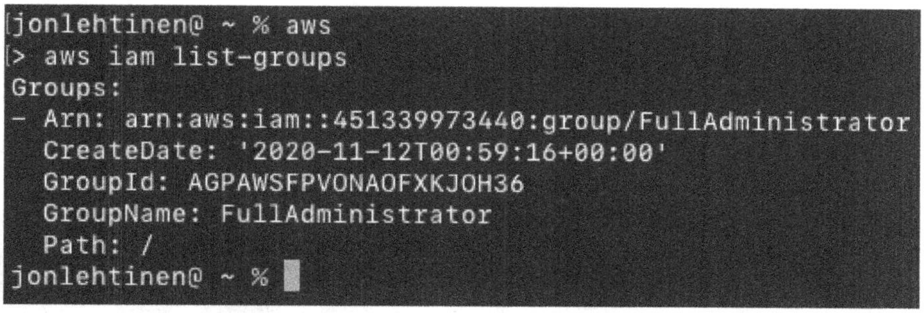

    ```
    [jonlehtinen@ ~ % aws
    > aws iam list-groups
    Groups:
    - Arn: arn:aws:iam::451339973440:group/FullAdministrator
      CreateDate: '2020-11-12T00:59:16+00:00'
      GroupId: AGPAWSFPVONAOFXKJOH36
      GroupName: FullAdministrator
      Path: /
    jonlehtinen@ ~ %
    ```

    Figure 2.25 – Previously created FullAdministrator group object

2. In fact, since we know that `redbeardidentityaccount` is our current non-root full administrator account, we will get the same output if we run a different command with some different parameters:

    ```
    $ aws iam list-groups-for-user --user-name
    redbeardidentity
    ```

Sure enough, we get this output:

```
[jonlehtinen@ ~ % aws
[> aws iam list-groups
Groups:
- Arn: arn:aws:iam::451339973440:group/FullAdministrator
  CreateDate: '2020-11-12T00:59:16+00:00'
  GroupId: AGPAWSFPVONAOFXKJOH36
  GroupName: FullAdministrator
  Path: /
[jonlehtinen@ ~ % aws iam list
[> aws iam list-groups-for-user --user-name redbeardidentity
Groups:
- Arn: arn:aws:iam::451339973440:group/FullAdministrator
  CreateDate: '2020-11-12T00:59:16+00:00'
  GroupId: AGPAWSFPVONAOFXKJOH36
  GroupName: FullAdministrator
  Path: /
jonlehtinen@ ~ %
```

Figure 2.26 – Demonstrating the group relationship with the current AWS CLI user

3. Of course, how does this group grant its members full administrative privileges to the AWS account? We can run some commands to explore the policy objects attached to the group to find out:

```
$ aws iam list-attached-group-policies --group-name
FullAdministrator
```

As we see in *Figure 2.27*, the AWS-managed `AdministratorAccess` policy is attached to this group:

```
[jonlehtinen@ ~ % aws
[> aws iam list-attached-group-policies --group-name FullAdministrator
AttachedPolicies:
- PolicyArn: arn:aws:iam::aws:policy/AdministratorAccess
  PolicyName: AdministratorAccess
jonlehtinen@ ~ %
```

Figure 2.27 – The attached policy objects for the FullAdministrator group

What is a managed policy, you may ask? A managed policy object is an AWS IAM authorization policy that is created and maintained by AWS and included natively with the AWS account. This is an example of how it is the relationship between these two distinct objects, and not necessarily the group object itself, that ultimately provides the identity outcome here.

Having confirmed that there is a group that will suit our needs, let's add the RBI_Admin user object to that group:

```
$ aws iam add-user-to-group --group-name
FullAdministrator --user-name RBI_Admin
```

4. No error message means the command was executed, but let's validate anyway by running the following:

```
$ aws iam list-groups-for-user --user-name RBI_Admin
```

And sure enough, we now see FullAdministrator listed among the groups that RBI_Admin belongs to:

```
[jonlehtinen@ ~ % aws
[> aws iam add-user-to-group --group-name FullAdministrator --user-name RBI_Admin
[jonlehtinen@ ~ % aws iam
[> aws iam list-groups-for-user --user-name RBI_Admin
Groups:
- Arn: arn:aws:iam::451339973440:group/FullAdministrator
  CreateDate: '2020-11-12T00:59:16+00:00'
  GroupId: AGPAWSFPVONAOFXKJOH36
  GroupName: FullAdministrator
  Path: /
jonlehtinen@ ~ %
```

Figure 2.28 – RBI_Admin added to and confirmed a member of the FullAdministrator group

Of course, being a member of the FullAdministrator group is of little use if the RBI_Admin user cannot sign in to the Management Console.

Creating and attaching a password

The next thing we will tackle is the creation and attachment of the password to the RBI_Admin user object. While you would think looking up commands involving password would lead you to the right command, the change-password command is only good for altering the password of the current user of the AWS CLI. As we are executing these commands under the redbeardidentity IAM user object profile for the RBI_Admin user object, that would not help us get RBI_Admin access to log on to the AWS Management Console.

The command we are looking for is actually `create-login-profile`, and just like within the AWS IAM dashboard, it provides options for either setting a temporary password that must be changed at first logon or just defining the password entirely. Let's make things simple and not force a change of password on the first login to the console:

```
$ aws iam create-login profile --user-name RBI_Admin --password
XXXXXXXXXX - --no-password-reset-required
```

The output confirms the creation of Management Console access:

```
[jonlehtinen@ ~ % aws
> aws iam create-login-profile --user-name RBI_Admin --password          --no-password-reset-required
LoginProfile:
    CreateDate: '2020-11-27T18:57:32+00:00'
    PasswordResetRequired: false
    UserName: RBI_Admin
jonlehtinen@ ~ %
```

Figure 2.29 – Creating RBI_Admin's Management Console password

We can verify that it worked by signing in to the Management Console with RBI_Admin and that password:

Figure 2.30 – Signing in to the Management Console with RBI_Admin

Now that we have access to the Management Console under the RBI_Admin user object, we could simply enable programmatic access from the Console. But for completeness' sake, let's also create the object's programmatic credentials from the CLI.

Creating and attaching the programmatic credentials

Finally, let's grant the RBI_Admin user object access to the AWS CLI creating its access key ID and secret access key. By combing the AWS CLI documentation at https://docs.aws.amazon.com/cli/latest/reference/iam/index.html, we can quickly find the relevant commands for listing, creating, and deleting access keys, which are (unsurprisingly) `list-access-keys`, `create-access-key`, and `delete-access-key`, respectively. Let's get right to it and create an access key for RBI_Admin by running the following:

```
$ aws iam create-access-key RBI_Admin
```

We immediately get the access key ID and secret access key values as the output:

```
Last login: Fri Nov 27 13:38:31 on ttys000

jonlehtinen@ ~ %
[jonlehtinen@ ~ % aws iam
> aws iam create-access-key --user-name RBI_Admin
AccessKey:
  AccessKeyId: AKIAWSFPVONACOFCK7WA
  CreateDate: '2020-11-28T17:21:48+00:00'
  SecretAccessKey:
  Status: Active
  UserName: RBI_Admin
jonlehtinen@ ~ %
```

Figure 2.31 – Successful creation of programmatic credentials using the CLI

> **Tip**
>
> **Capturing the secret access key upon creation**: Remember, the secret
> access key is only visible at creation time, unless you export the user object
> information, along with its credentials, as a CSV file during creation using the
> Management Console. If you lose the secret access key after creation, there is
> no recovery process. A new one will need to be issued.

We now have what we need to use the RBI_Admin user with our AWS account from the
CLI.

Using the new profile

Now that we have everything required to make the RBI_Admin user object as equally
functional as our other administrative accounts, we can create a new profile in our AWS
CLI configuration to use it. We will start by running the aws configure command
to populate the access key ID, secret access key, default region, and default output for the
profile under a new profile name of rbi_admin:

```
$ aws configure --profile rbi_admin
AWS Access Key ID [None]: AKIAWSFPVONACOFCK7WA
AWS Secret Access Key [None]: XXXXXXXXXXXXXXXXXXXXXXXXXXXXXXXX
XXXX
Default region name [None]: us-east-1
Default output format [None]: yaml
```

Then, we can test it out by issuing any number of AWS CLI commands to make sure it works. For example, let's list all of our access keys:

```
$ aws iam list-access-keys --profile rbi_admin
```

We get a list of all the access keys for the `rbi_admin` user object:

```
jonlehtinen@ ~ % aws iam list-access-keys --profile rbi_admin
AccessKeyMetadata:
- AccessKeyId: AKIAWSFPVONACOFCK7WA
  CreateDate: '2020-11-28T17:21:48+00:00'
  Status: Active
  UserName: RBI_Admin
jonlehtinen@ ~ %
```

Figure 2.32 – All access keys under the rbi_admin profile

As I am curious about whether we can see the keys for other users under our `rbi_admin` current profile, let's add an additional parameter to specify that we want to see all of the `redbeardbeardidentity` user object's access keys:

```
$ aws iam list-access-keys --user-name redbeardidentity
--profile rbi_admin
```

It looks like our full administrator permissions are allowing us to see keys on other user objects as well:

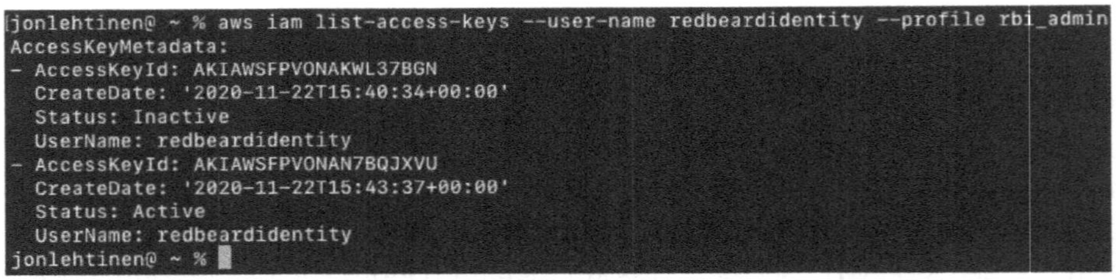

```
jonlehtinen@ ~ % aws iam list-access-keys --user-name redbeardidentity --profile rbi_admin
AccessKeyMetadata:
- AccessKeyId: AKIAWSFPVONAKWL37BGN
  CreateDate: '2020-11-22T15:40:34+00:00'
  Status: Inactive
  UserName: redbeardidentity
- AccessKeyId: AKIAWSFPVONAN7BQJXVU
  CreateDate: '2020-11-22T15:43:37+00:00'
  Status: Active
  UserName: redbeardidentity
jonlehtinen@ ~ %
```

Figure 2.33 – Viewing the access keys of other IAM user objects under the rbi_admin profile

With that, we have verified that our `rbi_admin` profile can exercise its administrative privileges the way we want it to from the CLI.

Scripting

Now, whereas entering all of these commands is arguably faster than doing it through the Management Console, it is tedious to enter several commands just to create a single IAM user object with full administrator privileges. Fortunately, this is where the power of the AWS CLI's programmatic capabilities is nearly limitless; rather than manually entering individual commands, how about we instead write a script that will string all the commands together for us?

On scripting best practices

The following example is designed to show how we can use scripting languages to accomplish the same series of tasks that we have already completed using ad hoc commands. You can use any scripting language you are comfortable with to accomplish your objectives. The important thing to realize is that by chaining AWS CLI commands in a script, you will be able to create complex arrangements of AWS objects and resources very quickly, using resources that you can comfortably edit and collaborate on using whatever toolsets and languages are optimal for your use case.

Let's take a moment to think about all of the individual commands we will need to run in order to create a new IAM user object with access to both the AWS CLI and Management Console, and that is also a member of the existing FullAdministrator group in our AWS account. By recalling what it took to accomplish this with the RBI_Admin user object, we can break the process down into a few steps:

1. Create the IAM user object.
2. Add that IAM user to the existing FullAdministrator group.
3. Create a login profile for the new IAM user object.
4. Create the access keys for the new IAM user object.

Bash is an approachable scripting language with plenty of searchable resources, so let's make a Bash script that will handle all of those steps for us with minimal interaction. Using a text editor or IDE, open a new file and save it as createiamuser.sh. To get started, let's capture the corresponding AWS CLI commands that will address those four things that we know we will need this script to do, and populate values for the username and initial password:

```
#!/bin/sh
aws iam create-user --user-name ScriptTestUser
aws iam add-user-to-group --group-name FullAdministrator
--user-name ScriptTestUser
aws iam create-login-profile --user-name ScriptTestUser
--password OurF1rstP@ssWord! --password-reset-required
aws iam create-access-key --user-name ScriptTestUser
```

Strictly speaking, this script would do the job. However, some critical components, such as the secret access key, would appear once in the shell window as the script runs. Additionally, this script would require editing for each new user object that we would need to create. Let's punch it up a bit with some quality-of-life improvements that will make it more useful:

```
#!/bin/sh
read -p ''Enter the username for the new IAM User Object: ''
username && touch ./$username
read -p ''Enter the initial password for AWS Management Console
Access: '' initialpassword && echo ''Your temporary AWS
Management Console password is $initialpassword'' > ./$username
aws iam create-user --user-name $username >> ./$username
aws iam add-user-to-group --group-name FullAdministrator
--user-name $username >> ./$username
aws iam create-login-profile --user-name $username --password
$initialpassword --password-reset-required >> ./$username
aws iam create-access-key --user-name $username >> ./$username
```

With these updates, we've added two user prompts at the beginning. The first will ask for the IAM user object's name. The script then takes the value that is entered and uses that value as the $username variable everywhere else in the script where the username is required. This makes the script more maintainable. The second prompt asks for a temporary password, which will be used for the first-time login to the Management Console.

The next major change is that the output of the script is now captured in a file named after the IAM user object that is being created. This makes it easy to hold onto the artifacts needed to set up a profile using this IAM user object in the AWS CLI or to deliver the results of the script to the designated principal who will be using that IAM object so that they can change their credentials and start their configuration.

Let's run the script and see the output. Please note that you may need to make the script executable before it can run:

```
$ chmod +x ./createiamuser.sh
$ ./createiamuser.sh
```

We are prompted for the IAM user object's name and the temporary password, and the script completes rather quietly:

```
[jonlehtinen@ Documents % ./createiamuser.sh
Enter the username for the new IAM User Object: ScriptTestUser
Enter the initial password for AWS Management Console Access: 0urF1rstP@ssWord!
jonlehtinen@ Documents %
```

Figure 2.34 – Command-line output of the createiamuser.sh script

However, there is now a file in the same working directory where we ran our script named for our new IAM user object. When we open it up, we see everything we need to use this IAM user just like RBI_Admin before it:

```
jonlehtinen@ Documents % cat ./ScriptTestUser
Your temporary AWS Management Console password is 0urF1rstP@ssWord!
User:
  Arn: arn:aws:iam::451339973440:user/ScriptTestUser
  CreateDate: '2020-11-28T19:27:03+00:00'
  Path: /
  UserId: AIDAWSFPVONACAUWRESDQ
  UserName: ScriptTestUser
LoginProfile:
  CreateDate: '2020-11-28T19:27:05+00:00'
  PasswordResetRequired: true
  UserName: ScriptTestUser
AccessKey:
  AccessKeyId: AKIAWSFPVONAPOCK5YWX
  CreateDate: '2020-11-28T19:27:06+00:00'
  SecretAccessKey:
  Status: Active
  UserName: ScriptTestUser
jonlehtinen@ Documents %
```

Figure 2.35 – The contents of the ScriptTestUser file generated by the createiamuser.sh script

This script is rudimentary and could be further refined. For example, it could take a feed from a separate list of users, generate a random string for the initial Management Console password, and create hundreds of users at once. However, advanced Bash scripting isn't the purpose of this chapter; the important thing to remember is that you can use nearly any scripting language in conjunction with the AWS CLI to manage complex arrangements of AWS resources very quickly.

Summary

Now that you have gone through this chapter, you have everything you need to begin using the AWS CLI effectively. The AWS CLI is a powerful administrative tool for managing your AWS account. Whereas this was definitely *not* a comprehensive review of every command, feature, and function available to us in the CLI, the basic concepts and examples you have learned will give you the tools you need to learn about and solve nearly any challenge or use case you may require the use of the AWS CLI for.

We will put that knowledge to good use in the next chapter, and every remaining chapter, of this book. Next, we will take a deep dive into IAM user accounts. This will include the types of accounts and the different administrative requirements that each has. We will also look more closely at passwords, password policy, programmatic credentials, and multifactor credentials. Moreover, we will learn how to manage each and every one of those things using the AWS CLI.

Questions

1. Is AWS CLI version 2 backward-compatible with AWS CLI v1?

2. What is the command to configure the default profile?

3. How would you configure an alternate profile using different programmatic credentials without impacting your default AWS CLI profile?

4. Name three ways to discover AWS CLI command syntax from the command line.

5. True or false: When importing a template that was generated using the `--generate-cli-skeleton` command, you cannot make any modifications to the template that was generated by that command.

6. What are the three ways to enable auto-prompt on a profile?

7. True or false: AWS CLI commands must be entered one at a time from the command line (this one is a trick question, so think carefully!).

8. What are the advantages of combining the AWS CLI with a scripting language?

Further reading

*   AWS CLI command reference: `https://docs.aws.amazon.com/cli/latest/index.html`

*   AWS CLI user guide: `https://docs.aws.amazon.com/cli/latest/userguide/cli-chap-welcome.html`

*   AWS API reference (RESTful interface for interacting with AWS resources): `https://docs.aws.amazon.com/general/latest/gr/aws-apis.html`

3
IAM User Management

Some of the most highly visible objects in **identity and access management (IAM)** are user accounts. Much of the discipline is centered on securing the credentials for those accounts, ensuring they have proper lifecycle management, and providing the governance to ensure that we can audit and document their proper use. And, of course, issues with accounts and passwords can also cause much user-experience friction in both enterprise and customer environments. All of those challenges are still with us in the cloud. In fact, it is arguable that the stakes for securely managing user accounts within cloud management backplanes are higher, as a loss of control there could have knock-on effects across dozens of apps and on critical enterprise infrastructure. With that in mind, let's take a look at **Amazon Web Services (AWS)** IAM user management.

In this chapter, we'll cover the following topics:

- What is an IAM user account?
- Managing and securing root IAM user accounts
- Managing and securing IAM user accounts
- Managing federated user accounts

Technical requirements

To get the most out of this chapter, you will need the following:

- An AWS account

- A workstation running the AWS **command-line interface (CLI)**

- A text editor or **integrated development environment (IDE)** to edit **JavaScript Object Notation (JSON)/YAML Ain't Markup Language (YAML)** files, such as **Microsoft Visual Studio Code (VS Code)**

What is an IAM user account?

In *Chapter 1, An Introduction to IAM and AWS IAM Concepts*, we introduced the foundational objects that AWS IAM uses to manage authentication and authorization to AWS resources under the context of AWS as an **infrastructure-as-a-service (IaaS)** platform. A principal—that is to say, a person or application that wants to access an AWS resource—will present itself using a known IAM object (such as an IAM user or a federated user) to AWS IAM. The principal validates their entitlement to assume that IAM object by confirming a shared secret, such as the IAM user object's password or access key ID and secret access key. By presenting the shared secret for the IAM user object, AWS IAM is able to authenticate the principal or determine who the principal is.

IAM user accounts are distinct user profiles managed by AWS IAM that distinguish and authenticate users within an AWS account. Using the AWS Management Console or the AWS CLI, we can perform lifecycle management on the accounts, manage their access to the Management Console, manage their programmatic access, and scope what they are allowed to do within the account by applying access policies to them.

These users are identified by a unique **Amazon Resource Name (ARN)**, which is the **universally unique identifier (UUID)** for that object within AWS. The format of that ARN is a combination of our account number and the user-friendly account name that we've used to identify the delegated accounts we've created so far. Here's the ARN for `redbeardidentity`:

```
arn:aws:iam::451339973440:user/redbeardidentity
```

Generally speaking, we typically interact with the user-friendly name, such as when signing in to the AWS Management Console, or even when executing many AWS CLI commands. However, some CLI commands and other AWS services require the specificity of the ARN for their purposes.

Principals

We keep referring to the actor behind the IAM user account as a principal instead of a person. Depending on your organization's needs, AWS IAM user accounts can map to applications, service accounts, bot process automation accounts, or all manner of additional use cases beyond an administrator accessing AWS resources with their account.

As such, users of IAM accounts can be any number of non-human entities. Not every IAM user account is used by an individual human being. Good IAM governance recommends that each user account used by an application be limited in its use to a single application. This limits potential damage should any given application's IAM user account be compromised, and facilitates access control and auditing of events to a specific application's account as well. Additionally, it is best practice to ultimately correlate each application's IAM user account to a named owner. However, this is a challenge that must be solved outside of AWS IAM.

When a principal authenticates itself as an AWS user account, the AWS IAM account is what it authenticates and authorizes to operate on AWS resources. It does not do this for the organizational structure (or lack thereof) of people using the account.

Additionally, some access to AWS is not done through user accounts but through assumed roles. In this use case, there is no user to refer to—at least not in the sense we've used that term so far when discussing AWS IAM. In those use cases, it is again a principal that accesses the AWS account, though through a distinct method. For the sake of consistency, we will refer to the actor that initiates the authentication event as the principal. This will be regardless of whether it is really a human being, a service account, or another application.

The *AWS IAM User Guide* further clarifies the role of a principal in a transaction by saying it is anyone *"authenticated as the AWS account root user or an IAM entity to make requests in AWS"* and *"as a best practice, do not use your root user credentials for your daily work"* (`https://docs.aws.amazon.com/IAM/latest/UserGuide/intro-structure.html`). We will see in the next section how and why the root user is a special class of account principals should avoid using.

Managing and securing root IAM user accounts

IAM user accounts are the basic units of accountability when a principal authenticates itself directly through AWS IAM, thus ensuring that those accounts are hardened is foundational to the security of the entire AWS account. However, before we begin on general account management and security, we need to address some peculiarities and best practices of a unique account type.

Differences between root user account and IAM user accounts

We've heard that repetition is key to learning. In both *Chapter 1*, *An Introduction to IAM and AWS IAM Concepts*, and *Chapter 2*, *An Introduction to the AWS CLI*, we created IAM users using both the AWS Management Console and the AWS CLI. The IAM users that were created were no less capable of doing anything inside of the AWS account by dint of them being members of the Full Administrators group. However, this still was an example of an AWS IAM security best practice. The AWS root account should not be used beyond creating other IAM user accounts.

Why is the AWS root account such a security risk? It is less that the AWS root account is inherently risky; it's more that the AWS root account—unlike any other IAM user account—cannot be deleted or replaced. It is intrinsically tied to the AWS account as the root account for that environment. Moreover, the root account cannot have its access scoped down at all; it will always be permanently endowed with a permissive, all-inclusive authorization policy. As such, increasing one's dependency on that account by building resources under it greatly increases certain risks, both in terms of the attack surface and because of the likelihood that a full administrator token or credential could be compromised by a malicious actor.

In comparison, even a full administrator IAM user account—if compromised—could have its credentials revoked, its group memberships adjusted to limit blast radius, or even authorization policies modified to alter what that account could do. There are no similar avenues for remediation with the root account. As such, there are some best practices that should be followed to ensure the root account does not become a security threat.

Managing the root account's password

When we set up our AWS account, we may not have understood the significance of that first password value we set. We're not merely setting up yet another password for yet another online service; that password is the security linchpin for a wealth of pay-per-use services.

AWS applies a service-wide password policy that it applies to all root accounts at signup to ensure a baseline level of security. However, we may be subject to additional regulatory requirements or organizational security policies that mandate enhanced password complexity. As with any other password, the root account's password should be changed on occasion. How frequently those changes occur depends upon your security requirements, but it shouldn't be left as the same value indefinitely.

> **Tip**
> The root account is only covered by AWS's own password policy. You have the option of creating a stricter password policy for your AWS IAM account, but those requirements will only apply to managed IAM user accounts. If you must comply with additional regulations or security policies, you will need to modify the root account password manually to comply with the policies required by your use case.

Unlike every other user, the root account is not administrated using the AWS IAM Dashboard. We can sign in as our root user, but we will not see the same information about password and access key age, last activity, or **multi-factor authentication (MFA)** enablement as we do with our other IAM users in the AWS IAM Dashboard.

This is demonstrated in the following screenshot:

	User name ▼	Groups	Access key age	Password age	Last activity	MFA
☐	RBI_Admin	FullAdministrator	✅ 7 days	8 days	7 days	Not enabled
☐	RBI_EC2	None	✅ 13 days	13 days	7 days	Not enabled
☐	RBI_S3	None	✅ 13 days	13 days	7 days	Not enabled
☐	redbeardidentity	FullAdministrator	✅ 13 days	23 days	Today	Not enabled

Figure 3.1 – Signed in as the root user, but no root account to be found

We find our password management options under the **My Security Credentials** section in the user menu, as illustrated in the following screenshot:

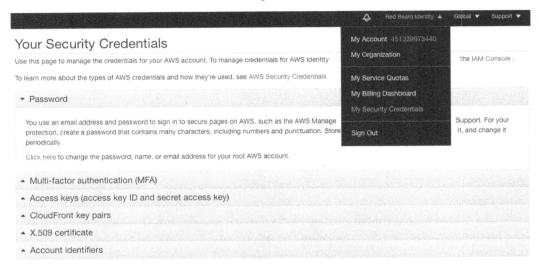

Figure 3.2 – Root account security-credential administration

It is here that we can update the root account's password based upon our own requirements or on your organization's security policy.

Do not create or use root account access keys

Similar to the password change options available to us in the root account's **My Security Credentials** form, we also have an option to administrate the root account's access keys. That said, there should *not* be any access keys listed to administrate under our root account, and not just because we never had occasion to create one. The form itself gives a warning as to why, as illustrated in the following screenshot:

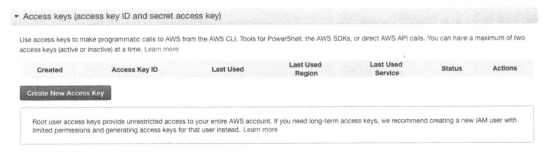

Figure 3.3 – Warning about root user access keys

We've already emphasized how we cannot limit the access of the root account using the authorization policy, unlike with other managed IAM user accounts. The existence of long-lived root credentials presents an unnecessary risk to the integrity of the environment when the same capabilities can be provided by an IAM user account with full administrator access. This is why we created the Full Administrator group and the IAM user accounts we configured for use with our AWS CLI profiles.

> **Important note**
>
> **Using the AWS CLI to administrate the root user**: As the root user is such a powerful user object, it may be tempting to try to automate the management of its credentials using the AWS CLI, to ensure some best practices (such as credential cycling) are followed. While there is an argument to be made for this kind of administration, consider other root account security best practices that would have to be contraindicated in order to even allow the root account AWS CLI access—namely, creating a long-lived access key ID and secret access key for the root account. Moreover, those credentials would reside on whatever system would be handling those administrative tasks, and would require protection there. There are specialized privileged account-credential vaulting tools that can facilitate these types of use cases. Notwithstanding, carefully consider the use case you are trying to solve and the security risks and trade-offs of issuing your root account programmatic credentials prior to undertaking such a venture.

Enabling MFA on the root account

When a user is authenticated, they provide proof or a piece of evidence that they are who they say they are. These proofs fall into the following three categories:

- **Something you know**: A shared secret that can be provided, which should only be known by the account being authenticated and the system doing the validation, such as a password

- **Something you have**: An item, token, or other object that was issued to the person being authenticated, which they can then present as evidence that they are the person who received that object, such as a certificate or a one-time-passcode token

- **Something you are**: Authentication through the validation of an inherent attribute or property of the user being authenticated, such as a biometric identifier (for example, a fingerprint)

MFA occurs when a user or subject of an authentication request provides proofs across two or more distinct authentication categories. For example, when signing in to a bank's website using our username and password (something you know) on a new device, we may be prompted to also enter a one-time code that will be sent to our cellphone number. Our receipt and entry of that code proves possession of the phone number the bank has on file with our account (something you have).

By compounding the different authentication methods, the level of assurance the bank has that we are who we say we are goes up. Additionally, it becomes much more complex for an attacker to impersonate us if we are asked to prove we are who we say we are in various ways. An attacker may steal our password in a data breach, but they will have a much more difficult time *also* stealing our phone. MFA is one of the most effective ways to secure an account.

As such, it should come as no surprise that it is a recommended best practice to apply MFA on the root account. In fact, if the root account on our AWS account does not have MFA enabled, the AWS IAM Dashboard displays a warning to any IAM user that is signed in.

We can enable MFA on the root account under the same **My Security Credentials** form where we addressed the other credentials we've examined, as follows:

1. Expand the **Multi-factor authentication (MFA)** section and click the **Activate MFA** button.

2. Here, we see the options available to us for our MFA devices. For some more detail on each of the MFA authenticator types, see the *MFA authenticators* callout shown next. We will use a **Virtual MFA device** with our root account. We select that option and click **Continue**, as illustrated in the following screenshot:

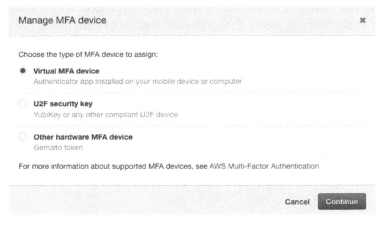

Figure 3.4 – Options for MFA credentials

3. The virtual authenticator we have on our device is Google Authenticator, but any **Time-based One-Time Passcode (TOTP)**-compliant virtual MFA application will work. A list of supported virtual MFA devices can be found at `https://aws.amazon.com/iam/features/mfa/?audit=2019q1`.

4. To enroll our device, we scan the **Quick Response (QR)** code with our TOTP app of choice and then enter two consecutive codes generated by that TOTP app. This ensures the pattern that seeded the number sequence was captured correctly.

    The process is illustrated in the following screenshot:

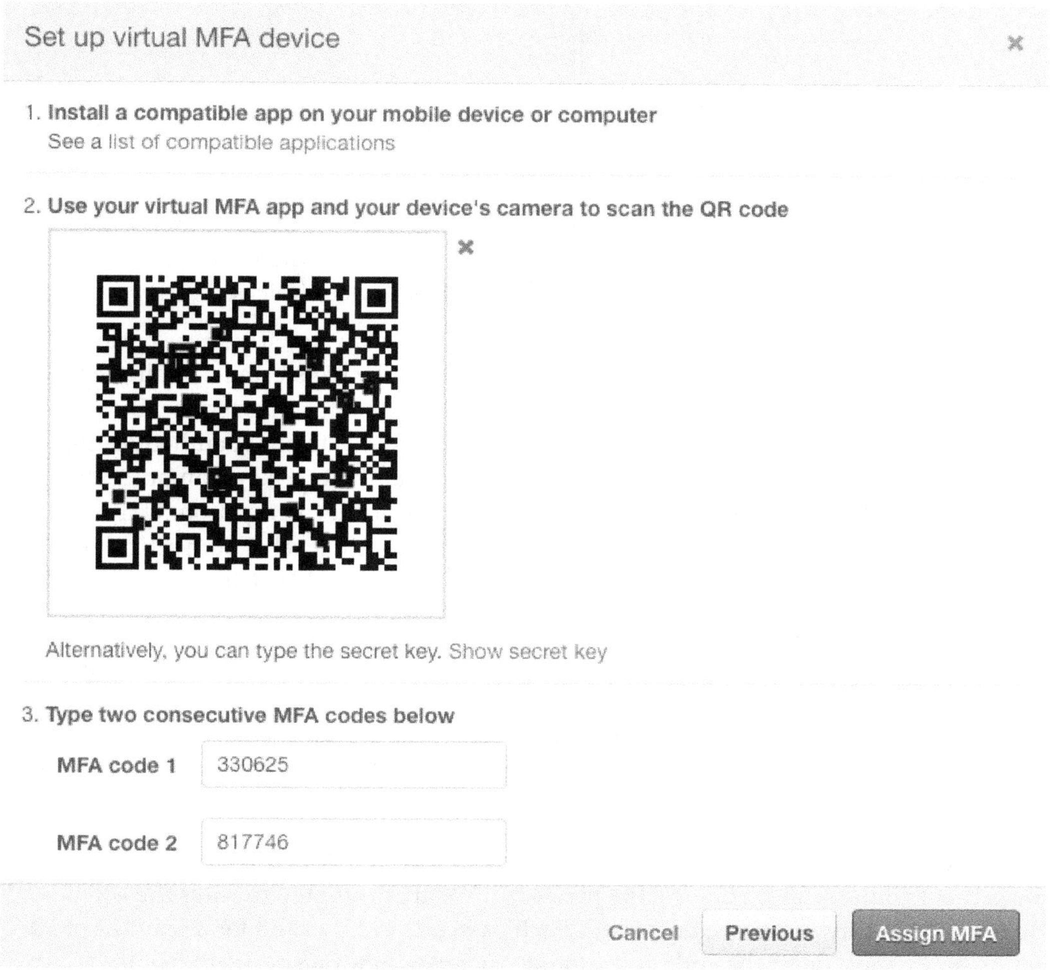

Figure 3.5 – TOTP authenticator configuration

5. If our sequential codes validate, we get a notification that our root account has been assigned a virtual MFA token. We can now see that specific authenticator, including its ARN, listed in the **Multi-factor authentication (MFA)** section of the **Your Security Credentials** form for the root account, as illustrated in the following screenshot:

▼ Multi-factor authentication (MFA)

Use MFA to increase the security of your AWS environments. Signing in to MFA-protected accounts requires a user name, password, and an authentication code from an MFA device.

Device type	Serial number	Actions
Virtual	arn:aws:iam::451339973440:mfa/root-account-mfa-device	Manage

Figure 3.6 – The root account's registered MFA credentials

AWS IAM supports three types of MFA devices on both the root account and IAM user accounts. This is a brief overview of each authenticator type:

*   The first is a virtual MFA device. This is a software token—usually in the form of a phone application—that will provide a one-time passcode. Microsoft Authenticator and Google Authenticator are popular virtual MFA device choices.

*   The next option is a **Universal 2nd Factor** (**U2F**) security key. These are hardware cryptographic tokens that authenticate the possessor by registering a public key corresponding to the device's private key with the authenticating service. By sending a signing statement signed with the device's private key at authentication time, the authenticating service can be confident that the U2F key is in possession of the person who registered it thanks to public/private key encryption. U2F security key MFA fulfills the requirements for the highest level of authentication assurance based on US Department of Commerce **National Institute of Standards and Technology** (**NIST**) 800-63-3 guidelines.

*   Finally, there are hardware one-time passcode tokens. These tokens cycle through a preset count of codes that the authenticating service validates by correlating them to the hardware token's registered serial number. These are used less frequently in light of the convenience of other tokens, but are popular in secured environments where **Universal Serial Bus** (**USB**) access or phones may not be permitted.

With MFA enabled on our root account, we have an extra step during sign-on to the Management Console after entering the password. We are prompted to enter the MFA code from the authenticator we registered. Each passcode is only valid for a window of 30 seconds, so we must enter the code and submit it before a new one cycles through.

The process is illustrated in the following screenshot:

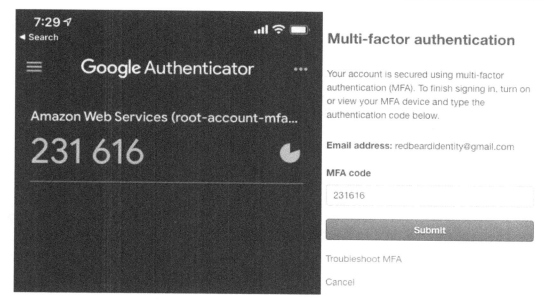

Figure 3.7 – TOTP code is now required on each root account login

This is now required with every login to the Management Console, which makes our root account—and, by extension, our AWS account—much more secure.

Now that we've hardened our root account, let's focus on best practices for securing normal IAM user objects.

Managing and securing IAM user accounts

Many of the same principles that apply to securing the root account apply broadly to individual AWS IAM user accounts. That said, as these are managed objects, they are subject to additional configurable security policies. Additionally, as we can use a delegated account to administer other delegated accounts, we can also use the CLI for some of these tasks, while doing the same for the root account would be ill advised.

IAM user lifecycle management

We have referred to user accounts as the most basic unit of accountability for AWS-managed users. However, as the complexity of the organization increases, it's less likely that administrators would provision and administrate IAM user accounts for their user base. Large organizations with complex AWS account structures rely on identity federation for user authentication into AWS. This relies on temporary security credentials and assumed roles for access. We will dive more deeply into this topic in the *Managing federated user accounts* section.

The lifecycle of a native IAM user account is dependent upon the manual (or scripted) processes available for creating, updating, and deleting them by the AWS account administrators. The specific business or identity lifecycle triggers that determine the circumstances in which the accounts are created will vary based on your organization's business and regulatory requirements. Regardless of what those specific requirements and lifecycle events are, there is a pattern to follow with these managed accounts, illustrated in the following diagram:

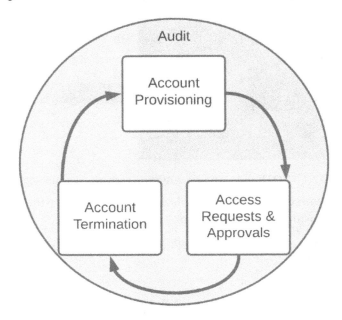

Figure 3.8 – Generic account lifecycle pattern

First, an established process should trigger the creation of the account. Such a process could be a service request, the onboarding of a new user, or nearly anything else; the important point is that the provenance of the account is known. Next, modifications to what that account can access or do also need to follow some sort of established process and—where appropriate—have approval logged. The goal here is to ensure changes that could give that account entitlements access beyond least privilege or in contradiction to separation-of-duties requirements will not be violated, and also to ensure that the user of the account cannot escalate their own privileges. Finally, the account should be terminated when no longer needed. Unmanaged accounts are a security risk.

Assuming that account belonged to an individual, it should be terminated when that individual leaves the organization or when the role that necessitated the account changes. All of these events take place on a background of continuous audit and review so as to ensure that controls are effective.

Password management

In this section, we will go over the password management options and best practices available to any given IAM user account using the Management Console and CLI. Then, we will look at the options available for us to apply account-wide password policies for all IAM users. **Password management policies** allow administrators to enforce minimum-complexity values, enforce credential cycling, and even require administrator intervention to re-issue a password or unlock an account. This next section will show you how to manage IAM user passwords and set password policies.

IAM user account password management

Let's take a look at the options we have for managing the `redbeardidentity` IAM user account password. We will start with the Management Console. The first thing we notice is that the IAM user accounts have a couple of additional options under **My security credentials** compared to the root account, thanks to them being managed objects. Not only is the layout different, but there are also additional credential types available for some specific AWS services such as **Keyspaces** and **CodeCommit**. More important than those differences are our options for making adjustments to how this form works for all user accounts.

An overview of the form is provided here:

Figure 3.9 – IAM user account credential management

> **Tip**
>
> **My security credentials** is a path to administrate the credentials of the IAM user account currently being used. We can also administrate those credentials by selecting that user account from the **User** section of the IAM Dashboard. If our user account has the appropriate authorization, we can select any user account and review or change its security credentials by opening the **Security credentials** tab on that user's **Summary** screen.

Let's look at how to apply a password policy next.

Applying a password policy to the AWS account

Figure 3.9 shown previously gives us a few indications about the password policy that applies to our AWS account. The password age and an admonition to regularly change that password appear above the **Change password** button, but there is no indication that we will be forced to change our password at any time.

If we feel these permissions are too lenient, or if we have security policy or regulatory purposes for doing so, we have the option to change the password policy that applies to all IAM user accounts within the AWS account.

Setting the password policy in the Management Console

We can review the current policy from the IAM Dashboard, under **Access management, Account settings**. Here, we see the default policy for our account, which is currently set to AWS' own baseline policy, as illustrated in the following screenshot:

▾ Password policy

A password policy is a set of rules that define the type of password an IAM user can set. Learn more

Password policy
This AWS account uses the following default password policy:

- Minimum password length is 8 characters
- Include a minimum of three of the following mix of character types: uppercase, lowercase, numbers, and ! @ # $ % ^ & * ( ) _ + - = [ ] { } | '
- Must not be identical to your AWS account name or email address

Change password policy

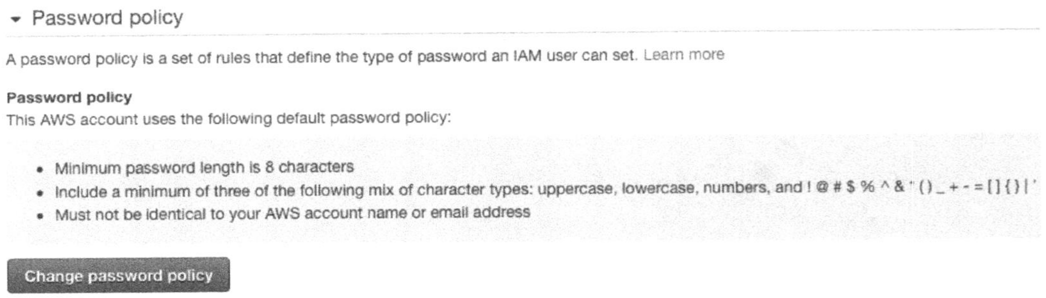

Figure 3.10 – The default IAM password policy

We can review the options available to us by clicking the **Change password policy** button. By doing so, we are brought to the **Set password policy** form, as illustrated in the following screenshot:

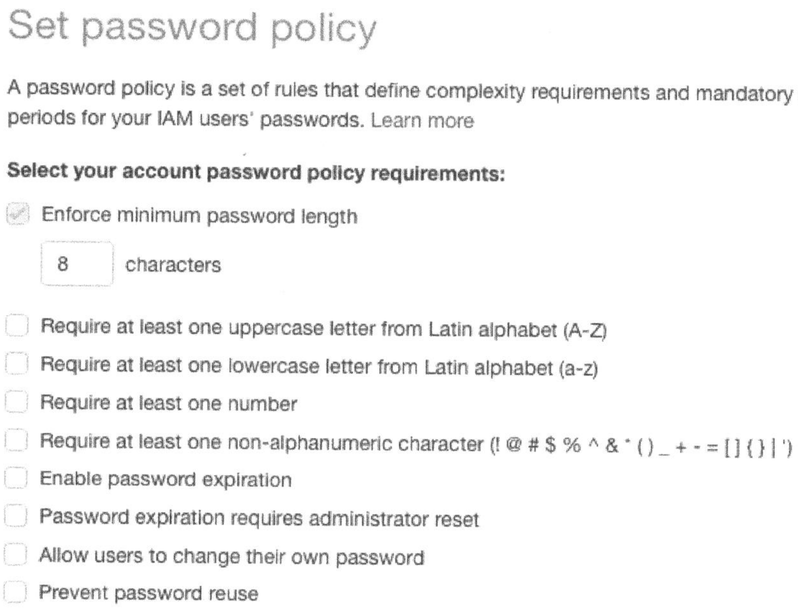

Set password policy

A password policy is a set of rules that define complexity requirements and mandatory rotation periods for your IAM users' passwords. Learn more

Select your account password policy requirements:

☑ Enforce minimum password length

> 8   characters

☐ Require at least one uppercase letter from Latin alphabet (A-Z)

☐ Require at least one lowercase letter from Latin alphabet (a-z)

☐ Require at least one number

☐ Require at least one non-alphanumeric character (! @ # $ % ^ & * ( ) _ + - = [ ] { } | ')

☐ Enable password expiration

☐ Password expiration requires administrator reset

☐ Allow users to change their own password

☐ Prevent password reuse

Figure 3.11 – Options available for creating an account password policy

On this page, we have options for the following:

- Minimum password length

- Requiring at least one uppercase letter

- Requiring at least one lowercase letter

- Requiring at least one number

- Requiring at least one non-alphanumeric character from a pre-determined list of values

- An option to force password expiration

- An option to force administrator intervention when a password has expired

- An option to allow users to change their own password

- An option that prevents previously used passwords from being re-entered

It's worth noting that the default password policy template does not start from a baseline of the default AWS service password policy. If we look through the options, we see that some things from our account's current policy—such as requiring alphanumeric characters—are not enforced. Each password policy is a blank slate.

> **Tip**
> The **Password expiration requires administrator reset** and **Allow users to change their own password** options are *not* contradictory, though they may appear to be. Requiring an administrator to reset an expired password means they will need to issue a new password to the user *only if the old one is allowed to expire*. A user should be able to change their own password before it ages 90 days, to avoid requiring administrator intervention.

For our account's new policy, we will enable each option. When we enable password expiration, we get a new option to enter a value for how many days until a password expires, and when we prevent password reuse, we can set the count of previous sequential passwords to not allow, as illustrated in the following screenshot:

Set password policy

A password policy is a set of rules that define complexity requirements and mandatory rotation periods for your IAM users' passwords. Learn more.

Select your account password policy requirements:

☑ Enforce minimum password length

> 8   characters

☑ Require at least one uppercase letter from Latin alphabet (A-Z)

☑ Require at least one lowercase letter from Latin alphabet (a-z)

☑ Require at least one number

☑ Require at least one non-alphanumeric character (! @ # $ % ^ & * ( ) _ + - = [ ] { } | ')

☑ Enable password expiration

> Expire passwords in 90   day(s)

☑ Password expiration requires administrator reset

☑ Allow users to change their own password

☑ Prevent password reuse

> Remember 5   password(s)

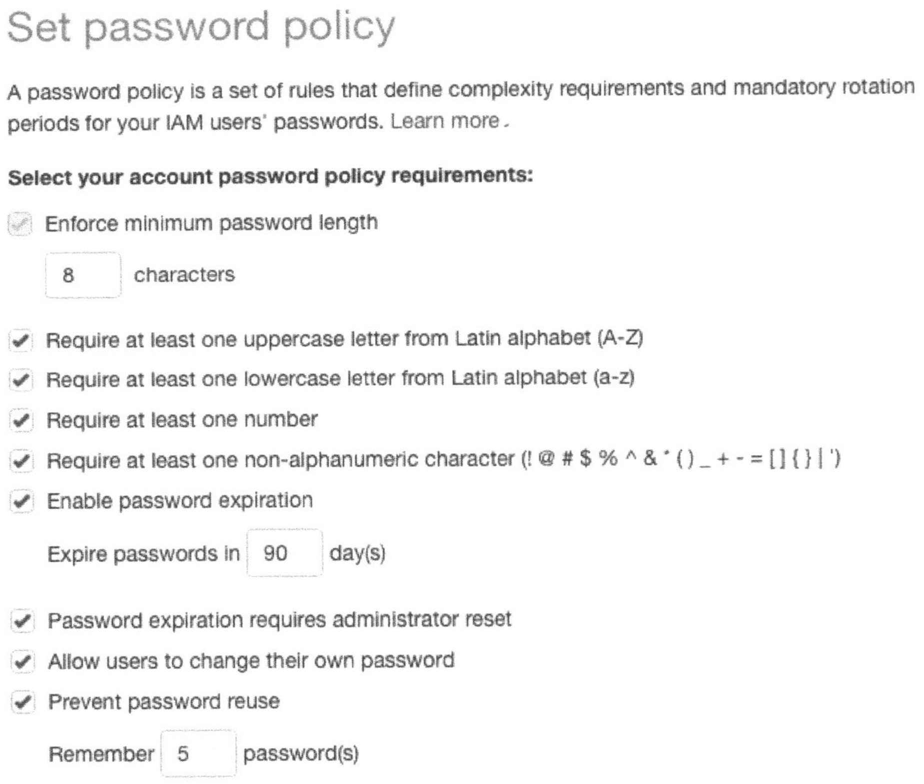

Figure 3.12 – Our AWS account's new custom password policy

Upon saving the changes, we can now see the new password policy reflected in the prior form within the IAM Dashboard, as illustrated in the following screenshot:

A password policy is a set of rules that define the type of password an IAM user can set. Learn more

Password policy
This AWS account uses the following custom password policy:

- Minimum password length is 8 characters
- Require at least one uppercase letter from Latin alphabet (A-Z)
- Require at least one lowercase letter from Latin alphabet (a-z)
- Require at least one number
- Require at least one non-alphanumeric character (! @ # $ % ^ & * ( ) _ + - = [ ] { } | ')
- Password expires in 90 day(s)
- Password expiration requires administrator reset
- Allow users to change their own password
- Remember last 5 password(s) and prevent reuse

Figure 3.13 – The new password policy is applied

We can make further piecemeal adjustments to the policy as needed by clicking **Change** and adjusting the settings on the password policy form. Alternatively, we can click **Delete** to revert to the default AWS policy.

Setting the password policy from the AWS CLI

Let's take a look at our new policy from the AWS CLI. From a terminal, run the following command:

```
$ aws iam get-account-password-policy
```

The output of that command will show the same values that we see in the IAM Dashboard, though some of the names for the options may be described differently. For example, HardExpiry, shown in the following screenshot, is really the **Password expiration requires administrator reset** option from the Management Console:

```
[jonlehtinen@ ~ % aws iam get-account-password-policy
PasswordPolicy:
    AllowUsersToChangePassword: true
    ExpirePasswords: true
    HardExpiry: true
    MaxPasswordAge: 90
    MinimumPasswordLength: 8
    PasswordReusePrevention: 5
    RequireLowercaseCharacters: true
    RequireNumbers: true
    RequireSymbols: true
    RequireUppercaseCharacters: true
jonlehtinen@ ~ %
```

Figure 3.14 – Reviewing IAM account password policy via CLI

Since most administrators are an overworked and cranky lot, we will eventually drop the HardExpiry requirement from our password policy. Let's run the following command to change that variable:

```
$ aws iam update-account-password-policy --no-hard-expiry
```

We can then confirm that the HardExpiry requirement was dropped by reviewing the policy once more, as illustrated in the following screenshot:

```
[jonlehtinen@ ~ % aws iam update-account-password-policy --no-hard-expiry
[jonlehtinen@ ~ % aws iam get-account-password-policy
PasswordPolicy:
    AllowUsersToChangePassword: false
    ExpirePasswords: false
    HardExpiry: false
    MinimumPasswordLength: 6
    RequireLowercaseCharacters: false
    RequireNumbers: false
    RequireSymbols: false
    RequireUppercaseCharacters: false
jonlehtinen@ ~ %
```

Figure 3.15 – Unexpected changes to the password policy

Unfortunately, only setting a single-parameter value set the remaining values to some values that will permit some weak passwords in our environment. We will need to run the update command with every variable populated, in order to prevent insecure values from being populated by default. Rather than type a several-hundred character length command and risk typos and frustration, we can instead opt to use a CLI skeleton and update the policy after editing it in an IDE or text editor. In the terminal, we enter the following command:

```
aws iam update-account-password-policy --generate-cli-skeleton
yaml-input
```

We get our YAML template for the password policy in return, as illustrated in the following screenshot:

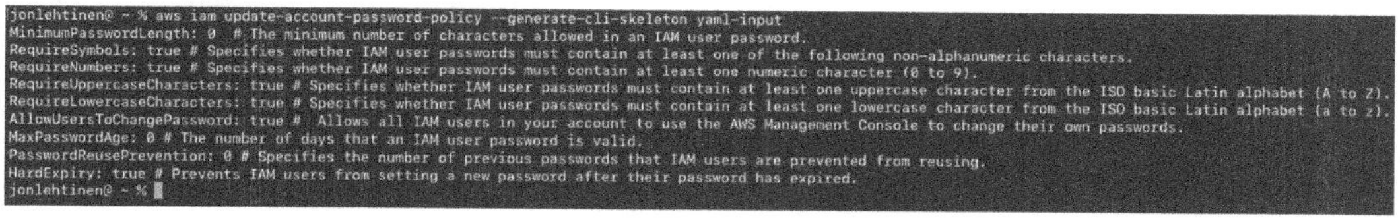

Figure 3.16 – Password policy template

After we create a file and populate it with our desired parameter values, we import it using the CLI. With that update executing successfully, we can view our updated policy with HardExpiry set to false, as illustrated in the following screenshot:

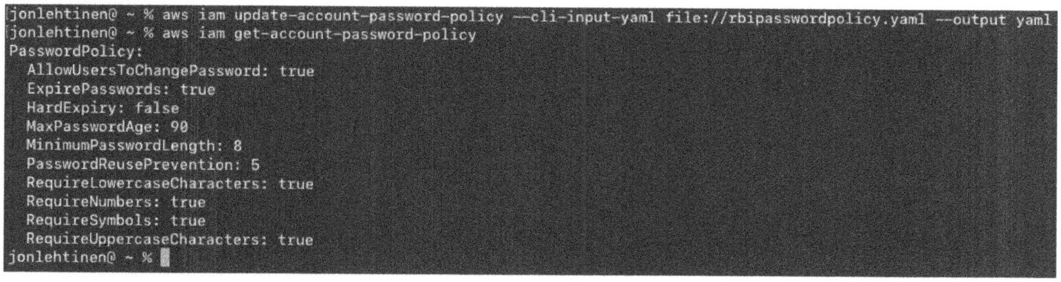

Figure 3.17 – Import of new password policy and review of updated policy

> **Tip**
>
> The aws iam update-account-password-policy command will execute successfully **with no arguments** from the CLI. This will disable all requirements around complexity, password age, and reset options. Be sure to always enter the command with all parameters included and set to the desired values.

Now that we know how to apply a password policy broadly to our account, let's look at passwords in relation to specific user accounts.

Password administration

Similar to our experience when administrating the root account password, we can use the AWS Management Console to change a password if we are logged in as that particular IAM user. The process is nearly identical to that of the root account, except that our IAM user account must comply with the password policy we applied to our AWS account.

An IAM user with the proper authorization policy can change the password on other user accounts using the Management Console. This is done within the IAM Dashboard. We will use our redbeardidentity user account to issue a new password to the RBI_S3 user account this way:

1. We select the user account from the list of users in the IAM Dashboard, and open that user's **Security credentials** tab.

2. There, we see information about that user account's console password—specifically, if it is currently enabled, and the date it was last used to log in. We click **Manage** to modify the password, as illustrated in the following screenshot:

Summary

User ARN	arn:aws:iam::451339973440:user/RBI_S3
Path	/
Creation time	2020-11-22 12:59 EST

Permissions	Groups	Tags (2)	Security credentials	Access Advisor

Sign-in credentials

Summary	• Console sign-in link: https://451339973440.signin.aws.amazon.com/console
Console password	Enabled (last signed in Yesterday) | Manage
Assigned MFA device	Not assigned | Manage
Signing certificates	None

Figure 3.18 – Information on another user account's console password status

3. A window appears, with options to change that user's access to the Management Console, set a new password, and to require the user to change their password on their next login to the console. We will opt for an autogenerated password and force that user account to change the password on their next login so that their password remains secret, as illustrated in the following screenshot:

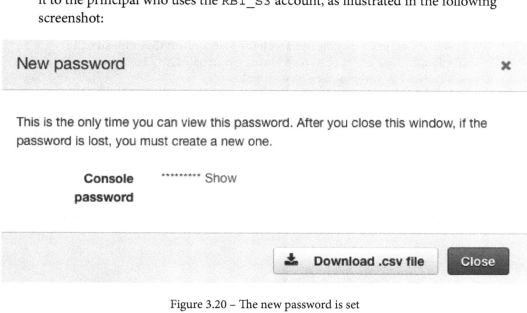

Figure 3.19 – Resetting the RBI_S3 user account's password

4. We are presented with the new password and warned that this will be our only chance to capture the password value. We can download the .csv file and deliver it to the principal who uses the RBI_S3 account, as illustrated in the following screenshot:

Figure 3.20 – The new password is set

We can also use the AWS CLI to reset IAM user account passwords, though we will need valid programmatic credentials and the appropriate access policy in order to do so. First, let's examine the current password for our redbeardidentity user account. From the terminal, we enter the following command:

```
$ aws iam user get-login-profile --user-name redbeardidentity
```

The Management Console access is referred to as the login profile within the CLI. When creating IAM users directly from the CLI, we must create a login profile for them if we intend for them to have access to the Management Console. Part of that creation process includes assigning a password to that login profile, as illustrated here:

```
[jonlehtinen@ ~ % aws iam list
[> aws iam get-login-profile --user-name redbeardidentity
LoginProfile:
  CreateDate: '2020-11-12T01:11:47+00:00'
  PasswordResetRequired: false
  UserName: redbeardidentity
jonlehtinen@ ~ %
```

Figure 3.21 – The redbeardidentity password status

This provides us information on our login profile, including its creation date and whether the associated password for that profile requires a reset.

> **Tip**
>
> AWS documentation on creating users with the Management Console and CLI is available at https://docs.aws.amazon.com/IAM/latest/UserGuide/id_users_create.html.

As a member of the Full Administrator group, the redbeardidentity account can reset the password for any IAM user account within our AWS account using the update-login-profile command, including its own. From the terminal, we run the following command:

```
$ aws iam update-login-profile --user-name redbeardidentity
--password dUpg09-vievut-tadziw --no-password-reset-required
```

If the new password value we entered complied with our AWS account's password policy, we will be able to use the new password value we defined with that command.

Whereas we are able to use this administrative command to update our own account's login profile, this is not a best practice. Setting aside that it would be unlikely for most AWS administrators to have access to this capability in the IAM service, directly altering the login profile on an account introduces additional complicating parameters, such as those focused on forcing a password reset or not. Nor does that command require the person executing it to demonstrate knowledge of the account's previous password—just because we have access to perform an administrative function does not mean we are authorized to do so.

Best practice for AWS CLI-based, self-service password resets is to use the following command:

```
$ aws iam change-password --old-password oldpasswordvalue
--new-password newpasswordvalue
```

There are fewer parameters, and it ensures whoever is changing the password can prove ownership by authenticating themselves with the old password prior to changing it to a new one.

We can use the update-login-profile command to issue temporary credentials to users whose passwords have expired. We enhance security by ensuring that only the account owner will know the final password after reset, as follows:

```
$ aws iam update-login-profile --user-name RBI_S3 --password
$TempP4ss --password-reset-required
```

When we sign in to the Management Console as the RBI_S3 user account using the temporary password, we are prompted to change our password, as illustrated in the following screenshot:

You must change your password to continue

AWS account 451339973440

IAM user name RBI_S3

Old password ··········

New password ····················

Retype new password ····················

Confirm password change

Sign in using root user email

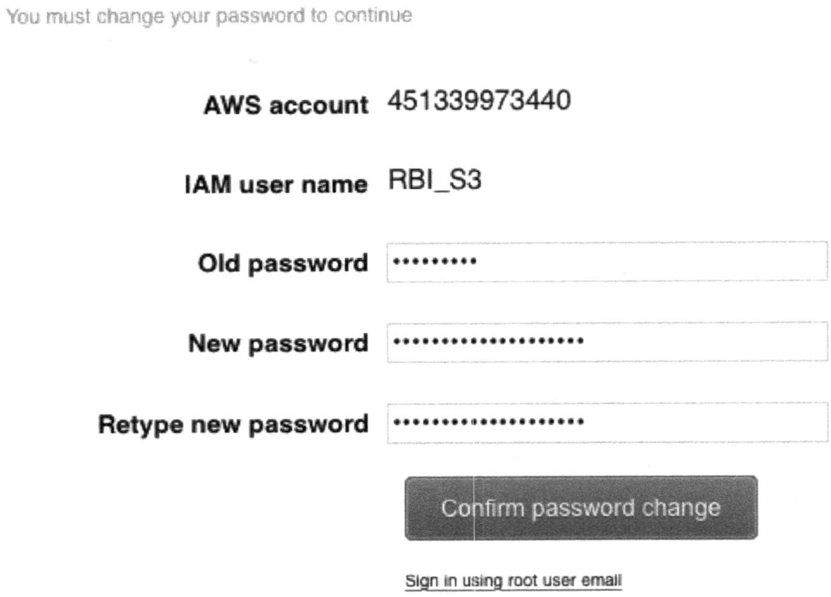

Figure 3.22 – The RBI_S3 user is prompted to change the temporary password

What happens if we run afoul of our password policy when attempting a self-service reset command? We can attempt to enter a password that fails to comply with our AWS account's password policy, but we get an error message, as illustrated in the following screenshot:

```
jonlehtinen@ ~ % aws iam change-password --old-password dUppo9-vimvut-tabziw --new-password 12345

An error occurred (AccessDenied) when calling the ChangePassword operation: User: arn:aws:iam::451
339973440:user/redbeardidentity is not authorized to perform: iam:ChangePassword on resource: user
 redbeardidentity with an explicit deny
jonlehtinen@ ~ %
```

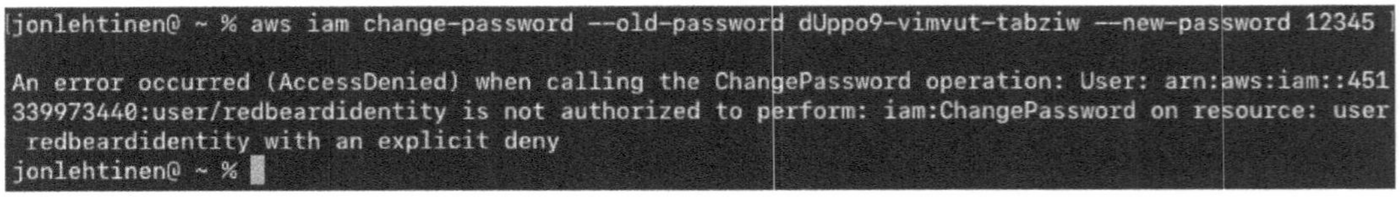

Figure 3.23 – Explicit deny due to failure to comply with password policy

The message indicates that we are not authorized to perform the operation on the resource with an explicit deny. This is because the password policy is acting as a sort of authorization policy for this specific transaction. By entering a password that fails its criteria, we are explicitly barred from executing the update.

Access key management

In this section, we will examine access key management best practices. Access keys are long-lived credentials used by principals to access AWS services programmatically. It is through these credentials that we access our AWS account with the AWS CLI, and how non-human principals such as service accounts and other applications leverage AWS services.

IAM user access key management in the Management Console

As with the Management Console password, access keys are administrated within the IAM Dashboard. Under the **Security credentials** tab for each IAM user, that user's access key IDs, their creation date, dates of last use, and status (either **Active** or **Inactive**) are available at a glance. This is also where existing access keys can be made inactive, or new ones created. The following screenshot shows the access key IDs listed under the `redbeardidentity` user account in our AWS account:

Access keys

Use access keys to make secure REST or HTTP Query protocol requests to AWS service APIs. For your protection, you should never share your secret keys with anyone. As a best practice, we recommend frequent key rotation. Learn more

Create access key

Access key ID	Created	Last used	Status	
AKIAWSFPVONAKWL37BGN	2020-11-22 10:40 EST	N/A	Inactive Make active	✖
AKIAWSFPVONAN7BQJXVU	2020-11-22 10:43 EST	2020-12-11 16:49 EST with iam in us-east-1	Active Make inactive	✖

Figure 3.24 – The redbeardidentity user account's current access keys

Each user account is limited to two access keys. In order to create a new access key, we will first need to delete one of them. We can tell by the last-used date on the inactive key that it was never used and thus will be safe to delete; no process or activity has ever used it, otherwise a date would be populated in that field. Conversely, we see a very recent last-used date on the `active` key, so deleting that one would have an adverse impact on our environment.

To delete the `inactive` key, we click the **x** on its line. We are immediately prompted to verify if we really want to delete that key, and a window reconfirming the user account the access key belongs to and its last-used date appear, to ensure we are not selecting the wrong key by mistake. As the information checks out we can confirm its deletion, as illustrated in the following screenshot:

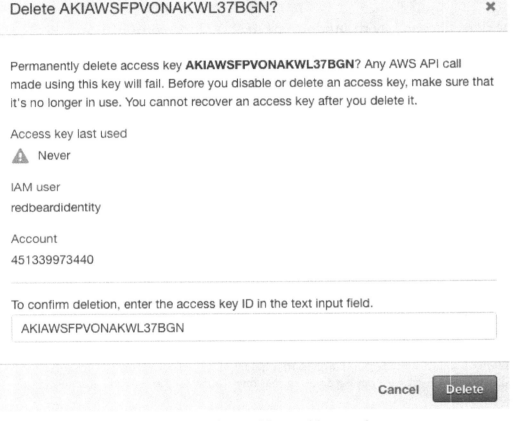

Delete AKIAWSFPVONAKWL37BGN?   ✖

Permanently delete access key **AKIAWSFPVONAKWL37BGN**? Any AWS API call made using this key will fail. Before you disable or delete an access key, make sure that it's no longer in use. You cannot recover an access key after you delete it.

Access key last used
⚠ Never

IAM user
redbeardidentity

Account
451339973440

To confirm deletion, enter the access key ID in the text input field.

AKIAWSFPVONAKWL37BGN

Cancel   Delete

Figure 3.25 – Confirming deletion of the access key

Now that the access key is gone, we can create a new key for our `redbeardidentity` user by clicking **Create access key**. A window pops up to show us our new access key ID and secret access key values, along with an option to download a `.csv` file with those values. As this is the only time the secret access key will be shared with us, we will opt to download the `.csv` file for future reference. If we were creating a new key for a user account owned by a different person, we would send the `.csv` file to them via a secure channel.

The process is illustrated in the following screenshot:

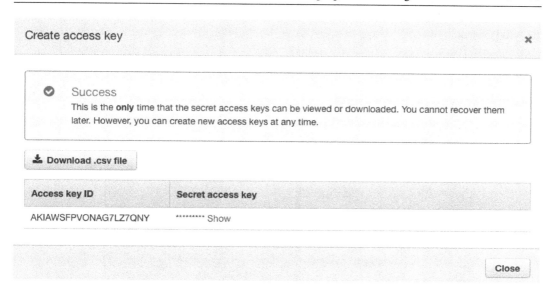

Figure 3.26 – New access key creation

The new access key is immediately active for our user account to use, as illustrated in the following screenshot:

Access key ID	Created	Last used	Status		
AKIAWSFPVONAN7BQJXVU	2020-11-22 10:43 EST	2020-12-11 16:49 EST with iam in us-east-1	Active	Make inactive	✖
AKIAWSFPVONAG7LZ7QNY	2020-12-12 15:35 EST	N/A	Active	Make inactive	✖

Figure 3.27 – The old and new access keys, both active and ready for use

Both access keys are now visible within the **Security Credentials** tab, with the new key showing as never having been used.

IAM user access key management in the AWS CLI

We can perform these same functions using the AWS CLI. First, we should view the access keys using the following command:

```
$ aws iam list-access-keys
```

This will provide a list of the keys associated with whatever user account was used to issue the command from the AWS CLI, which is the `redbeardidentity` user account in our case. Sure enough, that command returns the two access keys we had listed under that user account in the Management Console. In the following screenshot, note the creation dates on both keys indicate which is the newer one:

```
[jonlehtinen@ ~ % aws iam list-access-keys
AccessKeyMetadata:
- AccessKeyId: AKIAWSFPVONAN7BQJXVU
  CreateDate: '2020-11-22T15:43:37+00:00'
  Status: Active
  UserName: redbeardidentity
- AccessKeyId: AKIAWSFPVONAG7LZ7QNY
  CreateDate: '2020-12-12T20:35:37+00:00'
  Status: Active
  UserName: redbeardidentity
jonlehtinen@ ~ %
```

Figure 3.28 – The old and new access keys

In order to view the access keys that belong to a different user, we append a `--user-name` parameter to that command. We can view the key assigned to our other Full Administrator account, `RBI_Admin`, with the following command:

```
$ aws iam list-access-keys --user-name RBI_Admin
```

```
[jonlehtinen@ ~ % aws iam list-access-keys --user-name RBI_Admin
AccessKeyMetadata:
- AccessKeyId: AKIAWSFPVONACOFCK7WA
  CreateDate: '2020-11-28T17:21:48+00:00'
  Status: Active
  UserName: RBI_Admin
jonlehtinen@ ~ %
```

Figure 3.29 – Another user's access keys

As the process to create a new access key for the `redbeardidentity` account we are using in the CLI is the same process we would follow for creating a key for any other account, let's create a new access key for the `RBI_Admin` user account. From the terminal, we enter the following command:

```
$ aws iam create-access-key --user-name RBI_Admin
```

Out comes the new access key for that user account, complete with its secret access key. Similarly to when we create access keys in the Management Console, this will be the only time the secret access key will be revealed, so we'd better capture it. In the following screenshot, you can see where the secret access key appears:

```
[jonlehtinen@ ~ % aws iam create-acce
> aws iam create-access-key --user-name RBI_Admin
AccessKey:
  AccessKeyId: AKIAWSFPVONAJERMLWET
  CreateDate: '2020-12-12T21:00:45+00:00'
  SecretAccessKey:
  Status: Active
  UserName: RBI_Admin
jonlehtinen@ ~ %
```

Figure 3.30 – The new access key and secret access key

We now see two access keys listed under that account, including the one we just created, as illustrated in the following screenshot:

```
[jonlehtinen@ ~ % aws iam list-access-keys --user-name RBI_Admin
AccessKeyMetadata:
- AccessKeyId: AKIAWSFPVONACOFCK7WA
  CreateDate: '2020-11-28T17:21:48+00:00'
  Status: Active
  UserName: RBI_Admin
- AccessKeyId: AKIAWSFPVONAJERMLWET
  CreateDate: '2020-12-12T21:00:45+00:00'
  Status: Active
  UserName: RBI_Admin
jonlehtinen@ ~ %
```

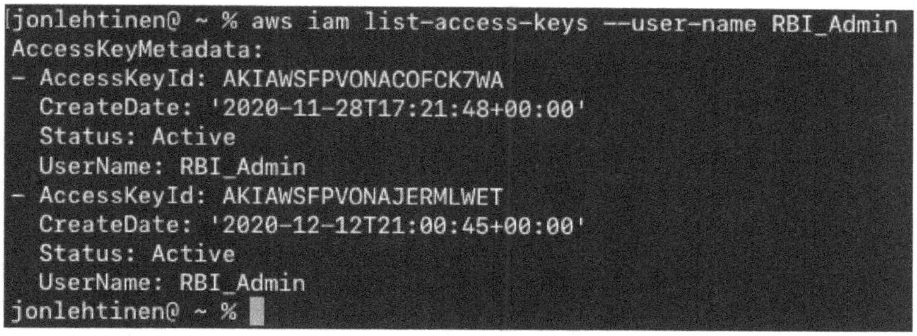

Figure 3.31 – The updated list of access keys for the RBI_Admin account

Unlike with the Management Console, that command does not include at-a-glance information on last-used datetime and service on the listed keys. If we wanted to know when these keys were last used, we could use the `aws iam get-access-key-last-used` command with the `--access-key-id` parameter to get that information.

Access key rotation

As they do with passwords, AWS considers it a best practice to rotate access keys at intervals. However, since access keys may be used by applications and infrastructure, casually removing an access key could have an impact that would not be felt until a process, script, or application eventually attempted to use those credentials. As such, the best practice for rotating access keys is slightly different from when changing a Management Console password.

Access key rotation using the Management Console

1. From the **User** section of the IAM Dashboard, select the user whose key needs to be rotated, and open their **Security credentials** tab. We will select the `redbeardidentity` user account for this example.

2. We want to rotate this user's only access key. We can see it is getting used frequently—and recently—by looking at the **Last used** date. We can also see what it was last used to do, which is useful for administrators who must rotate keys but who may not have full knowledge of what every user account and their access keys may be used for. This is illustrated in the following screenshot:

Access keys

Use access keys to make secure REST or HTTP Query protocol requests to AWS service APIs. For your protection, you should never share your secret keys with anyone. As a best practice, we recommend frequent key rotation. Learn more

Create access key

Access key ID	Created	Last used	Status	
AKIAWSFPVONAN7BQJXVU	2020-11-22 10:43 EST	2020-12-11 16:49 EST with iam in us-east-1	Active	Make inactive ✖

Figure 3.32 – The access key to be rotated with details on latest usage

3. Before we can replace the current key, we need to create its replacement. Once we do, we will have two active access keys assigned to this user account.

4. To make the rotation non-disruptive, we must now replace the old access key and secret key values with the new ones, everywhere they are referenced. For the `redbeardidentity` user account, this is simple; we will just rerun `aws configure` in the CLI to use the new keys. In more complex environments, the ease of discovering where these keys are referenced could be much more difficult. This is why the Management Console provides not just the last-used date, but also the last-used service and region where the command was issued. We see that that key was used against the IAM service in **us-east-1**. That makes sense as we've been running many IAM commands from the CLI, and the `redbeardidentity` user account's default region is set to **us-east-1**.

5. For our use case, we need to update our AWS CLI config to use the new keys, and then verify that the CLI works with those keys. After running `aws configure` and updating the default key values, we successfully run a command, as illustrated in the following screenshot:

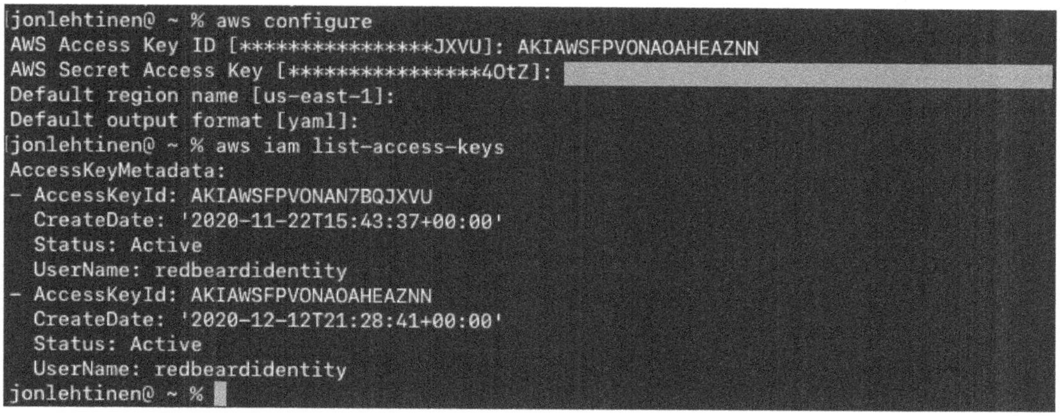

```
[jonlehtinen@ ~ % aws configure
AWS Access Key ID [****************JXVU]: AKIAWSFPVONAOAHEAZNN
AWS Secret Access Key [****************4OtZ]:
Default region name [us-east-1]:
Default output format [yaml]:
[jonlehtinen@ ~ % aws iam list-access-keys
AccessKeyMetadata:
- AccessKeyId: AKIAWSFPVONAN7BQJXVU
  CreateDate: '2020-11-22T15:43:37+00:00'
  Status: Active
  UserName: redbeardidentity
- AccessKeyId: AKIAWSFPVONAOAHEAZNN
  CreateDate: '2020-12-12T21:28:41+00:00'
  Status: Active
  UserName: redbeardidentity
jonlehtinen@ ~ %
```

Figure 3.33 – Updating AWS CLI and testing the new keys

6. After verifying that the new key works, we return to the user's **Security credentials** tab and disable the old access key. Why wouldn't we immediately delete it? Whereas in this example we are confident that the key was only used in one location, such certainty becomes more difficult in complex environments. Some functions that use the old key may only execute once a week or once a month, or less frequently than this, and it is advantageous to keep the old key around for a while to expedite recovery from a service outage.

    The following screenshot shows the old access key being disabled:

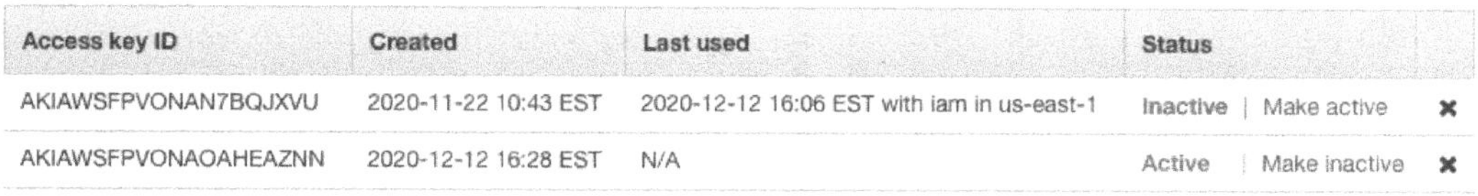

Access key ID	Created	Last used	Status		
AKIAWSFPVONAN7BQJXVU	2020-11-22 10:43 EST	2020-12-12 16:06 EST with iam in us-east-1	Inactive	Make active	✖
AKIAWSFPVONAOAHEAZNN	2020-12-12 16:28 EST	N/A	Active	Make inactive	✖

Figure 3.34 – Previous key disabled after successful testing of new key

> **Important note**
> Where's the last-used information for the new key? You may notice that the new key shows **N/A** for the **Last used** value. The values shown are pulled from CloudTrail audit records, which lag behind real-time events by a matter of minutes. We were confident in disabling the old key after doing functional tests using the new key in the AWS CLI.

7. Depending upon our requirements, once a sufficient amount of time has passed without incident, we can finally delete the old key.

Access key rotation using the AWS CLI

The process and guidance around key rotation are the same when using the AWS CLI; only the mechanics of execution differ:

1. We will once again be rotating the `redbeardidentity` user account's keys. If rotating the keys for a different user account, simply add the `--user-name` parameter to these commands. We will start by creating a new access key, as follows:

    ```
    $ aws iam create-access-key
    ```

2. Capture the access key and secret key values, as the secret key will not be revealed again.

3. Replace the old key with the new one. For this use case, this is a matter of running `aws configure` with the new-key values.

4. Once the new key has replaced the old one, wait a few days and then check if the old key has been used by running the following command:

    ```
    $ aws iam get-access-key-last-used --access-key-id
    AKIAWSFPVONAOAHEAZNN
    ```

    This will show the last date, time, service, and region where the specified access key was used, indicating if we missed replacing this key somewhere.

5. Assuming the key is no longer being used, we can go ahead and disable it, as follows:

    ```
    $ aws iam update-access-key --access-key-id
    AKIAWSFPVONAOAHEAZNN --status Inactive
    ```

6. We will leave the key in its `inactive` state for as long as is appropriate for our use case. In the event that we discover something that requires the key, we can revert it to an `active` state using the `update-access-key` command, remediate the issue, and return it to `inactive`. If there are no issues, we can finally delete the old key and the rotation is complete, as illustrated here:

    ```
    $ aws iam delete-access-key --access-key-id
    AKIAWSFPVONAOAHEAZNN
    ```

> **Tip**
>
> Rotating the programmatic credential this way will only work if the user has a single key defined. If they have multiple keys defined, they will get the following error:
>
> ```
> An error occurred (LimitExceeded) when calling the
> CreateAccessKey operation: Cannot exceed quota for
> AccessKeysPerUser: 2
> ```

Now that we've addressed programmatic access key rotation, let's move on to best practices for managing MFA credential management.

MFA credential management

This section addresses best practices for managing MFA credentials. MFA greatly enhances the security of IAM user account login by requiring additional authentication through a different channel aside from the user account's password.

IAM user account MFA credential management

We've already seen how to enable MFA on the root account earlier in this chapter. We can initiate that process on our own IAM user account the same way by selecting **My security credentials** from the **User** drop-down menu and clicking the **Assign MFA device** button. Alternatively, if our account has the appropriate authorization policy, we can open the **Security credentials** tab on our IAM user account in the IAM Dashboard and select **Manage** next to **Assigned MFA device**. That process is the same when administrating the MFA device for other IAM user accounts.

Interestingly, there is no intuitive account-wide setting to force IAM user accounts to use MFA for console logon, nor to enforce MFA credential enrollment if those accounts have Management Console access and a valid password, but no MFA device assigned. The best-practice method to solve this is to apply an authorization policy to the user accounts, which entitles them to manipulate their own security credentials and explicitly denies certain service interactions unless MFA authentication has occurred. We will dive into the details of authorization policy, including applying this policy, in *Chapter 4, Access Management, Policies, and Permissions*. As solving this use case outside of an authorization policy involves several antipatterns, we will focus on self-service MFA token administration use cases.

As we've already seen the process of assigning a virtual MFA device to our root account using the Management Console, we will not repeat the exercise again here.

We can also use the CLI to manage our user account's MFA device. You cannot do the same with the root account; it is explicitly forbidden (more information is available at `https://docs.aws.amazon.com/IAM/latest/UserGuide/id_credentials_mfa_enable_cliapi.html`). We can create hardware and virtual MFA devices using the CLI. As virtual MFA devices are much easier to procure and administrate, we will stick to using those:

1. First, we must create the MFA device. From a terminal, we run the following command:

    ```
    $ aws iam create-virtual-mfa-device --virtual-mfa-device-
    name googleauth1 --outfile ./redbeardidentitymfa.png
    --bootstrap-method QRCodePNG
    ```

    With this command, we give the virtual MFA device a name, designate a location for the QR code to be written on our local workstation, and designate the activation method for this device as a scanned QR code. If the command is successful, we will see the ARN of the virtual MFA device as the command's output in the CLI. It should be noted that the virtual device name value will be the name that appears for the account in the TOTP application on our phone. If we have several accounts in our TOTP application, we should be sure to name it something we would recognize so that we can readily recall which account it belongs to when we use the phone application.

    The aforementioned process can be seen in the following screenshot:

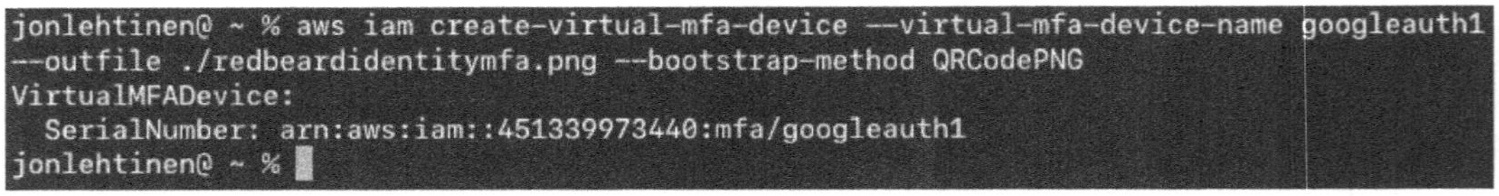

```
jonlehtinen@ ~ % aws iam create-virtual-mfa-device —virtual-mfa-device-name googleauth1
—outfile ./redbeardidentitymfa.png —bootstrap-method QRCodePNG
VirtualMFADevice:
  SerialNumber: arn:aws:iam::451339973440:mfa/googleauth1
jonlehtinen@ ~ %
```

Figure 3.35 – Successful creation of a virtual MFA device

We may alternatively use a different bootstrap method, and output a string called `Base32StringSeed`. That file and bootstrap provide the same information that can be scanned from the QR code, just in text form. That may be useful when using a desktop TOTP MFA application, but as we are using a mobile application, scanning the QR code is easier for us.

> **Important note**
>
> The output files are secrets: whichever method we use, both of the output files are considered secrets as sensitive as any other credential. A malicious actor with those files has much of what is needed to impersonate the MFA device attached to an IAM user. These files should be deleted after use or stored with the same care as for any other sensitive credential.

2. Next, we use our TOTP application to scan the QR code we generated in the previous command. This will add the account into the app and begin issuing the series of rotating codes we will need to complete the activation.

3. Next, we attach the virtual device to the IAM user account. From the terminal, run the following command:

```
$ aws iam enable-mfa-device --user-name redbeardidentity
--serial-number arn:aws:iam::451339973440:mfa/googleauth1
--authentication-code1 557439 --authentication-code2
380195
```

In this command, we proved possession of that specific virtual MFA device by entering two sequential codes from the TOTP application. If those codes match what is expected, there is no output in the terminal. The redbeardidentity user account now has an MFA device associated with it.

4. We can validate that the MFA device is associated with the account by signing in to the Management Console. After entering our console password, we are now prompted for a code to proceed, as illustrated in the following screenshot:

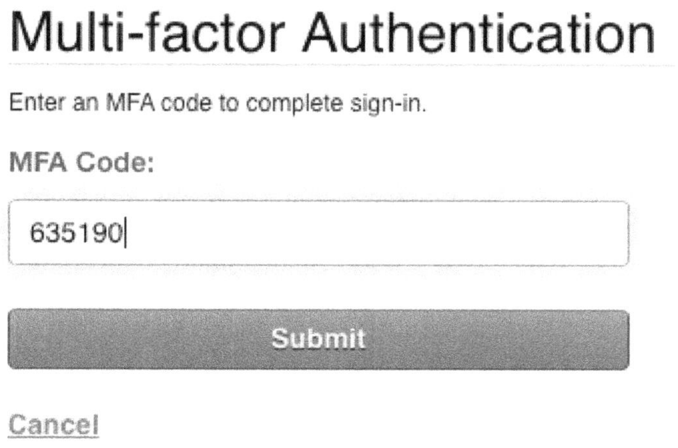

Multi-factor Authentication

Enter an MFA code to complete sign-in.

MFA Code:

635190|

Submit

Cancel

Figure 3.36 – MFA now enabled on the redbeardidentity user account

Removing MFA from a user account is simple. From the terminal, run the following command:

```
$ aws iam list-virtual-mfa-devices
```

This reveals all the virtual MFA devices created within our AWS account, along with information on who the device is assigned to. Here, we see three devices, each assigned to a different user account:

```
[jonlehtinen@ ~ % aws iam list-virtual-mfa-devices
VirtualMFADevices:
- EnableDate: '2020-12-09T00:17:06+00:00'
  SerialNumber: arn:aws:iam::451339973440:mfa/root-account-mfa-device
  User:
    Arn: arn:aws:iam::451339973440:root
    CreateDate: '2020-11-09T16:56:07+00:00'
    PasswordLastUsed: '2020-12-09T00:39:13+00:00'
    UserId: '451339973440'
- EnableDate: '2020-12-13T19:28:34+00:00'
  SerialNumber: arn:aws:iam::451339973440:mfa/rbis3
  User:
    Arn: arn:aws:iam::451339973440:user/RBI_S3
    CreateDate: '2020-11-22T17:59:25+00:00'
    PasswordLastUsed: '2020-12-13T18:13:27+00:00'
    Path: /
    UserId: AIDAWSFPVONACP5HVGP3C
    UserName: RBI_S3
- EnableDate: '2020-12-13T19:17:44+00:00'
  SerialNumber: arn:aws:iam::451339973440:mfa/googleauth1
  User:
    Arn: arn:aws:iam::451339973440:user/redbeardidentity
    CreateDate: '2020-11-12T01:11:46+00:00'
    PasswordLastUsed: '2020-12-13T19:22:54+00:00'
    Path: /
    UserId: AIDAWSFPVONALTHHLBKLK
    UserName: redbeardidentity
jonlehtinen@ ~ % █
```

Figure 3.37 – A list of the account's virtual MFA devices

Since we are not using the RBI_S3 user account, we will remove that one. Similar to the login profile, these devices cannot be deleted until they are first disassociated from the user account they are attached to. This is done with the following command:

```
$ aws iam deactivate-mfa-device --user-name RBI_S3 --serial-
number arn:aws:iam::451339973440:mfa/rbis3
```

If successful, there is no output. We can verify that the RBI_S3 user account no longer has the MFA device associated with it by listing the devices once more, as illustrated in the following screenshot:

```
[jonlehtinen@ ~ % aws iam deactivate-vir
[> aws iam deactivate-mfa-device --user-name RBI_S3 --serial-number arn:aws:iam::451339973440:m
fa/rbis3
[jonlehtinen@ ~ % aws iam list-virtual-mfa-devices
VirtualMFADevices:
- SerialNumber: arn:aws:iam::451339973440:mfa/rbis3
- EnableDate: '2020-12-09T00:17:06+00:00'
  SerialNumber: arn:aws:iam::451339973440:mfa/root-account-mfa-device
  User:
    Arn: arn:aws:iam::451339973440:root
    CreateDate: '2020-11-09T16:56:07+00:00'
    PasswordLastUsed: '2020-12-09T00:39:13+00:00'
    UserId: '451339973440'
- EnableDate: '2020-12-13T19:17:44+00:00'
  SerialNumber: arn:aws:iam::451339973440:mfa/googleauth1
  User:
    Arn: arn:aws:iam::451339973440:user/redbeardidentity
    CreateDate: '2020-11-12T01:11:46+00:00'
    PasswordLastUsed: '2020-12-13T19:22:54+00:00'
    Path: /
    UserId: AIDAWSFPVONALTHHLBKLK
    UserName: redbeardidentity
jonlehtinen@ ~ %
```

Figure 3.38 – The AWS account's MFA devices, with only two associated with user accounts

We now see that the device affiliated with the RBI_S3 user account is still listed, but this no longer includes user information when we execute the list-virtual-mfa-devices command. We can now delete it with the following command:

```
$ aws iam delete-virtual-mfa-device --serial-number
arn:aws:iam::451339973440:mfa/rbis3
```

Once again, if successful, there will be no output.

Thus far, we've addressed password policy and management, programmatic credential management, and MFA credential management for our own user accounts. With identity federation, authentication is delegated to an external identity provider, so we will not have the same credential management concerns. In fact, in certain federation use cases, we will not even have user accounts for our federated users at all. Let's take a closer look at federated user accounts on AWS IAM.

Managing federated user accounts

We've focused primarily on AWS IAM-managed user accounts in this chapter. Recall the distinction between a user account—referring to the AWS IAM user object, which a principal uses to identify itself to access AWS resources—and a principal, which is an end user of the system in a general sense. We've discussed at length how principals may use an AWS IAM-managed user account to access AWS resources; however, that is not the only way principals may do so.

Many organizations manage their own enterprise identities and would prefer to maintain control over the accounts and credentials that employees use when accessing business applications. Similarly, service providers or relying parties benefit from not needing to maintain an account's credentials. As we saw in the **Redbeard Identity (RBI)** example in *Chapter 1, An Introduction to IAM and AWS IAM Concepts,* the RBI organization would provision an account into various **software-as-a-service (SaaS)** providers, and Bob could use his RBI-issued username and password to sign in to those SaaS platforms using identity federation. Through identity federation, a service provider or relying party (such as a SaaS platform) delegates authentication of the principal to an external identity provider, as illustrated in the following diagram:

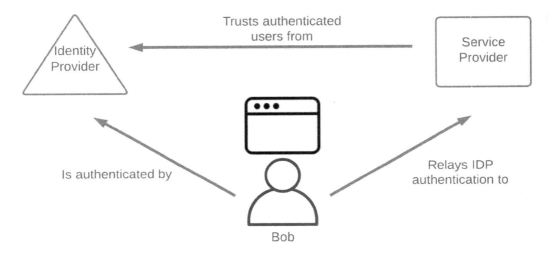

Figure 3.39 – Federated authentication relationships

AWS IAM supports federated authentication from an external identity provider. This allows organizations to retain lifecycle control of their accounts and ensure that access to the AWS account is revoked upon termination of the organization's user account. If the user cannot sign in to their work account, they cannot access AWS resources either.

Identity federation deals primarily with authentication; many service providers merely consume an authentication token from the identity provider and maintain no local record or account. In other cases, where the application has need of its own identity store and attributes for that federated user for certain application functions, the service provider might maintain a local, corresponding user record. That record can be created and updated in various ways. **Just-in-time** (**JIT**) provisioning creates new records and updates existing records each time a new authentication token arrives. Organizations can send flat files or data extracts from their authoritative sources to keep the SaaS user store and the organization's directories in sync. A standards-based method is federated identity provisioning through the **System for Cross-domain Identity Management** (**SCIM**), which offers a connector that provides RESTful **create, read, update, and delete** (**CRUD**) operations (where **REST** stands for **REpresentational State Transfer**) for maintaining user information in the directory. Suffice to say, whereas federated authentication is comparatively straightforward, managing remote identity stores remains a bit more scattershot.

What is interesting about the federated authentication use case on AWS IAM is that there are no AWS IAM user accounts involved—only the accounts managed by the external identity provider. There is no requirement to map a federated user to an AWS IAM-managed user account. If an authenticated federated user does not get correlated to an IAM user account, then which AWS IAM-managed entity is used to identify the federated user so that authorization may be applied to them?

At an exceedingly high level, when federating into an AWS account, the principal's authentication token from their **identity provider** (**IdP**) is exchanged through AWS **Secure Token Service** (**STS**) for a temporary credential, which is scoped to allow the principal to assume a role from AWS IAM. The authorization policies associated with that role determine what a federated user can and cannot do while their temporary credentials are valid.

The process is illustrated in the following diagram:

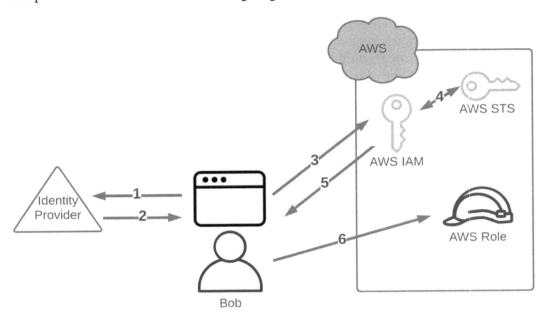

Figure 3.40 – Federated user assumes a role in an AWS account

We just introduced several topics and concepts in the preceding sentence that, while germane to federated user authentication and authorization, merit their own chapters in and of themselves. Don't worry for now if you don't fully understand the moving pieces; the important part is to understand how federated users are handled differently compared to AWS IAM users. *Chapter 9, Bringing Your Admins into to the AWS Backplane*, will delve deeply into this topic.

AWS Single Sign-On and federated users

AWS Single Sign-On (**AWS SSO**) is a relatively new service within AWS, aimed at solving complex AWS identity federation use cases. Large organizations often have several AWS accounts to manage and maintain ad hoc federated relationships between their existing IdP, and each of those AWS IAM instances grows into an increasing administrative burden with each new AWS account the organization spins up.

The following diagram illustrates this:

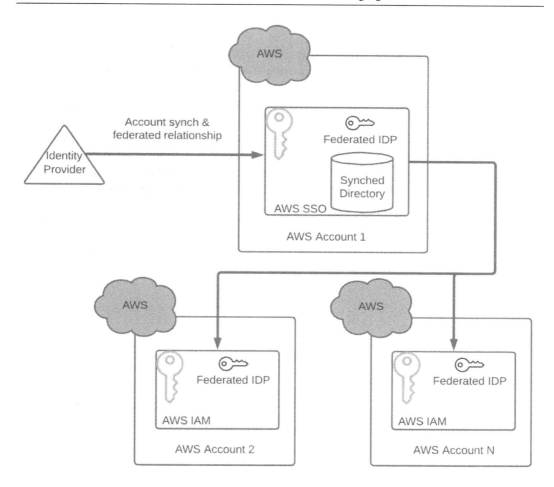

Figure 3.41 – AWS organizations and AWS SSO propagating IdPs in downstream AWS accounts

AWS SSO aims to solve this complexity by providing one connection, both for directory synchronization from an on-premises user store and as a single-service provider connection for all AWS accounts in use by the organization. The AWS SSO connection automatically registers the existing IdP as an IdP within the AWS IAM service of each downstream AWS account. This way, organizational administrators only need to tend the connection between their IdP and the main organizational AWS account, and every other AWS account included in their organization will be able to take advantage of delegated authentication without the per-account administrative overhead. This too will be examined in much more detail in later chapters, such as in *Chapter 9, Bringing Your Admins into the AWS Administrative Backplane*.

Summary

Now that you've made it through this chapter, you have a much better understanding of the best practices for administrating and securing AWS IAM-managed user accounts, including the root account. Additionally, you have learned why the root account merits extra consideration and why certain administrative functions are best left to managed IAM user objects. This chapter also increased your understanding of password, access key, and MFA device management within an AWS account, including how to perform those functions programmatically using the AWS CLI. Finally, you were introduced to what makes federated users different from AWS IAM users, in order to ensure you had a complete understanding of how principals use both to interact with AWS services.

Now that we have discussed managing our AWS IAM users and the various ways we can authenticate them, it is time to turn our attention to controlling what they can do within an AWS account afterward. This is access management and authorization. In the next chapter, we will take a look at AWS authorization policies, including the various types, how they are written, and what they can do.

Questions

1. What is a principal?

2. How does an IAM user account differ from a root account?

3. Why is not considered a best practice to use access keys with a root account?

4. What is MFA, and how does it improve account security?

5. What kind of multifactor authenticators can be used with any IAM user, including a root account, to access the Management Console?

6. Describe how federated users access AWS resources, and how that differs from AWS IAM users.

4
Access Management, Policies, and Permissions

The Access Management model of AWS is based on policies. At a high level, we can use these policies to determine what an AWS identity object or resource can do to or with a resource or service within an AWS account. Of course, this quickly becomes very complicated once we must apply and manage policies across the multiple places where they may have been applied. We may also need to customize the existing access policies within an AWS account, or even create new policies from scratch to accommodate our own use cases. This chapter does not represent a complete compendium of knowledge regarding the complexities of access management in AWS, but it will introduce you to the foundational concepts required to understand and solve several common authorization use cases.

With this knowledge as your foundation, you will be able to strategically select further areas of study based on your own personal or professional requirements.

The chapter will cover the following topics:

- What is access management?
- Introducing the AWS access policy types
- The anatomy of an AWS JSON policy document
- Exploring the AWS policy types
- Governance

Technical requirements

To get the most out of this chapter, you will need the following:

- An AWS account
- A workstation running the AWS CLI
- A text editor or IDE to edit JSON/YAML files, such as Microsoft Visual Studio Code

What is access management?

As the two words that make up the "AM" of IAM, **access management** represents one of the core functions of IAM as an enabling technology. Access management covers two things, the first of which is the validation that a request comes from a legitimate source. To frame that in AWS IAM terms, it means that it can provide the shared secrets affiliated with their IAM user account to prove the request is valid. This is the **authentication** side of access management; we dealt with how to authenticate AWS IAM user accounts in depth in *Chapter 3, IAM User Management*.

The second function of access management is to make sure that the request itself is authorized. This is to say that there is nothing about the request, such as the target of the action, or the location of the object or requestor, or anything else that runs afoul of the rules that apply to that specific request and requestor, which then determine what they should and should not be able to do. This is the **authorization** component of access management. In AWS IAM, requests are authorized by evaluating several different policy types to determine what an IAM user object or resource can and cannot do when they make a request.

Introducing the AWS access policy types

We've mentioned the word policy before. In an organizational, regulatory, or legal setting, a policy represents the rules, patterns, and guidance meant to steer a decision-making process. In the context of IAM, a policy is how things such as business logic, security controls, and compliance requirements are translated into an access management system, such as AWS IAM. Within AWS IAM, policy are objects that specifically spell out the permissions of a principal or resource they are attached to. This can be seen in the following diagram:

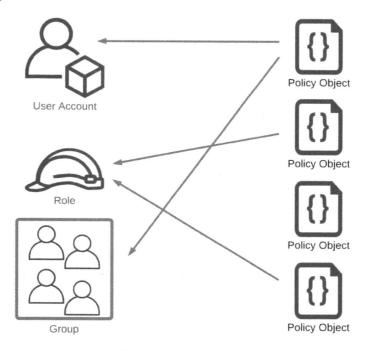

Figure 4.1 – An example of policy objects that can apply to one or more AWS objects

Access policies can apply to IAM objects, as shown in the preceding diagram. They can also apply to specific AWS objects, such as S3 buckets, or even across multiple AWS accounts under the management of an AWS Organization.

In some regards, an AWS access policy can be thought of as a "mix and match" system containing infinitely combinable access control snippets. Depending on the type of policy used, a given policy can exist within AWS IAM, but it will only affect an action if it is associated with an object. A single policy object may be associated with a single object, no objects at all, or thousands of objects. A single AWS resource may have a single policy applied to it, or as many as can be accommodated by the AWS IAM service quotas (not that the latter would be a desirable administrative situation).

There are six high-level policy types that are used in AWS IAM that we will discuss in detail momentarily. These policy types are as follows:

- Identity-based policies, which include managed policies and inline policies
- Resource-based policies
- Permissions boundaries
- Organizations service control policies
- Access control lists
- Session policies

In order to understand the nuances between these six main policy types, we will also introduce some additional policy classifications that are based on how a policy is created and managed, and how that policy becomes associated with an AWS identity or resource. We will discuss those additional policy sub-classifications in the context of their relationship to the six major policy types. These additional policy classifications are as follows:

- Managed policies, which include the following:

  i) AWS-managed policies

  ii) Customer-managed policies
- Inline policies

These policy classifications are more of a mechanical description of the policy, indicating how a given policy is managed and attached to the AWS resources they affect. The main policy types describe what kind of policy they are and where they are applied in the AWS IAM policy evaluation flow.

Next, let's take a look at how these policies are constructed and evaluated using AWS's JSON policy language.

The anatomy of an AWS JSON policy document

Most, but not all, policy types are written and stored as JSON documents within AWS. These include identity-based policies, resource-based policies, permissions boundaries, organizational service control policies, and session policies. Access control lists use a distinct syntax, depending on the service where it is being applied:

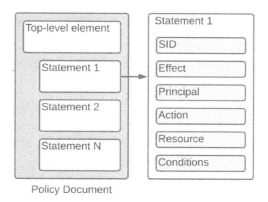

Figure 4.2 – A policy document, its components, and the elements of its statements

Let's take a look at the preceding diagram. Here, we can see a logical example of a policy document and its constituent components. Within the main document, there is a **Top-level element** that contains policy-wide information, followed by one or more statements. Each of these statements describes a permission or entitlement through the information contained within its individual elements.

We can see these elements within the JSON of a policy document by examining any of the existing policies available to us within our AWS account. Let's take a look at the one that we are already familiar with – the AdministratorAccess policy:

```
{
    ''Version'': ''2012-10-17'',
    ''Statement'': [
        {
            ''Effect'': ''Allow'',
            ''Action'': ''*'',
            ''Resource'': ''*''
        }
    ]
}
```

Though this is a powerful policy, it is comparatively simple. We can see the Version and Statement top-level elements, and the elements of the statement contained within. Without instructions, we could likely infer what this policy does and how those asterisks operate, given how we've used it so far. Notice that no principal element is explicitly defined. This is relevant to the type of policy that this is. The reason why there is no explicit principal element will be explained as we dive deeper into the six main policy types in the *Exploring the AWS policy types* section.

Defining JSON policy document elements

Let's take a closer look at what elements are within the JSON policy structure. Please note that unless specified otherwise, all the items are elements within the statement section of the policy document:

- `Version` (Policy top-level element): This defines the version of the policy language that's used when writing the policy document. AWS' documentation indicates that a best practice is to always use the latest version, which is currently `2012-10-17`. Please note that this version refers to the version of the policy language used within the policy document, *not* the version of the policy document itself. The `Version` element is required:

  `''Version'': ''2012-10-17''`

- `Id` (Policy top-level element): A unique ID for the policy document. This is an optional element. Best practice recommends ensuring that a unique value is used for this element if it's used since certain services require it to be unique. Setting the `Id` value to a **Universally Unique Identifier (UUID)** is recommended:

  `''Id'': ''cd3ad3d9-2776-4ef1-a904-4c229d1642ee''`

- `Statement` (Policy top-level element): This is the main element of the policy that houses the stuff that describes a permission. The top-level statement can house a single statement or multiple statements within itself. This is required. Note the differing bracket and braces requirements required due to JSON notation:

  i) For a single statement: `''Statement'': {...}`

  ii) For multiple statements: `''Statement'': [{...},{...},{...}]`

- `Sid`: Sid stands for "statement ID," and it lets us optionally apply a unique identifier or version to our policy statement. Each Sid must be unique within a policy document:

  `''Sid'': ''1''`

- `Effect`: This is the "verb" of the policy that determines what a principal subject to the policy can or cannot do. It may be set to either `Allow` or `Deny`. By default, AWS denies access to resources unless explicitly allowed by the policy:

  `''Effect'':''Allow''`

- `Principal`: This is the "subject" of the policy statement and determines who or what should or should not be allowed access to a resource. As several things may act as a principal during an action, the formatting for the subject of this element varies, depending on the specific use case, policy, and principal in play.

   i) **AWS accounts**: If an account is named as the principal in a statement, authorization to access the resources in that account will be delegated to all the identity objects in the account named in the principal element, assuming those objects in the account named in the principal element have policies granting them access to those resources:

   `''Principal'': { ''AWS'': ''451339973440'' }`

   ii) **IAM users**: These are specific IAM user accounts:

   a) For a single user: `''Principal'': { ''AWS'': ''arn:aws:iam:: 451339973440:user/redbeardidentity'' }`

   b) For multiple users: `''Principal'': {''AWS'': [''arn:aws:iam::4 51339973440:user/RBI_EC2'',''arn:aws:iam::451339973440:u ser/RBI_S3'']}`

   iii) **Federated web identity users**: These are federated users that are authenticated by a social identity provider such as Google, Facebook, or Cognito:

   `''Principal'': { ''Federated'': ''accounts.google.com'' }`

   iv) **Federated SAML users**: Users federated from an external SAML identity provider registered within the account's AWS IAM service:

   `''Principal'': { ''Federated'': ''arn:aws:iam:: 451339973440:saml-provider/rbi_idp'' }`

   v) **IAM roles**: An assumed role within the account:

   `''Principal'': { ''AWS'': ''arn:aws:iam::451339973440:r ole/rbi_s3_admin_role'' }`

   vi) **Assumed-role sessions**: Assumed role sessions are unique identifiers for a role that may be assumed by several different principals. By assigning a unique name to the assumed role session, we can track activity to a distinct principal's activity, though many principals may be entitled to assume a specific role:

   `''Principal'': { ''AWS'': ''arn:aws:sts::451339973440 :assumed-role/rbi_s3_admin_role/rbi_account2_unique_ session_name'' }`

vii) **AWS services**: Other AWS services can assume IAM roles through something called **service roles**. These service roles perform functions within the AWS account on behalf of a principal. These service roles are governed by inline resource-based policies called **trust policies**. Trust policies define the principal as the AWS service assuming the service role:

a) For a single service: `''Principal'': {''Service'': {''eks.amazonaws.com''}}`

b) For multiple services: `''Principal'': {''Service'': [''eks.amazonaws.com'',''codebuild.amazonaws.com'']}`

viii) **Anonymous users**: Some AWS services allow resource-based policies to accommodate public access through a wildcard operator in the `Principal` value. Consider a use case such as allowing anyone to view images hosted in an S3 bucket. This is not considered a best practice and should be tempered with something such as a condition to preclude unfettered access to the role:

`''Principal'': ''*''`

- `NotPrincipal`: This is an exclusionary version of the principal element. If a policy using `NotPrincipal` includes an allow action, the principals named in the `NotPrincipal` element would be explicitly denied access, and all other principals, *including anonymous principals*, could be granted access. If a policy using `NotPrincipal` included a deny action, the principal's names in the `NotPrincipal` element would potentially be allowed access to the resource if another policy explicitly granted them access to the resource, whereas every other principal would be excluded. As this gets confusing very quickly, a best practice is to limit use of `NotPrincipal` for use cases where there is no other option.

- `Action`: This defines which capabilities of a service are either allowed or denied by the policy document. Each AWS service has their own collection of actions. In a policy document, the actions of a service are prefixed by the service's name, such as `ec2:AllocateHosts` or `ecs:CreateCluster`. Wildcards may be used to indicate all the actions on all services (such as what we've seen in the `AdministratorAccess` policy), or they may be used after the initial service prefix to indicate full access or exclusion to just the actions available within that specific service, such as `ec2:*`. The wildcard can also be used to subdivide actions within a service. For example, `ec2:Create*` will address all the create actions available within the EC2 service.

- NotAction: The exclusionary version of the Action element. The actions listed under a NotAction, in combination with an Allow Effect, are explicitly denied to the principals listed in the policy. If a Deny Effect is used, all actions except those listed under the NotAction element are explicitly denied, though the principal will only be able to take the actions listed under NotAction if a different statement or policy explicitly allows access.

- Resource: This is the object or wildcard group of objects that the statement is determining access to. How these are specified may vary from service to service, but generally, they may be explicitly named by **Amazon Resource Name (ARN)**, a list of ARNs, or a wildcard indicating every resource matching the ARN up to a certain path. The resource element may also use a **policy variable**, which is a placeholder value that is replaced and evaluated during runtime based on the request context.

- NotResource: The exclusionary version of the Resource element that explicitly matches on every resource except those listed under that element. When used with a Deny Effect, it explicitly denies access to all resources not listed. It is not recommended to use NotResource with the Allow Effect as it would grant permissions to all resources not listed in the NotResource element.

- Condition: The Condition element places requirements on the policy taking affect. The condition may be simple or complex. Condition also supports policy variables.

> **Tip**
> Both policy variables and all the available operations available within the Condition element are tremendously useful when building maintainable policy objects, though they would require their own chapters to do them justice. For more information, please check out the *Further reading* section.

Now that we have examined the structure of the JSON policy documents and seen how each of the elements work, let's look at the policy types available to us in AWS in detail.

Exploring the AWS policy types

There are more than a few policy objects available within AWS. Every request and action within an AWS account is evaluated against these policies at execution time. Since that is a lot of moving parts determining permissions, let's take a look at the six major policy types and how they are used.

Identity-based policies

Identity-based policies are the policies that determine what an identity object can do. These policies are JSON documents that spell out the user, group, or role that can perform the action, the resources that those actions can be performed on, and the conditions under which those actions are valid. These identity-based policies are better understood by some further categorization into three additional policy types, which we will now explore in greater detail.

AWS managed policies

AWS IAM comes prepopulated with several hundred policy objects. They are not natively used in a new AWS account; they are simply available for use to facilitate good access management practices. These policies are created and maintained by AWS. They cannot be edited by any of our AWS IAM users, nor by our root account, as they are considered standalone policies. This means they have their own unique ARN, and that ARN *doesn't* reference any AWS account number. AWS managed policies are globally referenceable.

These policies are advantageous to use as they are maintained by AWS. This means that as the services, methods, and capabilities change within the platform itself, the function and access control boundaries provided by managed policies will be adjusted accordingly. Given the rate of feature introduction in AWS, we can save ourselves significant administrative burden by sticking to these policies.

We can review the AWS managed policies from the Management Console by going to the **Policies** section under **Access Management** on the IAM Dashboard. The AWS managed policies have an orange icon next to them to distinguish them from other policy types. Additionally, they are clearly labeled in the **Type** column of the policy list:

		Policy name ▾	Type	Used as	Description
○	▶	AWSSystemsManagerChange...	AWS managed	None	Provides access to AWS resources managed or used by the AWS Syste...
○	▶	AWSThinkboxAssetServerPolicy	AWS managed	None	This policy grants the AWS Portal Asset Server the necessary permission...
○	▶	AWSThinkboxAWSPortalAdmin...	AWS managed	None	This policy grants AWS Thinkbox's Deadline software full access to multi...
○	▶	AWSThinkboxAWSPortalGatew...	AWS managed	None	This policy grants the AWS Portal Gateway machine the necessary permi...
○	▶	AWSThinkboxAWSPortalWorke...	AWS managed	None	This policy grants the Deadline Workers in AWS Portal the necessary per...
○	▶	AWSThinkboxDeadlineResourc...	AWS managed	None	Grants permissions required for the operation of AWS Thinkbox's Deadlin...
○	▶	AWSThinkboxDeadlineResourc...	AWS managed	None	Grants permissions required to create, destroy, and administer AWS Thin...
○	▶	AWSThinkboxDeadlineSpotEve...	AWS managed	None	Grants permissions required for AWS Thinkbox's Deadline Spot Event Pl...
○	▶	AWSThinkboxDeadlineSpotEve...	AWS managed	None	Grant permissions required for an EC2 instance running AWS Thinkbox D...
○	▶	AWSTransferConsoleFullAccess	AWS managed	None	Provides full access to AWS Transfer via the AWS Management Console
○	▶	AWSTransferFullAccess	AWS managed	None	Provides full access to AWS Transfer Service.
○	▶	AWSTransferLoggingAccess	AWS managed	None	Allows AWS Transfer full access to create log streams and groups and pu...
○	▶	AWSTransferReadOnlyAccess	AWS managed	None	Provide readonly access to AWS Transfer services.
○	▶	AWSTrustedAdvisorReportingS...	AWS managed	None	Service Policy for Trusted Advisor Multi-account Reporting

Figure 4.3 – AWS managed policies listed in the IAM Dashboard

AWS designed many of these policies based on common use cases and job functions. We've already taken advantage of one such managed policy when we set about complying with the best practice guidance around not using the root account, by creating a full administrator IAM user account to use instead. Earlier, we created the **FullAdministrator** IAM group to manage admin access to our account. Rather than write our own access policy for a super user, we attached an existing managed policy that would handle this for us:

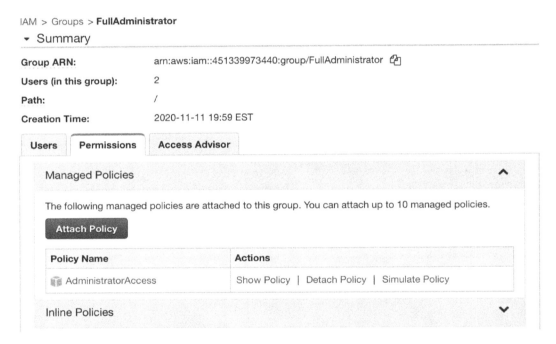

Figure 4.4 – The AdministratorAccess AWS managed policy attached to the FullAdministrator group

We can also review AWS managed policy objects using the CLI. As there is no icon to help indicate which of the policies listed are managed when using the `list-policies` command, we can use a parameter to ensure only AWS managed policies are returned instead:

```
$ aws iam list-policies --scope AWS
```

As there are just under 800 of these policies, it still returns more information than we can comfortably scrub through on screen. However, they are all AWS managed policies:

```
- Arn: arn:aws:iam::aws:policy/AWSDirectConnectReadOnlyAccess
  AttachmentCount: 0
  CreateDate: '2015-02-06T18:40:08+00:00'
  DefaultVersionId: v4
  IsAttachable: true
  Path: /
  PermissionsBoundaryUsageCount: 0
  PolicyId: ANPAI23HZ27SI6FQMGNQ2
  PolicyName: AWSDirectConnectReadOnlyAccess
  UpdateDate: '2020-05-18T18:48:22+00:00'
- Arn: arn:aws:iam::aws:policy/AmazonGlacierReadOnlyAccess
  AttachmentCount: 0
  CreateDate: '2015-02-06T18:40:27+00:00'
  DefaultVersionId: v2
  IsAttachable: true
  Path: /
  PermissionsBoundaryUsageCount: 0
  PolicyId: ANPAI2D5NJKMU274MET4E
  PolicyName: AmazonGlacierReadOnlyAccess
  UpdateDate: '2016-05-05T18:46:10+00:00'
- Arn: arn:aws:iam::aws:policy/AWSMarketplaceFullAccess
  AttachmentCount: 0
  CreateDate: '2015-02-11T17:21:45+00:00'
  DefaultVersionId: v3
  IsAttachable: true
  Path: /
  PermissionsBoundaryUsageCount: 0
  PolicyId: ANPAI2DV5ULJSO2FYVPYG
  PolicyName: AWSMarketplaceFullAccess
  UpdateDate: '2018-08-08T21:13:02+00:00'
- Arn: arn:aws:iam::aws:policy/aws-service-role/ClientVPNServiceRolePolicy
  AttachmentCount: 0
  CreateDate: '2018-12-10T21:20:25+00:00'
  DefaultVersionId: v5
  IsAttachable: true
:
```

Figure 4.5 – Managed policies, as viewed through the CLI

We can confirm they are AWS managed policies since their ARN does not include an AWS account number. This indicates these objects exist outside of our own or anyone else's account, though they are available for us to use and reference.

Customer managed policies

A **customer managed policy** is a policy object that is not managed by AWS, so it's being created by, and thus the responsibility of, the AWS account owner. These are referred to as "customer managed" because AWS is an infrastructure as a service platform. Whereas we may think of ourselves in terms of administrators and owners, from the perspective of AWS, we are their customers. If AWS does not manage something in an environment, then the customer does.

Customer managed policies are useful for addressing an organization's unique access requirements that are not satisfied by any of the AWS managed policies. As the name implies, the onus of maintaining these policies is with the customer in light of continuous AWS service iteration, which may introduce new services and functionality that are not accounted for by the original policy. If we decide to create our own policies, it is considered a best practice to use an existing AWS policy as a base template and make our modifications as needed.

Customer managed policies are viewed and administrated the same way as AWS managed ones are. In the Management Console, they are labeled **Customer managed** under **Type** and are bereft of the icon the AWS managed policies have:

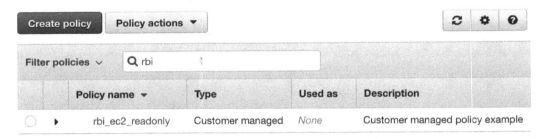

Figure 4.6 – Customer managed policy in the Management Console

We can also review these policy objects using the CLI. Using the `list-policies` command, we can apply a different parameter to limit the scope to returning only customer managed policies:

```
$ aws iam list-policies --scope Local
```

This gives us all the policies we created ourselves within our AWS account:

```
jonlehtinen@~ % aws iam list-pol
> aws iam list-policies --scope Local
Policies:
- Arn: arn:aws:iam::451339973440:policy/rbi_ec2_readonly
  AttachmentCount: 0
  CreateDate: '2020-12-21T17:38:40+00:00'
  DefaultVersionId: v1
  IsAttachable: true
  Path: /
  PermissionsBoundaryUsageCount: 0
  PolicyId: ANPAWSFPVONAERAJADVZD
  PolicyName: rbi_ec2_readonly
  UpdateDate: '2020-12-21T17:38:40+00:00'
- Arn: arn:aws:iam::451339973440:policy/selfServiceSecurityCredentialManagement
  AttachmentCount: 0
  CreateDate: '2020-12-13T18:11:43+00:00'
  DefaultVersionId: v1
  IsAttachable: true
  Path: /
  PermissionsBoundaryUsageCount: 0
  PolicyId: ANPAWSFPVONAFZIZEFRRT
  PolicyName: selfServiceSecurityCredentialManagement
  UpdateDate: '2020-12-13T18:11:43+00:00'
jonlehtinen@~ %
```

Figure 4.7 – All customer managed roles in our AWS account

Another difference between these policies and those managed by AWS is the ARN. Customer managed policies include the AWS account in the ARN and are only available locally within that AWS IAM instance. Both AWS managed and customer managed policies are considered managed policies. Managed policies exist as their own objects within AWS IAM. They have their own ARN, they may be iterated upon, and they can be attached to many, or no, identity or resource objects to apply authorization.

Both AWS managed and customer managed policies may be referred to as managed policies. This is to distinguish them from the next policy variation that's included in identity-based policies.

Inline policies

In contrast to managed policies, inline policies are applied directly to an object. An **inline policy** functionally becomes part of the object itself; there is no longer a relationship between a resource and a policy object that determines what that resource's permissions are – there is now an innate property of that object that determines this:

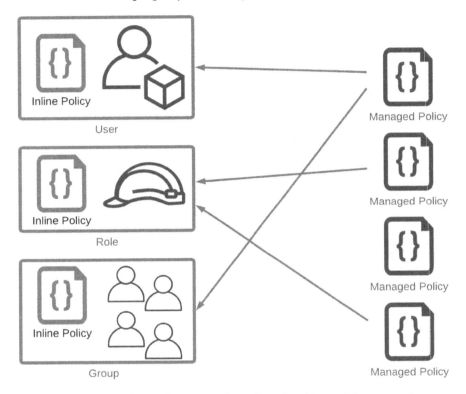

Figure 4.8 – Inline policies innately apply to the object, while managed
policies are associated with objects

Inline policies exist specifically in the context of the single object that they are attached to, meaning a specific inline policy applied to one object will not impact any other inline policy, nor the access of any other identity object. An object with an inline policy may still have managed policies applied to it, though administrators will need to be cognizant of how a given inline policy could adversely impact the expected functionality granted by a managed policy. Inline policies are not the most efficient way to administrate authorization in AWS at scale. They are best used for narrow, purpose-built policies that will only apply to a single identity.

Let's say we've decided that we will never have enough AWS administrators who would merit the level of access that the FullAdministrator group grants its members, and as such, we would like to delete that group. We still want the redbeardidentity user account to retain its super user capabilities; we are just eliminating that group as an avenue for any IAM user account to do so. If we wanted to ensure that only the redbeardidentity user account had super administrator capabilities *as an innate characteristic of its object*, we can create an inline policy to do so. Let's do just that:

1. From the AWS Management Console, open the FullAdministrator group. We need to examine the AWS managed policy that gives the FullAdministrator group its privileges to replicate them via an inline policy on the redbeardidentity user account, so we will click **Show policy** next to that managed policy:

Show Policy   ⊠

```
{
  "Version": "2012-10-17",
  "Statement": [
    {
      "Effect": "Allow",
      "Action": "*",
      "Resource": "*"
    }
  ]
}
```

Cancel

Figure 4.9 – AdministratorAccess policy document

2. This pops up a window with the JSON policy document on it. For now, we just need to capture this document as it will become the basis for our own inline policy.

3. Next, we will hop over to the user management side of the IAM Dashboard and open the `redbeardidentity` user record. The first tab shows us all the permissions that have been applied to that user, with the `AdministratorAccess` managed policy included in the list as being inherited from a group. We can even see that JSON again by expanding the policy:

Figure 4.10 – Policies applied to redbeardidentity

4. Click **Add inline policy** to go to the policy creation form. We can either manually sort through the services and actions by using the visual editor, or we can select the **JSON** tab and write the policy document directly. Regardless of how we choose to edit the policy, since we know the exact policy whose permissions we are trying to recreate for this account, we can follow a best practice and base our custom inline policy off that existing AWS managed policy. We can do this by selecting **Import managed policy** and selecting **AdministratorAccess** from that list to import it:

Figure 4.11 – Policy creation form prior to importing the existing AdministratorAccess policy

5.   The permissions and the JSON for that AWS managed policy now appear in our policy creation form. Next, we give the policy a name and review the permissions it provides. If everything checks out, we hit the **Create policy** button:

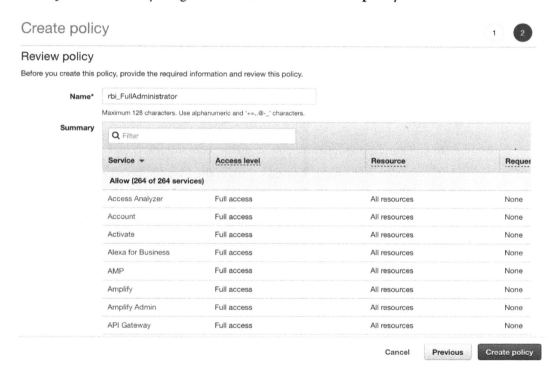

Figure 4.12 – Reviewing and naming our new inline policy

6. Now, back on the `redbeardidentity` user account information screen, we can see our new inline policy included among the policies that apply to that user account. Take note of the new policy's name and how we gave it a prefix to help make finding it easier:

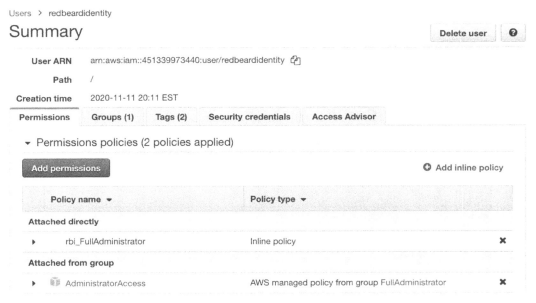

Figure 4.13 – The inline and group-inherited policies on the redbeardidentity user account

7. Finally, we will remove `redbeardidentity` from the `FullAdministrators` group. If the new inline policy is functioning as we expect it should, the account's access should not change thanks to the new inline policy:

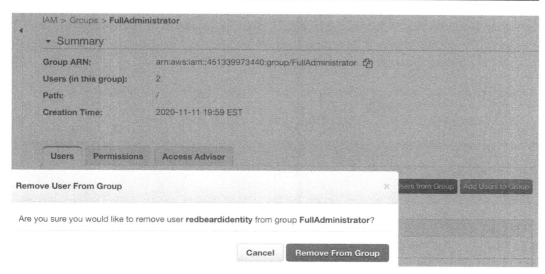

Figure 4.14 – Account has been removed from the FullAdministrator group

Even though we created that inline policy in the policy editor, named it, and saved it, it will not appear among our AWS account's listed policies. We named the policy `rbi_FullAdministrator`. Using the CLI, we can list all our account's local policies to see if it shows up:

```
$ aws iam list-polices --scope Local
```

We only get a single unrelated policy in response:

```
jonlehtinen@~ % aws iam list-policies --scope Local
Policies:
- Arn: arn:aws:iam::451339973440:policy/rbi_ec2_readonly
  AttachmentCount: 0
  CreateDate: '2020-12-21T17:38:40+00:00'
  DefaultVersionId: v1
  IsAttachable: true
  Path: /
  PermissionsBoundaryUsageCount: 0
  PolicyId: ANPAWSFPVONAERAJADVZD
  PolicyName: rbi_ec2_readonly
  UpdateDate: '2020-12-21T17:38:40+00:00'
jonlehtinen@~ %
```

Figure 4.15 – All managed policies in the account

This reinforces the concept that unlike managed policies, inline policies are truly part of the object they are applied to. We can find the new policy that we created by looking at the specific object where that policy was applied:

```
$ aws iam list-user-policies --user-name redbeardidentity
```

This shows us our policy, embedded within the user that it is part of:

```
jonlehtinen@~ % aws iam list-user-policies --user-name redbeardidentity
PolicyNames:
- rbi_FullAdministrator
jonlehtinen@~ %
```

Figure 4.16 – Inline user policies on the redbeardidentity account

If we imagine an organization that were to apply a standard inline policy to each of its user objects, that organization would functionally manage a unique instance of that standardized user policy for each user object. Comparatively, if that organization attached a standard managed policy to each of its user objects, it would only manage a single policy that applies to every user object. If there are use cases where it would be desirable to use and maintain inline policies for certain unique objects, it is considered a best practice to take advantage of the centralized administration that managed policies offer.

So, to recap: identity-based policies are policy objects that apply to identity objects, such as users, groups, and roles. There are various policy types that can contribute to identity-based policies, such as managed policies and inline policies. Managed policies that are created and maintained by AWS are AWS managed, and account-specific managed policies that are created and administrated by an account administrator are customer managed. Managed policies can apply to several identity objects at once, whereas inline policies are unique to the object where they are written and applied.

Next, we will look at resource-based policies, which are policy statements that are applied directly to resources.

Resource-based policies

The second of the main six AWS IAM access management policy types is resource-based policies. **Resource-based policies** are a lot like identity-based policies in that they are JSON documents that describe what principals can perform actions, and what actions those principals are allowed to do. The difference between these resource policies and identity policies is that resource policies describe the full access policy as it relates to that resource, including the principals allowed to perform the actions. Identity policies *imply* who the principal of the policy is by virtue of the policy being attached to them.

To help us visualize this, look at the following diagram. To aid with readability, we've omitted the extra JSON formatting found in real policy documents and just left the relevant policy structures in the diagram. Here, we can see two of our IAM user accounts: `redbeardidentity` and `RBI_S3`. The `redbeardidentity` account has an identity policy attached to it that gives it full administrative access to all the resources within our account. The `RBI_S3` account has no identity policies applied to it; outside of authenticating to the tenant, it cannot do much. The policy on the right implies what the principals are through its attachment to a user account. A principal is not specifically enumerated inside that policy document:

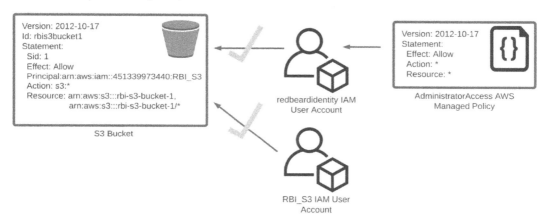

Figure 4.17 – Identity policies versus resource policies

If we look to the left, we can see that the `rbi-s3-bucket-1` S3 bucket has a resource policy applied to it. This policy specifically names the principals that can act on the resource, as well as the specific actions that principal can take on that resource. Despite having no identity policies applied to it, the bucket's resource policy entitles the `RBI_S3` user account to perform all S3 functions on the bucket object itself, as well as all objects contained within that bucket. The `redbeardidentity` user is also entitled to do the same, but that entitlement comes from the AWS managed identity policy that is attached to the user account.

We can see this in action by attempting to list the objects in the rbi-s3-bucket-1 S3 bucket. We can verify that the resource policy has been applied to that bucket through the Management Console by going to the **Permissions** section of that bucket and reviewing the bucket policy. Here is what we will find for the rbi-s3-bucket-1 S3 bucket:

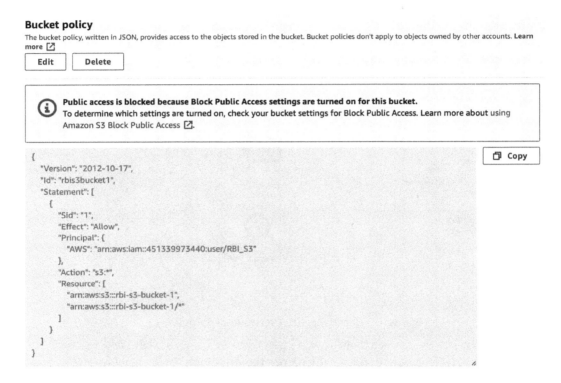

Bucket policy

The bucket policy, written in JSON, provides access to the objects stored in the bucket. Bucket policies don't apply to objects owned by other accounts. **Learn more** ☑

Edit   Delete

ⓘ **Public access is blocked because Block Public Access settings are turned on for this bucket.**
To determine which settings are turned on, check your bucket settings for Block Public Access. Learn more about using Amazon S3 Block Public Access ☑.

☐ Copy

```
{
    "Version": "2012-10-17",
    "Id": "rbis3bucket1",
    "Statement": [
        {
            "Sid": "1",
            "Effect": "Allow",
            "Principal": {
                "AWS": "arn:aws:iam::451339973440:user/RBI_S3"
            },
            "Action": "s3:*",
            "Resource": [
                "arn:aws:s3:::rbi-s3-bucket-1",
                "arn:aws:s3:::rbi-s3-bucket-1/*"
            ]
        }
    ]
}
```

Figure 4.18 – The rbi-s3-bucket-1 resource policy

Using the AWS CLI, we can validate that both the redbeardidentity and RBI_S3 accounts can successfully list that bucket's contents. First, run the list-objects-v2 command using the default profile (which executes under redbeardidentity), and then again with the --profile RBI_S3 parameter to force use of the RBI_S3 account:

```
[jonlehtinen@ ~ % aws s3api list-objects-v2 --bucket rbi-s3-bucket-1
Contents:
- ETag: '"8b90b6ee7d41e2ad8a14876c1620aa0d"'
  Key: HeadshotQuarantine2020.png
  LastModified: '2020-11-24T00:13:55+00:00'
  Size: 4304386
  StorageClass: STANDARD
- ETag: '"4ef64056a7e5209fd3c2fb64d6b4a8a2"'
  Key: headshot_jul2019.jpeg
  LastModified: '2020-11-24T00:04:21+00:00'
  Size: 1115
  StorageClass: STANDARD
[jonlehtinen@ ~ % aws s3api list-objects-v2 --bucket rbi-s3-bucket-1 --profile RBI_S3
Contents:
- ETag: '"8b90b6ee7d41e2ad8a14876c1620aa0d"'
  Key: HeadshotQuarantine2020.png
  LastModified: '2020-11-24T00:13:55+00:00'
  Size: 4304386
  StorageClass: STANDARD
- ETag: '"4ef64056a7e5209fd3c2fb64d6b4a8a2"'
  Key: headshot_jul2019.jpeg
  LastModified: '2020-11-24T00:04:21+00:00'
  Size: 1115
  StorageClass: STANDARD
jonlehtinen@ ~ %
```

Figure 4.19 – Both user accounts can list objects in the bucket

Both accounts could execute these commands, despite the differing policy mechanisms that granted them both access to do so.

In this situation, the `redbeardidentity` user account was able to access that S3 bucket, despite not being a named principal within the resource policy. That user account received that permission from the inline permission granting full administrator access to all the resources within the AWS account earlier. If we wanted to, we could amend that bucket's resource policy to explicitly deny the `redbeardidentity` user account, which would sufficiently override the permissions granted by the inline identity policy for that single bucket resource.

So far, the identity-based policy types and resource policies we have looked at have followed some fairly intuitive access management logic because they are additive, meaning that they grant a principal access to perform actions. The next policy type, permissions boundaries, introduces a different, yet complementary, access management paradigm.

IAM permissions boundaries

The next major policy type is **IAM permissions boundaries**. True to their name, they are policies that define the maximum boundary of permissions for an identity object, regardless of whatever identity or resource policy may counter-indicate this. A permissions boundary does not grant access to anything; it merely sets the limits of the access that the identity object could theoretically have. This is useful for use cases where an organization would like to constrain certain critical administrative functions from subsets of administrative users.

Let's take a look at the RBI_EC2 user account in the Management Console. This was a user account we created and attached a managed policy to that entitled it to have full access to the EC2 service within our account. We saw this option to enable an IAM permissions boundary when we created new IAM user accounts within the AWS Management Console, though we can add permissions boundaries to existing users as well:

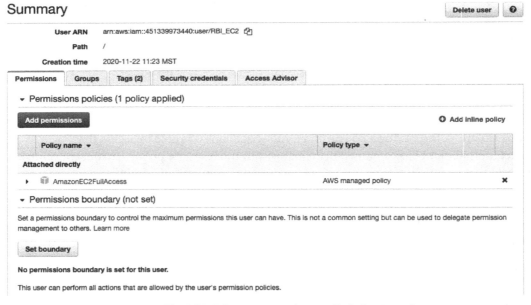

Figure 4.20 – The RBI_EC2 account and its applied identity policies

This user account has a single identity policy applied to it. Let's use this as an opportunity to demonstrate how permissions boundaries work, and why we would consider using them in light of their functionality.

First, we will set a boundary from within the Management Console. Follow these steps:

1. Open the user account from the IAM Dashboard, as shown in the preceding screenshot.

2. Select the **Set boundary** button.

3. IAM permissions boundaries use managed policies to define the access boundary it will apply for the user. We can either use an existing AWS managed policy or create our own to define the boundary. For the purposes of this demonstration, let's simply select the `AmazonEC2FullAccess` AWS managed policy and use that as our permission boundary. Clicking the **Set boundary** button applies the boundary to the user object.

   We can now see the permission boundary applied to the `RBI_EC2` user account:

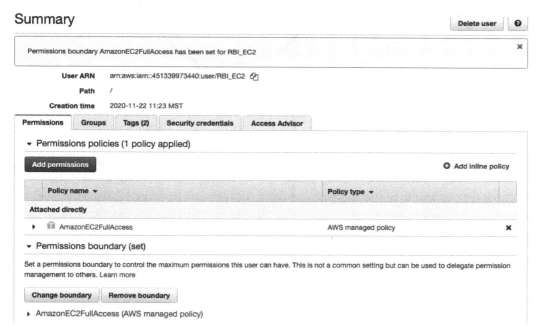

Figure 4.21 – Permission boundary applied to the RBI_EC2 user account

So, how has what we've done here altered this account's access? Strictly speaking, we have not modified its access in any way the principal would notice. All we have done is constrain what this principal can do in the future, if it were given additional access. The following diagram may help us visualize the impact of the permissions boundary more easily:

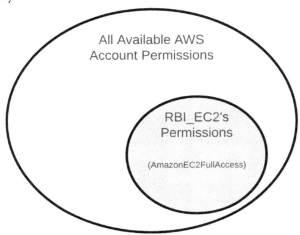

Figure 4.22 – Venn diagram of RBI_EC2's permissions versus all available permissions

In the preceding diagram, we can see how the `AmazonEC2FullAccess` identity policy gave the `RBI_EC2` user account a subset of all available account permissions available. In theory, we could grow that inner circle by attaching additional identity policies to that user account, up to granting it full access to everything that can possibly be done with every service inside of the account, similar to how the `redbeardidentity` user account has full access to every service. If we diagrammed out the permissions of `redbeardidentity` against all the available permissions of the account, the result would be two perfectly overlapping circles:

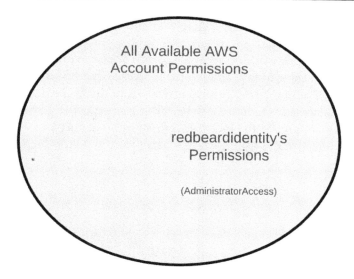

Figure 4.23 – Venn diagram of redbeardidentity's permissions versus all available permissions

Now, let's give `RBI_EC2` an identity policy, just like what `redbeardidentity` has. If `RBI_EC2` did not have a permissions boundary, our Venn diagram would be three overlapping circles. However, since we put a permissions boundary scoped to the `AmazonEC2FullAccess` managed policy on `RBI_EC2`, that account's functional access will be limited to what is within the scope of that permissions boundary, regardless of the more-permissive identity policy it now has:

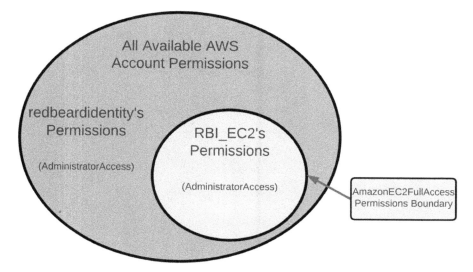

Figure 4.24 – RBI_EC2 account constrained by a permissions boundary,
despite having AdministratorAccess

We can see the permission boundary in action by attempting to do something that is within scope of the identity policy but prohibited by the permissions boundary. Since we know that the AdministratorAccess policy grants access to everything, we can safely assume that we should be able to take action on any service that is inside the permissions boundary. We still need to know exactly what is allowed in the AmazonEC2FullAccess policy in order to validate that the boundary is working as expected. Fortunately, we can. View the policy summary and/or the JSON document in the same place where we applied the permissions boundary to the RBI_EC2 user account. The policy document is as follows:

```
{
    ''Version'': ''2012-10-17'',
    ''Statement'': [
        {
            ''Action'': ''ec2:*'',
            ''Effect'': ''Allow'',
            ''Resource'': ''*''
        },
        {
            ''Effect'': ''Allow'',
            ''Action'': ''elasticloadbalancing:*'',
            ''Resource'': ''*''
        },
        {
            ''Effect'': ''Allow'',
            ''Action'': ''cloudwatch:*'',
            ''Resource'': ''*''
        },
        {
            ''Effect'': ''Allow'',
            ''Action'': ''autoscaling:*'',
            ''Resource'': ''*''
        },
        {
            ''Effect'': ''Allow'',
            ''Action'': ''iam:CreateServiceLinkedRole'',
            ''Resource'': ''*'',
```

```
        ''Condition'': {
            ''StringEquals'': {
                ''iam:AWSServiceName'': [
                    ''autoscaling.amazonaws.com'',
                    ''ec2scheduled.amazonaws.com'',
                    ''elasticloadbalancing.amazonaws.com'',
                    ''spot.amazonaws.com'',
                    ''spotfleet.amazonaws.com'',
                    ''transitgateway.amazonaws.com''
                ]
            }
        }
    }
]
}
```

The first statement indicates that any `ec2` command will be allowed on any resource. As such, we will use an `ec2` command to test something that will be within the scope of the permissions boundary. The final statement indicates that only a very narrow subset of `iam` commands are enabled by this policy. We can use something simple, such as `aws iam list-users`, to test an action that should be out of bounds.

Let's run the `aws ec2-describe-addresses` command under both the RBI_EC2 and redbeardidentity user accounts from the CLI. This command simply describes the elastic IP addresses in use within the EC2 service. If there are none in use (which is the case with our account), a successful response will show an empty list of addresses:

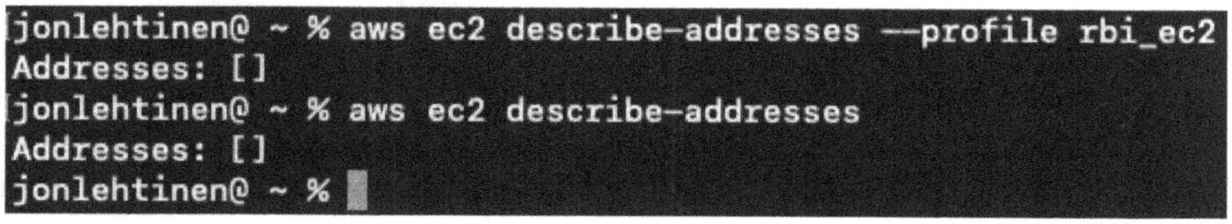

```
jonlehtinen@ ~ % aws ec2 describe-addresses --profile rbi_ec2
Addresses: []
jonlehtinen@ ~ % aws ec2 describe-addresses
Addresses: []
jonlehtinen@ ~ %
```

Figure 4.25 – Both accounts successfully executing the ec2 command

Here, we can see that running that command first under the RBI_EC2 account's profile and again under the default redbeardidentity account both completed successfully. Next, let's try aws iam list-users:

```
An error occurred (AccessDenied) when calling the ListUsers operation: User: arn:aws:iam::451339973440:
user/RBI_EC2 is not authorized to perform: iam:ListUsers on resource: arn:aws:iam::451339973440:user/
jonlehtinen@ ~ % aws iam list-users
Users:
- Arn: arn:aws:iam::451339973440:user/RBI_Admin
  CreateDate: '2020-11-27T17:12:40+00:00'
  PasswordLastUsed: '2020-11-27T19:01:15+00:00'
  Path: /
  UserId: AIDAWSFPVONAMULEKBAT7
  UserName: RBI_Admin
- Arn: arn:aws:iam::451339973440:user/RBI_EC2
  CreateDate: '2020-11-22T18:23:15+00:00'
  Path: /
  UserId: AIDAWSFPVONACROBSEOSS
  UserName: RBI_EC2
- Arn: arn:aws:iam::451339973440:user/RBI_S3
  CreateDate: '2020-11-22T17:59:25+00:00'
  PasswordLastUsed: '2020-12-13T18:13:27+00:00'
  Path: /
  UserId: AIDAWSFPVONACP5HVGP3C
  UserName: RBI_S3
- Arn: arn:aws:iam::451339973440:user/redbeardidentity
  CreateDate: '2020-11-12T01:11:46+00:00'
  PasswordLastUsed: '2021-01-02T20:17:25+00:00'
  Path: /
  UserId: AIDAWSFPVONALTHHLBKLK
  UserName: redbeardidentity
jonlehtinen@ ~ %
```

Figure 4.26 – RBI_EC2's permissions boundary blocks an action permitted by
an attached identity policy

Predictably, the redbeardidentity account succeeded. However, the RBI_EC2 account failed. Though both accounts share the same identity policy, RBI_EC2 cannot exceed its permissions boundary.

Finally, let's look at how permissions boundaries interact with resource-based policies. For this, we will update the resource-based policy that we applied to the rbi-s3-bucket-1 S3 bucket, so that both the RBI_S3 and RBI_EC2 user accounts are named as the principals who can perform operations on the bucket and its objects. The resource-based policy now looks like this:

```
{
    ''Version'': ''2012-10-17'',
    ''Id'': ''rbis3bucket1'',
    ''Statement'': [
        {
            ''Sid'': ''1'',
```

```
            ''Effect'': ''Allow'',
            ''Principal'': {
                ''AWS'': [
                    ''arn:aws:iam::451339973440:user/RBI_EC2'',
                    ''arn:aws:iam::451339973440:user/RBI_S3''
                ]
            },
            ''Action'': ''s3:*'',
            ''Resource'': [
                ''arn:aws:s3:::rbi-s3-bucket-1'',
                ''arn:aws:s3:::rbi-s3-bucket-1/*''
            ]
        }
    ]
```

Let's see what happens when each of these users attempts to list the bucket's contents from the CLI, similar to how we did earlier in this chapter when we first introduced resource-based policies:

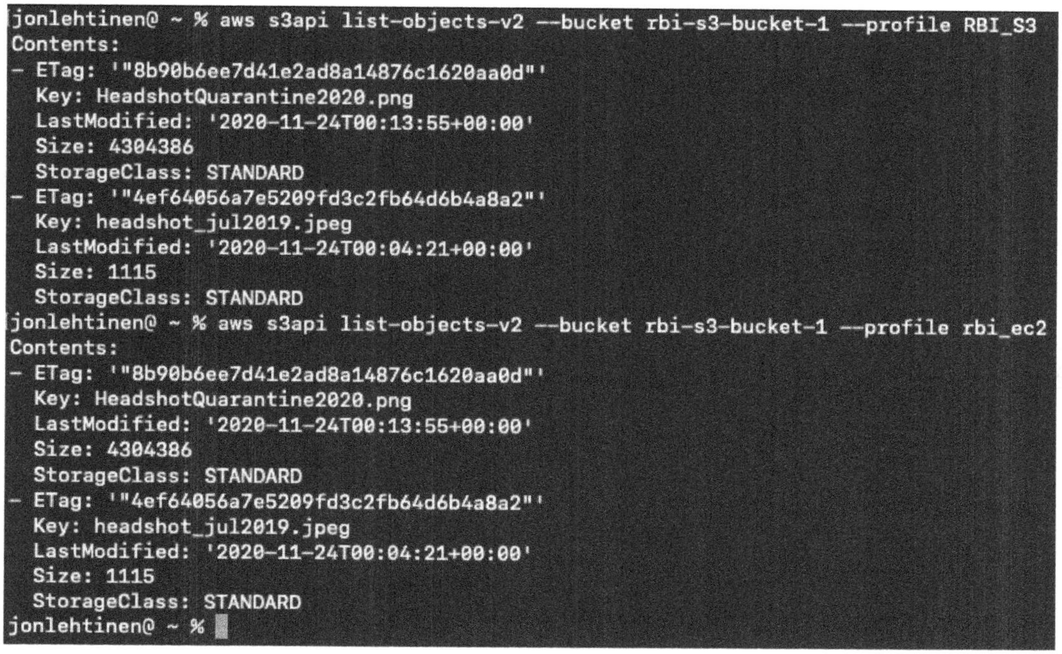

```
jonlehtinen@ ~ % aws s3api list-objects-v2 --bucket rbi-s3-bucket-1 --profile RBI_S3
Contents:
- ETag: '"8b90b6ee7d41e2ad8a14876c1620aa0d"'
  Key: HeadshotQuarantine2020.png
  LastModified: '2020-11-24T00:13:55+00:00'
  Size: 4304386
  StorageClass: STANDARD
- ETag: '"4ef64056a7e5209fd3c2fb64d6b4a8a2"'
  Key: headshot_jul2019.jpeg
  LastModified: '2020-11-24T00:04:21+00:00'
  Size: 1115
  StorageClass: STANDARD
jonlehtinen@ ~ % aws s3api list-objects-v2 --bucket rbi-s3-bucket-1 --profile rbi_ec2
Contents:
- ETag: '"8b90b6ee7d41e2ad8a14876c1620aa0d"'
  Key: HeadshotQuarantine2020.png
  LastModified: '2020-11-24T00:13:55+00:00'
  Size: 4304386
  StorageClass: STANDARD
- ETag: '"4ef64056a7e5209fd3c2fb64d6b4a8a2"'
  Key: headshot_jul2019.jpeg
  LastModified: '2020-11-24T00:04:21+00:00'
  Size: 1115
  StorageClass: STANDARD
jonlehtinen@ ~ %
```

Figure 4.27 – The resource-based policy on the bucket explicitly allows
RBI_EC2 to perform all operations

As the `RBI_EC2` user has no S3 service capabilities listed within its permissions boundary, we may expect that the permissions boundary would block it from taking action on this S3 resource. However, it is important to remember the primary distinction between resource-based policies and the identity-based policies that use managed policy objects. More specifically, identity-based policies implicitly refer to a principal through its association with an identity object. This is also true of permissions boundaries. Conversely, a resource-based policy explicitly names a principal or principals, along with the entitlements that principal has. For this example, though `RBI_S3` was outside of its permissions boundary, it was explicitly permitted full access to the `rbi-s3-bucket-1` S3 bucket inside that bucket's resource-based inline policy.

Next, we will look at service control policies, which contain different type of restrictive policy objects that behave similarly to permissions boundaries but apply at the AWS account level, as opposed to the individual objects or user accounts within the account.

Service control policies

Many organizations use several AWS accounts. In addition to the chronic challenges of aggregating accounts that have been discovered by shadow IT or mergers and acquisitions, organizations may choose to deliberately divide their accounts among business units. An emerging best practice is to issue AWS accounts on a per-application basis, to preclude resource competition within an account and reduce the risk of unauthorized or unintentional service impacts caused by access control failures.

AWS Organizations is a service that aims to help organizations manage all their AWS accounts. We briefly touched on AWS Organizations in *Chapter 3, IAM User Management*, specifically in the context of how that service assisted in managing federated users. In the context of access control, AWS Organizations allows us to define a management AWS account and organizational units where subordinate member accounts may be placed. It is here that we can apply service control policies in order to define the maximum permissions available within these subordinate AWS accounts. In this regard, **service control policies** can be thought of permissions boundaries that are applied to entire AWS accounts:

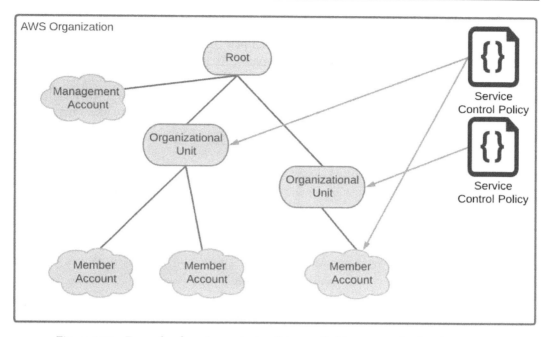

Figure 4.28 – Example of service control policies applied to an organizational structure

Service control policies may be applied at various levels within the organizational structure. A policy applied at an organizational unit level will affect all the member accounts listed beneath that OU. SCPs applied directly to the member account will apply only to the member account. Similar to other policy types, we can apply the same policy object to various levels within the OU structure; the impact this has on the member account's maximum possible permissions will be a combination of policies.

Service control policies use the same JSON document structure as other policy types. It is important to remember that SCPs do not innately grant permissions; just like permissions boundaries on identity objects, they merely set the maximum possible permissions. An additional policy will still be required to grant principals access to resources within each member AWS account.

The next policy type we will examine, access control lists, will be familiar to anyone with a background in network firewall management, though their use is not constrained to only network resources in AWS.

Access control lists

Access control lists determine whether an external principal can access a resource within an AWS account. Unlike the other policy types, ACLs do not use the JSON policy language. Different AWS services use different formats for their ACLs; for example, S3 uses an XML-formatted document, whereas several of the network-oriented services such as AWS **Virtual Private Cloud** (**VPC**) use something that looks more like a network allow/deny list, as shown here:

Figure 4.29 – A permissive ACL for a VPC subnet allowing anonymous principals from the greater internet

Curiously, the AWS IAM documentation indicates that ACLs "cannot be used to control access for a principal within the same account" (`https://docs.aws.amazon. com/IAM/latest/UserGuide/access_policies.html#policies_acl`, yet something every AWS neophyte quickly learns is how to open the appropriate ports on their EC2 instances using something like a network security group. While this may seem contradictory, it is important to remember transactions such as SSHing into an EC2 instance or connecting to a web server from the broader internet are not done under the context of an AWS IAM user account. SSHing to an EC2 instance is a form of local authentication to that specific EC2 instance. The ACL evaluates the inbound connection as an anonymous principal for the purposes of AWS IAM evaluation, while additional authorization occurs within the EC2 machine that is serving content.

The last policy type we will explore is session policies. These policies are used to limit the scope of access of temporary credentials that are issued through an AWS Security Token Service.

Session policies

In order to understand session policies, we must briefly discuss how federated authentication works. You will recall from *Chapter 3, IAM User Management*, that federated users do not map to an AWS IAM user account object, but rather they assume an AWS IAM role. They assume that role by exchanging the signed authentication token that was provided by their federated identity provider, which authenticated them with the AWS Security Token Service:

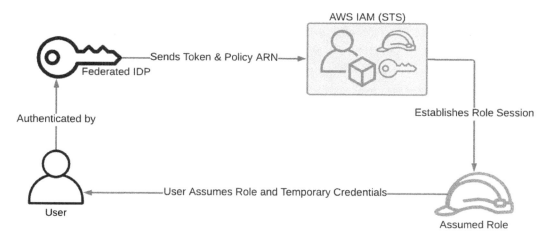

Figure 4.30 – A principal receives temporary credentials governed by a session policy

A session policy is included in that call from the IDP to the Security Token Service, which will constrain the scope of the temporary credentials issued by AWS STS. This session policy will be in addition to any other policy type that may be applied to the role identity object, or any inline resource policies that may be actioned upon by the principal while using those temporary credentials.

With so many different policy types in play in AWS, what is the order of operations for determining the final level of authorization when policy types interact? Fortunately for us, there is a consistent policy evaluation flow to ensure consistent access evaluation.

Policy evaluation

Now that we've looked at all the policy types available to us within AWS, the question becomes, how do each of these policy types interact with each other, and is there some sort of order or operations in play among them for processing requests? Fortunately, there is an overall pattern of evaluation logic that we can follow to see how an action is evaluated. Unfortunately, there are plenty of exceptions and nuances – more than can be reasonably detailed here. Whereas it is still valuable to understand how requests are generally assessed by AWS IAM, it is always prudent to review the service-specific documentation for unique behaviors.

First, there are a few overarching rules:

- The AWS root account has full access by default.

- Requests from all other principals are denied by default.

- For identity-based policies and resource-based policies, an explicitly enumerated allow statement will override the default deny.

- Organizational service control policies, session policies, and permissions boundaries may override an explicitly enumerated allow statement in an identity- or resource-based policy.

- An explicitly enumerated deny statement overrides any other allow statement.

Since that is as clear as mud, let's approach policy evaluation as a process diagram:

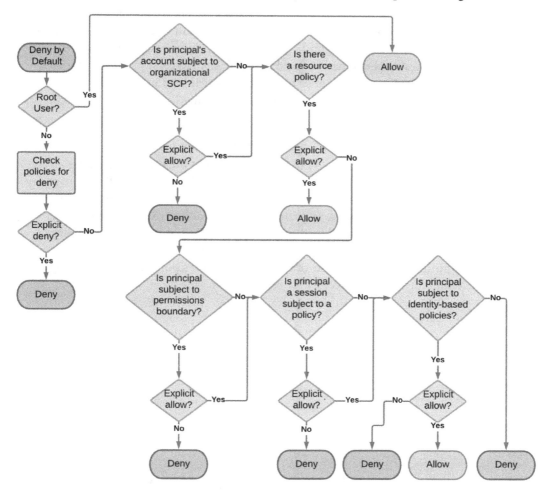

Figure 4.31 – Policy evaluation process diagram

Every request for every resource gets evaluated through this flow diagram every time a request is made. Additionally, each evaluation step could have several policies applied, with each policy containing several statements and conditions. With so many moving parts, it becomes important to double check that policies do what we expect them to do. Fortunately, there are tools available to help us with our governance challenges.

Governance

We've spent the last several pages detailing the mechanics of access management and authorization. It may seem tautological as to why we would want to enact a sound access management policy; we want to protect our AWS resources. However, there are also legal and regulatory requirements that we need to fulfill, such as least privilege, evidence of events, and audit. We will now look at a couple tools available to us to fulfill the governance requirements that come with access management.

Access Analyzer

Access Analyzer is a feature of AWS IAM that helps highlight potential weaknesses in existing authorization policy. As we've seen over the course of this chapter, there are many inputs, options, and places where a policy change could have unintended consequences for access control. As we start intertwining additional AWS accounts, and perhaps even AWS accounts not owned or managed by our own organization, it becomes increasingly important (and difficult) to visualize those potential access leaks. Access Analyzer facilitates this by automating the analysis of our AWS account's policies and providing a report on potential findings that necessitate remediation:

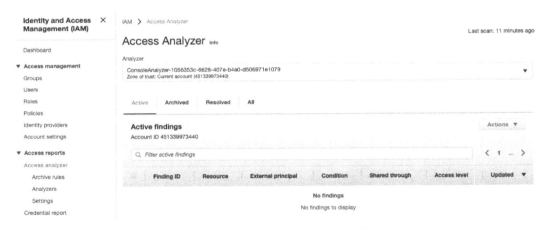

Figure 4.32 – Access Analyzer shows no issues within our account

In the preceding screenshot, we did not find any findings that required remediation. This was likely because our AWS account was not connected to any others, and sparsely populated with IAM objects.

Least-privilege is an essential component of good governance, but we also need administrative logging to tie actions to principals. Fortunately, there is a dedicated service which provides that event monitoring and logging.

AWS CloudTrail

AWS CloudTrail is the audit tool available for administrators to review actions that have been taken within the AWS account. It provides detailed event logging on a per-principal basis for all actions, changes, and events caused by that specific principal:

CloudTrail > Event history

Event name	Event time	User name	Event source	Resource type	Resource name
ConsoleLogin	January 10, 202...	redbeardidentity	signin.amazonaws.com	-	-
DeleteRole	January 09, 202...	redbeardidentity	iam.amazonaws.com	AWS::IAM::Role	rbi_ec2_lambda_s3_fu...
DetachRolePolicy	January 09, 202...	redbeardidentity	iam.amazonaws.com	AWS::IAM::Policy, AW...	arn:aws:iam::aws:polic...
DetachRolePolicy	January 09, 202...	redbeardidentity	iam.amazonaws.com	AWS::IAM::Policy, AW...	arn:aws:iam::aws:polic...
DeleteInstancePr...	January 09, 202...	redbeardidentity	iam.amazonaws.com	AWS::IAM::InstanceP...	rbi_ec2_lambda_s3_fu...
RemoveRoleFrom...	January 09, 202...	redbeardidentity	iam.amazonaws.com	AWS::IAM::InstanceP...	rbi_ec2_lambda_s3_fu...
GenerateServiceL...	January 09, 202...	redbeardidentity	iam.amazonaws.com	-	-
AttachRolePolicy	January 09, 202...	redbeardidentity	iam.amazonaws.com	AWS::IAM::Policy, AW...	arn:aws:iam::aws:polic...
AddRoleToInstan...	January 09, 202...	redbeardidentity	iam.amazonaws.com	AWS::IAM::InstanceP...	rbi_ec2_lambda_s3_fu...
CreateInstancePr...	January 09, 202...	redbeardidentity	iam.amazonaws.com	AWS::IAM::InstanceP...	arn:aws:iam::4513399...
AttachRolePolicy	January 09, 202...	redbeardidentity	iam.amazonaws.com	AWS::IAM::Policy, AW...	arn:aws:iam::aws:polic...
CreateRole	January 09, 202...	redbeardidentity	iam.amazonaws.com	AWS::IAM::Role, AWS...	rbi_ec2_lambda_s3_fu...
ConsoleLogin	January 09, 202...	redbeardidentity	signin.amazonaws.com	-	-
CreateAccessKey	January 07, 202...	redbeardidentity	iam.amazonaws.com	AWS::IAM::AccessKey...	AKIAWSFPVONAAHOJ...
DeleteAccessKey	January 07, 202...	redbeardidentity	iam.amazonaws.com	AWS::IAM::AccessKey...	AKIAWSFPVONALCDN...

Figure 4.33 – A record of events, timestamps, impacted resources, and
the principal responsible in CloudTrail

As we can see, we can review all the actions taken by the `redbeardidentity` user account. All actions, whether they're performed in the console or on the CLI, are logged. CloudTrail is a powerful governance and diagnostic tool for providing attestation to auditors that events have taken place, such as account termination.

Now that we have learned about the various policy types, how they interact, how they are evaluated, and how they are audited, we now have everything we need to control access to the resources within our AWS accounts.

Summary

Now that you've made it through this chapter, you should be familiar with the basics of AWS access management. Though we've reviewed the high-level components we need in order to be conversant on this topic and move forward, it is prudent to be cautious when it comes to access management and entitlements. Many security incidents that stem from excessive entitlements having been applied in an environment in the name of expediency. As such, consider this chapter a primer on learning *how* to learn more deeply about this topic, and consider the access management challenges that will surface throughout the remainder of this book.

The next chapter will see us shift from purely focusing on AWS IAM and looking at AWS as an infrastructure as a service offering. AWS Cognito is a service designed to offer applications simplified identity services, including user management, authentication, and authorization. Whereas we will reference many of the topics we introduced at the beginning of the book, this chapter represents the first service we will examine where AWS is acting as a platform as a service.

Questions

1. What is access management?
2. What is a policy document?
3. What is a statement within a policy document?
4. How many values are available for the `Effect` element and what are they?
5. Name the six major policy types available in AWS.
6. Describe why permissions boundaries and service control policies do not actually grant access to anything.
7. What tools are available to assist with access management audit and governance for AWS IAM and what do they do?

Further reading

- `Condition - AWS User Guide`
- `Variables - AWS User Guide`

5
Introducing Amazon Cognito

So far, we have approached identity for AWS in the context of managing authentication and authorization to AWS resources within an AWS account. We've examined the primary service that governs that access, known as AWS IAM, and seen how user accounts are managed, how their credentials are administrated, and how authorization policies are applied. Most of these use cases focus on using AWS in the context of an Infrastructure as a Service platform.

Amazon Cognito is, above all, a service for applications, with documentation and examples targeted at application developers. In fact, many of the use cases attempt to solve certain use cases by offering reference implementations that further enmesh the application architecture into AWS. This is what we mean when we say that Amazon Cognito offers identity services for AWS in the context of **Platform as a Service (PaaS)** and that AWS IAM handles identity for AWS as **Infrastructure as a Service (IaaS)**. Regrettably, this laser focus on application and platform integration also makes Cognito slightly less intuitive at first glance, even compared to the overly complex AWS IAM service.

By the end of this chapter, you will understand what each part of Cognito does and how they interoperate, as well as how they optionally interact with other AWS services.

The following topics will be covered in this chapter:

- What is Amazon Cognito?
- Amazon Cognito use cases
- Creating an Amazon Cognito user pool
- Exploring the hosted UI
- Creating an Amazon Cognito identity pool

Technical requirements

To get the most out of this chapter, you will need the following:

- An AWS account
- A workstation running the AWS CLI
- A text editor or IDE to edit JSON/YAML files, such as Microsoft Visual Studio Code

What is Amazon Cognito?

Amazon Cognito provides identity management, user authentication, and authorization for web applications. Amazon Cognito is a service that externalizes the components for application user identity management and authorization for application developers who do not wish to manage those items within the context of their own application. The Amazon Cognito service, within a given AWS account, can accommodate several distinct collections of user accounts, called **pools**. While Amazon Cognito is an identity service, it is distinct from AWS IAM in terms of its purpose and functionality. However, there are use cases and design patterns where Amazon Cognito and AWS IAM interact:

Figure 5.1 – The Amazon Cognito service from within the Management Console

Like nearly everything else in AWS, Amazon Cognito can be fully configured using both the Administrative Console and the AWS CLI. As shown in the preceding screenshot, Amazon Cognito offers two main features: **user pools** and **identity pools**. At a high level, the primary distinction between the two is that user pools address identity management and authentication use cases for applications, whereas identity pools address authorization to access AWS resources for application users.

In a nutshell, we could say that Cognito is a service that handles IAM applications through two components. The first is user account management and authentication through user pools, while the second is application authorization management to AWS resources through identity pools. As things are seldom that simple, let's examine each of these features more closely.

Amazon Cognito user pools

Amazon Cognito has two main components that make up the service. The first is user pools. A **user pool** is a managed directory for application accounts, and it offers a full suite of application user management and security features, including the following:

- New user sign-up

- New account verification workflows, such as phone number and email verification

- Prepackaged and customizable logon forms

- Federated authentication using social **identity providers** (**IDPs**), including Google, Apple, Facebook, and Login with Amazon, as well as support for any other standards-based SAML and OIDC identity provider

- Strong authentication with support for multifactor credentials

- Built-in security tools, including checks for compromised credentials and account takeover protection

Unlike AWS IAM, where there is no corresponding user account object and which only provides references to an assumed role when delegating user authentication to a federated provider, every user within an Amazon Cognito user pool is a user account object. Regardless of whether it is the Amazon Cognito user pool or a federated provider handling the authentication, that authentication is mapped to a user account within the user pool.

User pool tokens

Amazon Cognito uses standard OpenID Connect flows for its user authentication and default attribute schema. **OpenID Connect** (**OIDC**) is an identity layer that's built upon the OAuth2 authorization framework. This standard extends the functionality of an OAuth2 authorization server to include authentication tokens and claims about end users using REST-like transactions across security domains.

Once a user has been authenticated, Amazon Cognito provides user pool tokens to the application, which can then be used to mediate access to application-side resources. These user pool tokens are simply standard OIDC tokens. In this scenario, the Amazon Cognito user pool acts as the IDP and authorization server, while the application acts as the resource server in a standard OIDC transaction.

The following diagram shows an example of a flow between an application that uses Amazon Cognito user pools for authentication and user management:

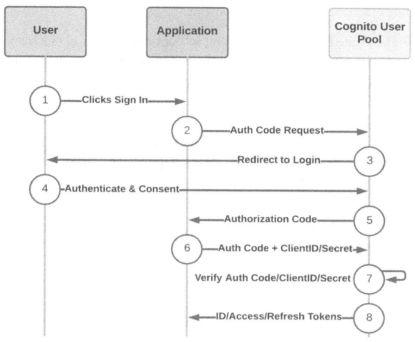

Figure 5.2 – Example of an authorization code flow with a proof of key for code exchange through an
Amazon Cognito user pool

This is an example of an **authorization code plus proof of key for code exchange (PKCE)**
flow, which is one of several different standardized flows available in the OpenID Connect
standard. While there are key differences between each available flow, their optimal
use cases, and the tokens they provide afterward, the authorization code flow is a good
reference to examine as it can provide all the available token types.

> **Tip**
>
> For additional information on the OpenID Connect standard, its available
> flows, and its best practices for implementation, please visit the OpenID
> Foundation at `https://openid.net`. For more information about
> each of these flows within an Amazon Cognito user pool, see `https://`
> `aws.amazon.com/blogs/mobile/understanding-amazon-`
> `cognito-user-pool-oauth-2-0-grants/`.

In the preceding diagram, we can see the following:

1. An unauthenticated user with an existing account clicks a sign-in link for a website that uses an Amazon Cognito user pool for user management and authentication.

2. The application's Amazon Cognito registered client generates a random `code_verifier` that it uses to generate a cryptographically generated `code_challenge`.

3. The website redirects the user's browser and `code_challenge` to the user pool's `/authorize` endpoint to make an authorization request.

4. The user pool redirects the user's browser to the Amazon Cognito login page to challenge the user for their credentials.

5. The user enters their credentials and (optionally) consents to the Amazon Cognito user pool sharing the information stored within the user pool directory, along with the application the user is trying to access.

6. The user pool stores `code_challenge` and redirects the user back to the application with a short-lived, one-time-use authorization code. This authorization code is the short-lived proof that the user provided their credentials.

7. The application sends the authorization code and `code_verifier`, which it created in *Step 2*, to the Amazon Cognito user pool authorization server. The authorization code is what the application's client exchanges for tokens once the authorization server validates the client using `code_verifier` and `code_challenge`.

8. The user pool verifies `code_verifier` and `code_challenge`. If they cryptographically correspond, then the authorization server has validated that the client can make this request for user information from the Amazon Cognito user pool as the authorization server.

9. The authorization server responds to the application with a signed **ID token** that attests that the authorization server validated the user's credentials and that they were valid. That ID token also contains embedded information about that user from the user pool directory, called **claims**. It also sends an **access token**, which the application may use on behalf of the user to access application resources protected by the authorization server. Finally, it may also send a **refresh token**, which the application may exchange for new ID, access, and refresh tokens without requiring additional user interaction.

Whereas it may not be essential to understand every OIDC flow if we want to use Amazon Cognito, it would be prudent to at least understand the basics of OAuth2 and OIDC. The application and authorization models for Amazon Cognito user pools are built upon the OpenID Connect specification.

Next, we will look at the second half of Cognito; that is, its identity pools.

Amazon Cognito identity pools

The second component of Amazon Cognito is its **identity pools**. Application developers can use identity pools to bridge application and user access to AWS resources. Similar to how federated AWS IAM users obtain temporary credentials, identity pool users may also take advantage of the AWS Security Token Service to obtain temporary credentials. Just like those federated AWS IAM users, there is no corresponding AWS IAM user account for these Amazon Cognito identity pool users. Access comes through an AWS IAM trust policy that mediates the Amazon Cognito identity pool users' access.

An identity pool requires a federated identity provider of some sort to provide the identities for its pool. That provider could be an Amazon Cognito user pool, a social provider such as Facebook or Twitter, a standards-based SAML2 or OIDC identity provider, or a mixture of them all.

Whereas user pools are self-contained platforms, identity pools should be incorporated into an application. The model behind identity pools is authorizing access to AWS resources that an application may use, with the application itself being hosted on the AWS platform. Identity pools are a developer-centric offering, with sample code and SDKs to facilitate their adoption.

Amazon Cognito identity pools can be a difficult concept to understand as they can seem redundant to other AWS identity services that address authorization to AWS resources, except that identity pools are purpose-built for applications that have been *designed* for deployment on AWS. The IaaS nature of AWS means we can solve a specific business or application use case in multiple ways, whereas the PaaS nature of Amazon Cognito offers a prescriptive pattern for solving that same use case. As we examine the use cases for both user pools and identity in the next section, the purpose-built functionality of Amazon Cognito (compared to the other AWS IAM patterns) will become easier to understand.

Amazon Cognito use cases

There are several common deployment patterns and use cases that Amazon Cognito accommodates. While each of these patterns may involve different Amazon Cognito, AWS IAM, or other app and AWS service components, they all share the same underlying purpose: to facilitate application identity services on applications deployed on AWS. Let's examine a few of these use cases and patterns and see how the different Amazon Cognito components come into play for each one.

User authentication for application access

The simplest design pattern to accommodate when using Amazon Cognito is fully externalized user account management and authentication. In this pattern, the Cognito user pool acts as the IDP and user store for the application:

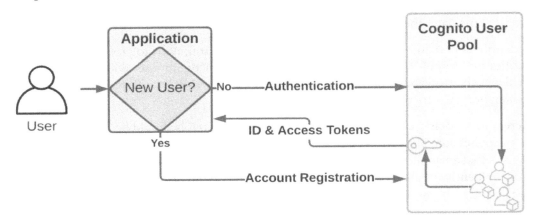

Figure 5.3 – Application authentication and user management with a user pool

Applications can take advantage of Amazon Cognito's hosted account management, sign-up, and verification process to register new user accounts in the user pool if a user does not have credentials to access the application. Once the user registers their account and sets up their credentials, the app looks to Amazon Cognito as its identity provider. Using a standards-based authentication flow, the application receives confirmation from Amazon Cognito that the user has been authenticated by being issued a signed, Amazon Cognito-provided ID token and access token.

This model can also be extended to look at external, federated identity providers for user authentication:

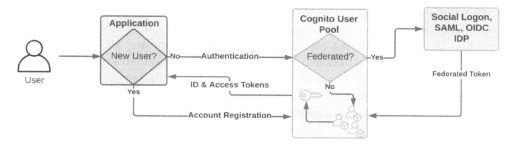

Figure 5.4 – App authentication and user management, including a federated provider

Under this model, the only significant difference is the option to change where the user is authenticated, as well as where their credentials are managed. The Amazon Cognito user pool may look to the locally managed accounts within its directory or to a federated provider at user authentication time; the application continues to only respect Amazon Cognito as its source of user identity. Though federated users may delegate authentication to an identity provider, they still exist as records within the user pool, and any information that's presented to the application about those users will come from Amazon Cognito using attributes from that pool.

The next model builds upon this one but adds authorization into the mix.

User authentication and authorization for access to application resources

This model is similar to the previous one, with one key exception: the application that uses Amazon Cognito for its identity provider also acts as a resource server. In addition to the user authentication and registration features, the application will also use the Amazon Cognito user pool as its authorization server in an OAuth2 pattern:

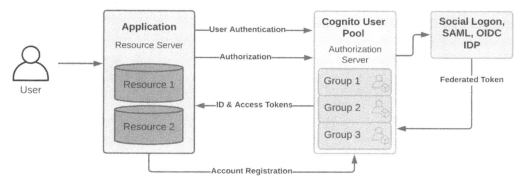

Figure 5.5 – Application using Amazon Cognito user pool tokens for scoped access to server-side resources

In the preceding diagram, just as in *Figure 5.4*, the application looks to Amazon Cognito for user authentication, new user sign-up, and account verification. The Amazon Cognito user pool may optionally look to a federated identity provider for user authentication. However, this time, the application also needs those users to be able to access either or both of its available resources once they have been authenticated to the application. Amazon Cognito user pools can facilitate this through group assignments within its user directory. We can assign entitlements, such as **members of group 1 may access resource 1**, that map to scopes that Amazon Cognito (as the authorization server) and the application (as the resource server) use to limit the access that's granted by the access token that's issued at user authentication time.

> **Important Note**
>
> Amazon Cognito user pools leverage OAuth2 concepts heavily. The IETF RFC may be useful for bootstrapping familiarity with the terms, roles, and mechanics of the framework. It is available here: `https://tools.ietf.org/html/rfc6749`. If you would like more information on the mechanics of calling the OAuth2 endpoints available in Amazon Cognito, you may read more at `https://docs.aws.amazon.com/cognito/latest/developerguide/cognito-userpools-server-contract-reference.html`. Finally, for more information on configuring application servers as resource servers for use with Amazon Cognito user pools, see the documentation available here: `https://docs.aws.amazon.com/cognito/latest/developerguide/cognito-user-pools-define-resource-servers.html`.

With all this talk of scopes, resource servers, and authorization servers, you would think that APIs would be in the mix somewhere in this model. To be fair, they could be as resources available on the application/resource server. However, since leaving API endpoints exposed is not a security best practice, AWS recommends pairing the Amazon Cognito user pool with AWS API Gateway:

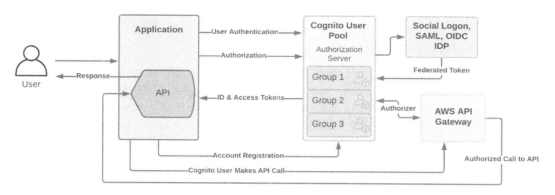

Figure 5.6 – Amazon Cognito as an authorizer for AWS API Gateway

In this relationship, on top of all the previously established relationships and services between Amazon Cognito user pools and the application, we must also establish a relationship between the user pool and the Amazon API Gateway service, so that Amazon API Gateway knows that Amazon Cognito can act as an authorization server for the application APIs that it proxies. Just as in the earlier variant of this pattern, user groups within the Amazon Cognito user pool map to scopes that determine which APIs the access token is permitted to call. This pattern is very platform-specific to the AWS services that are referenced in it. The documentation on implementing it is available here: `https://docs.aws.amazon.com/apigateway/latest/developerguide/apigateway-integrate-with-cognito.html`.

The next pattern drops the Amazon API Gateway in favor of Amazon Cognito identity pools.

User authentication and access to AWS services exposed through an application

The next pattern is the first to include both Amazon Cognito identity pools and a use case where an application and its users would have need to access AWS resources. Let's imagine we have an application that uses Amazon Cognito for its identity management, and part of that app's functionality includes file uploads and downloads from a repository, which is really just an S3 bucket:

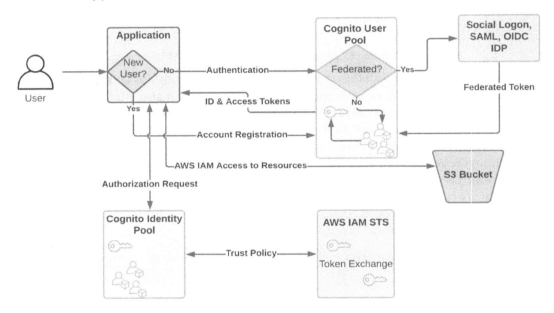

Figure 5.7 – Architecture with identity pools and a token exchange to access AWS resources

All the capabilities and features of Amazon Cognito user pools remains the same as they did previously. What's new here is that the application calls the identity pool to exchange the user's Amazon Cognito user pool tokens for AWS IAM credentials, which will allow them to access AWS resources in accordance with a predefined trust policy. The application delegates authorization to AWS IAM. Once the user assumes those credentials – more specifically, a role that was defined and scoped to allow access from that identity pool – they can interact with the AWS resource through the application.

Though this pattern is common, in the next section, we will see that an Amazon Cognito identity pool does not require a user pool to operate.

Federated user authentication and access to AWS services exposed through an application

This pattern is nearly identical to the previous one, except it sidesteps the use of Amazon Cognito user pools entirely in favor of only depending on federated identity providers. This may be confusing as an Amazon Cognito user pool may also use a federated identity provider for user authentication; the key difference is that a user pool provides authentication and user management for the application that looks to it for those things, regardless of whether the user pool itself federates to an external IDP. On the other hand, an Amazon Cognito identity pool is indifferent to the identity provider the application looks to for user authentication. Identity pools only care about the tokens that are issued by the authoritative identity provider since those attributes can be used to determine authorization and entitlement mapping:

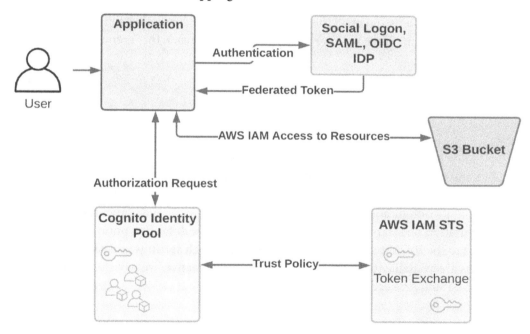

Figure 5.8 – Amazon Cognito identity pool and an external federated provider governing app access to AWS resources

Many of the same mechanics from the previous architecture are in play here as well, except this time, in lieu of an Amazon Cognito user pool token, the token from the federated provider is used to identify the user, collect critical attribute mappings, and align the user to the appropriate trust policy that governs the assumed role they will get during token exchange.

These common architectures may help us conceptualize how Amazon Cognito user and identity pools can be used to solve application identity challenges, but they do not give us a sense of what options are available for us within these services. In the next section, we will learn how to create an Amazon Cognito user pool and get a better sense for the service.

Creating an Amazon Cognito user pool

We will create an Amazon Cognito user pool using the Management Console. To do this, follow these steps:

1. Go to the Amazon Cognito service within the Management Console.

2. Click the **Manage User Pools** button.

3. This takes us to a listing of all the user pools that have currently been set up inside our AWS account. Since we have not configured any inside this account, it should be empty:

User Pools | Federated Identities
Your User Pools

Create a user pool

You have no user pools. Click here to create a user pool.

Figure 5.9 – Our empty list of Cognito user pools

4. Click the **Create a user pool** button to start creating our first pool. We will immediately be prompted for a pool name and given options for either reviewing the default configuration recommended by AWS or stepping through the configuration one step at a time. Selecting the **Review defaults** option will simply skip us to the **Review** page, so let's select **Step through settings** and see what options are available to us. Since we are not overly creative, we will call our first pool `rbipool` and proceed:

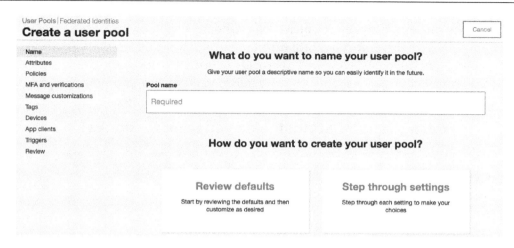

Figure 5.10 – Create a user pool screen

5. First, we must review the attribute settings for our user pool. These attribute settings will determine the username, required attributes, and any potential alias attributes we want our users to have. Let's leave the default identifier of username in place so that our users can choose their own usernames rather than being forced to identify via email or phone number. We should also note, and leave enabled, the case insensitivity option for the username so that users won't need to worry about issues such as browser autocapitalization during signup, causing them to lose access to their accounts:

User Pools | Federated Identities

Create a user pool   Cancel

Name
Attributes
Policies
MFA and verifications
Message customizations
Tags
Devices
App clients
Triggers
Review

> You can't change the sign-in and attribute options on this page after you've created your user pool. Make sure that you've decided on the settings that you want.

How do you want your end users to sign in?

You can choose to have users sign in with an email address, phone number, username or preferred username plus their password. Learn more.

○ **Username** - Users can use a username and optionally multiple alternatives to sign up and sign in.
  ☐ Also allow sign in with verified email address
  ☐ Also allow sign in with verified phone number
  ☐ Also allow sign in with preferred username (a username that your users can change)

○ **Email address or phone number** - Users can use an email address or phone number as their "username" to sign up and sign in.
  ○ Allow email addresses
  ○ Allow phone numbers
  ○ Allow both email addresses and phone numbers (users can choose one)

You can choose to enable case insensitivity on the username input for the selected sign-in option. For example, when this option is selected, the users can sign in using either "username" or "Username".

☑ (Recommended) Enable case insensitivity for username input

Figure 5.11 – Defining the user pool's identifying attribute

6. Next, we can optionally select which of the standard attributes will be required at signup. Each of these attributes will exist for every user record within the user pool, though users will only be required to provide information at signup for the ones we check here. Let's balance useful directory information against user privacy and only add a few, such as **given name**, **family name**, and **phone number**:

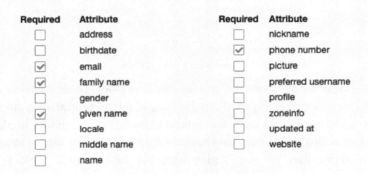

Figure 5.12 – Selecting the attributes required at signup

7. We also have the option to define custom attributes. This is an opportunity to add either string or number value attributes that our app would require but are not natively included in the Cognito/OpenID profile attribute schema. Let's define a custom attribute for costcenter and continue:

Do you want to add custom attributes?

Enter the name and select the type and settings for custom attributes.

Type	Name	Min value	max value	Mutable
number	custom:costcenter	0	1000	✓

Add another attribute

Back   Next step

Figure 5.13 – Defining a custom attribute

8. The next step covers password policies. The first option specifies password complexity requirements. As these default values look good, we can leave them at their default values:

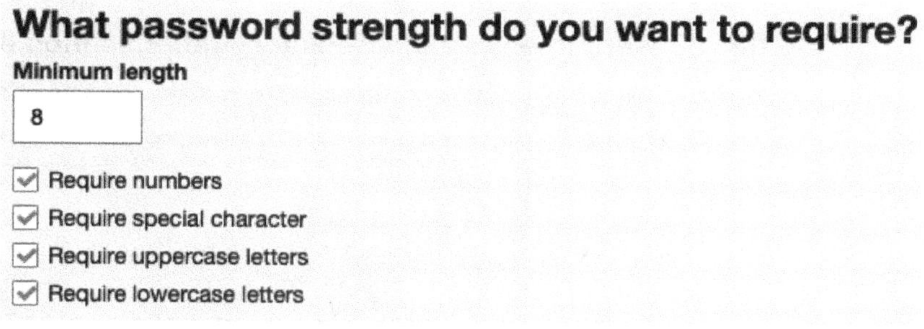

What password strength do you want to require?

Minimum length

> 8

☑ Require numbers
☑ Require special character
☑ Require uppercase letters
☑ Require lowercase letters

Figure 5.14 – Password policy for the user pool

9. On the same page, we can opt to only allow administrators to create user accounts, either through manual creation or through an import process, or allow users to sign themselves up. Enterprise use cases would likely select the first option, while consumer use cases would select the second. Since we are merely kicking the tires on a user pool today and not deploying for an enterprise use case, let's **Allow users to sign themselves up**:

Do you want to allow users to sign themselves up?

You can choose to only allow administrators to create users or allow users to sign themselves up. Learn more.

◯ Only allow administrators to create users
◉ Allow users to sign themselves up

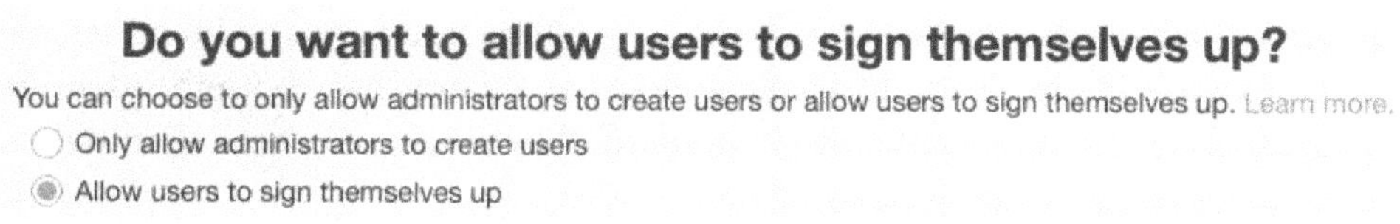

Figure 5.15 – Options for user self-signup

10. The final setting under the **Policies** setting defines the expiration period for temporary passwords when they're not being used. It is a good idea to terminate temporary passwords if they have not been used after some time as they represent a vector for account takeover. This is especially true in consumer use cases where there are no forcing functions requiring a user to interact with the application within a certain timeframe. We can leave this at 7 days and move on:

How quickly should temporary passwords set by administrators expire if not used?

You can choose for how long until a temporary password set by an administrator expires if the password is not used. This includes accounts created by administrators.

Days to expire

> 7

Figure 5.16 – Temporary password expiration value

11. The next section deals with multi-factor authentication and account recovery options. Let's choose to enforce MFA and select the **Time-based One-time Password** option:

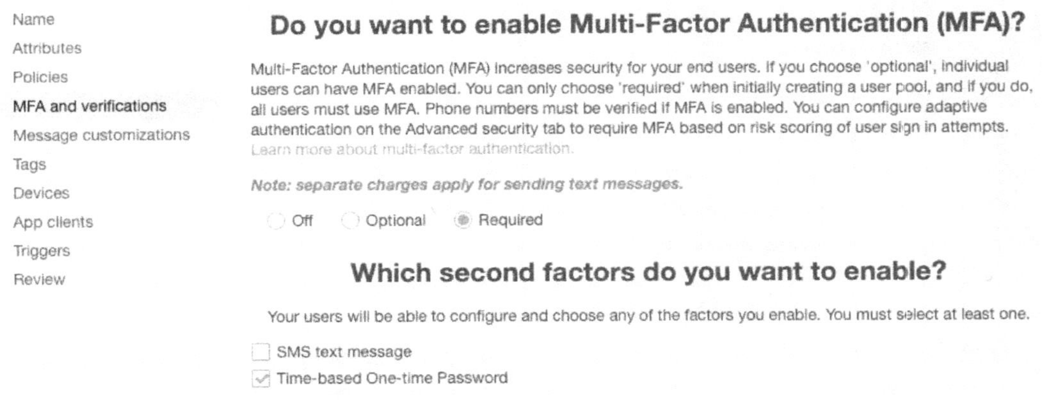

Name

Attributes

Policies

MFA and verifications

Message customizations

Tags

Devices

App clients

Triggers

Review

Do you want to enable Multi-Factor Authentication (MFA)?

Multi-Factor Authentication (MFA) increases security for your end users. If you choose 'optional', individual users can have MFA enabled. You can only choose 'required' when initially creating a user pool, and if you do, all users must use MFA. Phone numbers must be verified if MFA is enabled. You can configure adaptive authentication on the Advanced security tab to require MFA based on risk scoring of user sign in attempts. Learn more about multi-factor authentication.

Note: separate charges apply for sending text messages.

◯ Off   ◯ Optional   ◉ Required

Which second factors do you want to enable?

Your users will be able to configure and choose any of the factors you enable. You must select at least one.

☐ SMS text message

☑ Time-based One-time Password

Figure 5.17 – MFA options on the user pool

12. Still on the **MFA and verifications** page, we can select how we want to allow users to recover their accounts if they lose their passwords. There are many options available here. Best practice recommends keeping the account recovery channel and the MFA channel separate. Had we allowed the phone number as an identifier, a malicious actor in possession of a stolen phone would now have both the identifier and second factor authenticator for an account; they would only need to initiate account recovery to reset the password.

If account recovery also runs through the phone, it becomes the single channel that's required for an account takeover. As we require our user pool accounts to set a unique username and they are using TOTP for MFA, we can simply rely on email recovery. However, we will want to confirm at signup that the user is in possession of the email address, so we will want to enforce verification prior to allowing them to log in after signup:

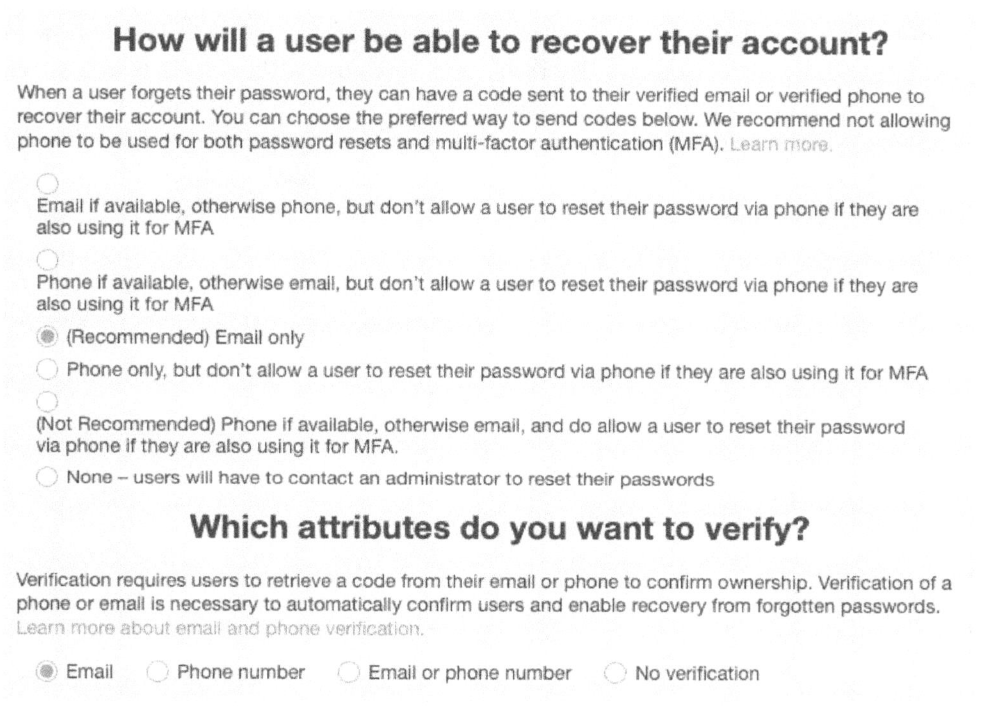

How will a user be able to recover their account?

When a user forgets their password, they can have a code sent to their verified email or verified phone to recover their account. You can choose the preferred way to send codes below. We recommend not allowing phone to be used for both password resets and multi-factor authentication (MFA). Learn more.

○ Email if available, otherwise phone, but don't allow a user to reset their password via phone if they are also using it for MFA

○ Phone if available, otherwise email, but don't allow a user to reset their password via phone if they are also using it for MFA

◉ (Recommended) Email only

○ Phone only, but don't allow a user to reset their password via phone if they are also using it for MFA

○ (Not Recommended) Phone if available, otherwise email, and do allow a user to reset their password via phone if they are also using it for MFA.

○ None – users will have to contact an administrator to reset their passwords

Which attributes do you want to verify?

Verification requires users to retrieve a code from their email or phone to confirm ownership. Verification of a phone or email is necessary to automatically confirm users and enable recovery from forgotten passwords. Learn more about email and phone verification.

◉ Email   ○ Phone number   ○ Email or phone number   ○ No verification

Figure 5.18 – Account recovery and email verification settings

13. Since we did not include SMS for any portion of our user pool configuration, we do not need to allow the wizard to create the appropriate AWS IAM role, which will let Cognito send SMS messages. If we had chosen SMS, this would be required. We are now ready to move on:

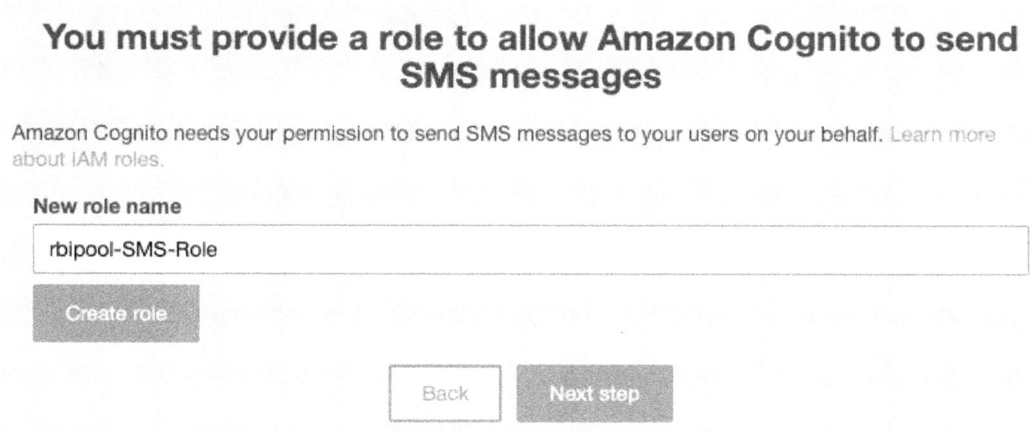

You must provide a role to allow Amazon Cognito to send SMS messages

Amazon Cognito needs your permission to send SMS messages to your users on your behalf. Learn more about IAM roles.

New role name

rbipool-SMS-Role

Create role

Back   Next step

Figure 5.19 – SMS IAM role prompt

14. As we will be sending verification emails to our users, we need to configure how we will be sending those messages. There are options to use AWS **Simple Email Service (SES)** or Amazon Cognito natively. Each option has its own limitations, namely that Amazon SES requires that we register and confirm an email within the Amazon SES service to select Amazon SES as the sender for the user pool. Additionally, we must apply for production access to Amazon SES if we wish to send to more addresses than just the one we initially registered with within Amazon SES. As that seems like more work than I am willing to put into my verification emails for my first user pool, I will defer to the second option; that is, letting Amazon Cognito can handle it natively:

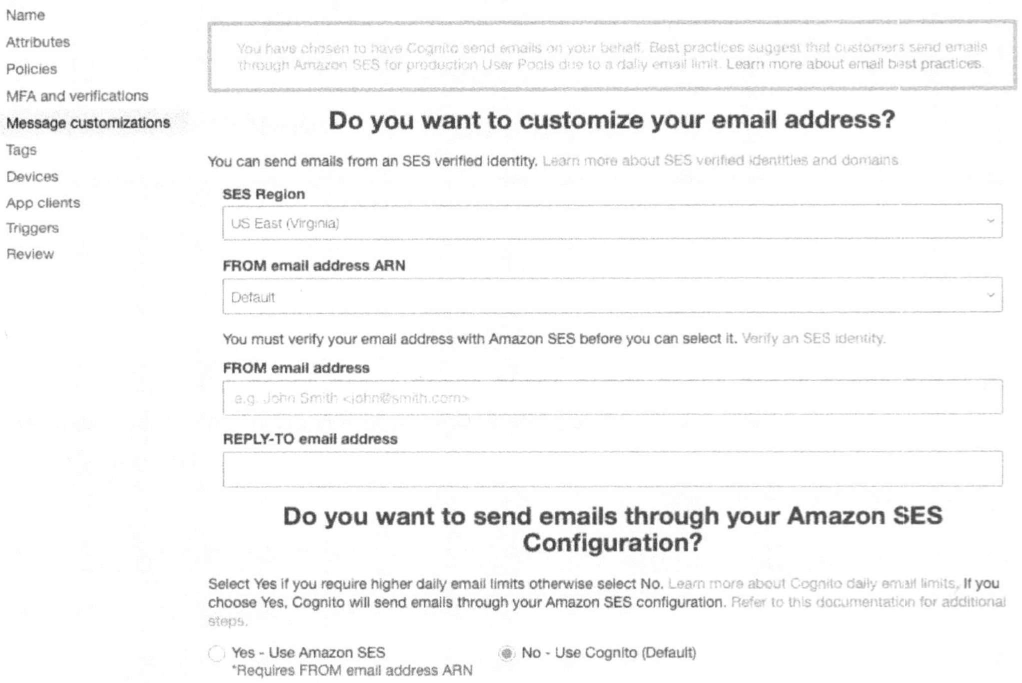

Figure 5.20 – Message sender and SES configuration options

Using Amazon Cognito also has its limitations, such as only a limited number of messages can be sent per day. Real organizations are encouraged to apply for production action to Amazon SES if they anticipate a significant number of signups. Now, we can move on.

15. Next, still on the **Message customizations** page, we have the option to customize the email verification message that will be sent to the new users that sign up. We can either have the users retrieve and enter a code or click a confirmation link. Let's reduce user friction at signup and have them click a link:

Do you want to customize your email verification messages?

You can choose to send a code or a clickable link and customize the message to verify email addresses. Learn more about email verification.

Verification type

◯ Code   ◉ Link

Email subject

Your verification link

Email message

Please click the link below to verify your email address. {##Verify Email##}

You can customize the message above, but it must include the "{####}" placeholder, which will be replaced with the link.

Figure 5.21 – Customizing the verification link template for the user pool

16. Finally, we have the option to customize both the SMS and email templates that are used for user invitation. If we are content with the default values, we can move on to the next section:

Do you want to customize your user invitation messages?

SMS message

Your username is {username} and temporary password is {####}.

You can customize the message above and include HTML tags, but it must include the "{username}" and "{####}" placeholder, which will be replaced with the username and temporary password respectively.

Email subject

Your temporary password

Email message

Your username is {username} and temporary password is {####}.

You can customize the message above and include HTML tags, but it must include the "{username}" and "{####}" placeholder, which will be replaced with the username and temporary password respectively.

Figure 5.22 – Customizing user invitation messages

17. Like most things in AWS, user pools support tags. We can define them here. Let's create one that we think may be useful and move on:

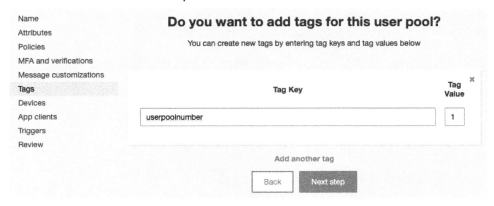

Figure 5.23 – User pool tagging

18. The next section gives us some options around remembering our user's devices. This feature allows the user pool to affiliate the device with the user's profile, and optionally allow us to opt to bypass the MFA prompt on known devices. The **Always** option associates any device that's used at logon time with the user's profile. **User Opt In** gives the user a choice in the matter and is the optimal balance of user experience and security. Let's select the **User Opt In** option and move on:

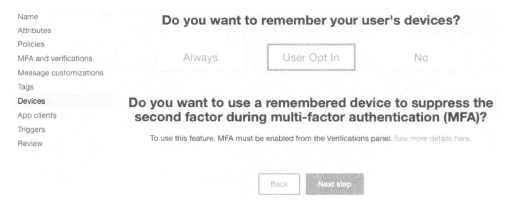

Figure 5.24 – Device association options for the user pool

19. The next step lets us define an app client. An **app client** is the OpenID Connect client that an application can use to execute one of the OIDC user authentication flows against this user pool, in its capacity as an identity provider. We will need to create an app client to make this user pool useful for forcing authentication to access a website, so let's select **Add an app client**:

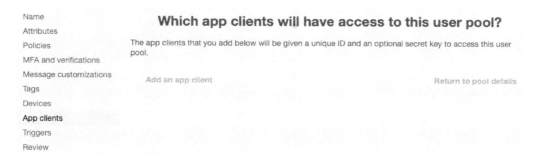

Name

Attributes

Policies

MFA and verifications

Message customizations

Tags

Devices

App clients

Triggers

Review

Figure 5.25 – Option to add a client to the user pool

20. This opens up a bunch of options, but we don't need to manipulate many values here. Let's name the client `rbiclient1` and leave the remaining values in their default positions:

App client name

rbiclient1

Refresh token expiration

30   days and   0   minutes

Must be between 60 minutes and 3650 days

Access token expiration

0   days and   60   minutes

Must be between 5 minutes and 1 day. Cannot be greater than refresh token expiration

ID token expiration

0   days and   60   minutes

Must be between 5 minutes and 1 day. Cannot be greater than refresh token expiration

☑ Generate client secret

Auth Flows Configuration

☐ Enable username password auth for admin APIs for authentication (ALLOW_ADMIN_USER_PASSWORD_AUTH)   Learn more.

☑ Enable lambda trigger based custom authentication (ALLOW_CUSTOM_AUTH)   Learn more.

☐ Enable username password based authentication (ALLOW_USER_PASSWORD_AUTH)   Learn more.

☑ Enable SRP (secure remote password) protocol based authentication (ALLOW_USER_SRP_AUTH)   Learn more.

☑ Enable refresh token based authentication (ALLOW_REFRESH_TOKEN_AUTH)   Learn more.

Security configuration

Prevent User Existence Errors   Learn more.

○ Legacy
◉ Enabled (Recommended)

Set attribute read and write permissions

Cancel   Create app client

Figure 5.26 – Application client configuration

Here, we can see that the refresh token is fairly long-lived by default, with an option to extend it to up to 10 years. The best practice for refresh token longevity is to find a balance for longevity that benefits the user experience and the risk appetite for not requiring reauthentication, for however long that period may be. Depending on the data classification or the regulatory environment that this app operates in, this could be much shorter than 30 days. Access tokens and ID tokens are much shorter-lived by comparison, with a default validity period of 1 hour.

There are additional options for various Auth Flows. These are AWS- and Amazon Cognito-specific capabilities that are designed to accommodate local application authentication mechanisms. While application developers may be content reinventing the wheel with such things, identity practitioners strongly advise adherence to standards-based protocols. The basic OpenID Connect flow is sufficient for our use case. Details on implementing the other authentication flows can be found in the Amazon Cognito documentation here: `https://docs.aws.amazon.com/cognito/latest/developerguide/amazon-cognito-user-pools-authentication-flow.html#amazon-cognito-user-pools-server-side-authentication-flow`.

Finally, we can toggle the attribute read/write permissions. By default, the application client is entitled to read and write to the user's record in Cognito. If we wanted to limit what this client could do, we could remove its access to individual attributes, scopes (which determine access to collections of attributes), or even read/write access entirely. We are content with the defaults, so let's create our client and click on the **Next step** button.

21. Next, can define certain AWS Lambda triggers upon certain events. Since we have no Lambda functions defined to associate with any of these triggers, we can move on to the review.

22. Here, we can see an overview of the choices we've made so far. We can go back and make edits or create the pool if we are confident in what we have. Let's create the pool:

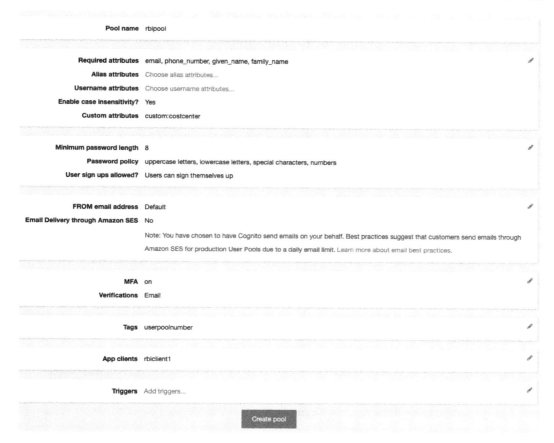

Figure 5.27 – Reviewing our user pool

23. Our user pool now appears in our list of user pools:

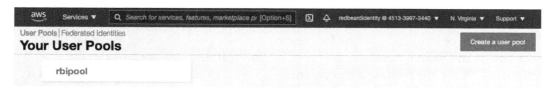

Figure 5.28 – Our newly created user pool

24. Now that we have our user pool, we need to configure the domain that our user pool will use as it acts as our identity provider for any applications that we choose to point to it. We can do this by clicking on `rbipool`, which we just created, and selecting **Domain name** beneath **App Integration** from the left-hand menu on the page:

What domain would you like to use?

Type a domain prefix to use for the sign-up and sign-in pages that are hosted by Amazon Cognito. The prefix must be unique across the selected AWS Region. Domain names can only contain lower-case letters, numbers, and hyphens. Learn more about domain prefixes.

This domain is available.

Amazon Cognito domain
Prefixed domain names can only contain lower-case letters, numbers, and hyphens. Learn more about domain prefixes.

Domain prefix

| https:// | rbipool | .auth.us-east-1.amazoncognito.com | Check availability |

Your own domain
This domain name needs to have an associated certificate in AWS Certificate Manager (ACM).☐ You also need the ability to add an alias record to the domain's hosted zone after it's associated with this user pool. Learn more about using your own domain.

Domain name

| auth.redbeardidentity.com |

AWS managed certificate

| redbeardidentity.com (arn:aws:acm:us-east-1:451339973440:certificate/f78e5e3d-01c4-49d1-a2b3-983edecb429e) | ▼ |

| Cancel | Save changes |

Figure 5.29 – Selecting an available Amazon Cognito domain or choosing our own domain for our user pool

We must select an **Amazon Cognito domain** name to use with our authorization server. To keep things consistent, let's use `rbipool` as the subdomain. We must then check the availability of that subdomain by hitting the **Check availability** button. If it is available, we can proceed.

There were a lot of steps and options for creating our user pool. However, consider what a user pool is and what it can do. A user pool is a full-featured identity provider for applications. When the process is recontextualized through that lens, it makes sense that it is complicated. Finish this setting clicking on the **Save changes** button.

Of course, an identity provider is not worth much without an identity store being behind it. We will populate our user pool in the next section.

Populating users in a user pool

Now that we have our user pool, we need some users to populate it. Users can enter our user pool in three different ways. Since we enabled self-registration, users will be able to sign up at login time to whatever applications we use with this user pool. We can also either manually create accounts from the user pool management console or bulk import users. Let's look at the user management options available to us in our user pool:

1. Using the menu on the left of the page, click **Users and groups** beneath **General settings**.

2. This page would normally present a list of users in our pool. Since we have none, it is looking pretty bare. We will create our first user by using the **Create user** button:

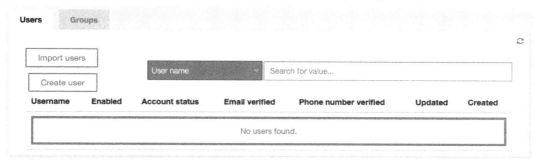

Figure 5.30 – The empty user management screen inside our user pool

3. Upon hitting the **Create user** button, we will get a form with the required attributes we configured for our user pool, along with a few administrative options:

Create user

Username (Required)

rbiuser1

☑ Send an invitation to this new user?

☐ SMS (default)   ☑ Email

Temporary password

Phone Number

+1804212

☑ Mark phone number as verified?

Email

redbeardidentity+1@gmail.com

☑ Mark email as verified?

Create user

Figure 5.31 – Manually creating a user for the user pool

As this is our first user, we will assign it the username of `rbiuser1`. We have the option to invite the user to activate their account, as well as set a temporary password. We can also define their phone number and email values, and optionally preemptively mark them as validated without forcing the user to go through the registration process. When we are satisfied with our selections, we can create the user.

4. Upon returning to the **Users** and **Groups** screen, we will see the `rbiuser1` account, along with some account information there. Note that both the email and phone number have been verified, and that the password has been set to force a change on the next logon:

Figure 5.32 – The new user in our user pool

5. If we go to our inbox, we will see an invitation message from Amazon Cognito that tells us what our username and temporary password are:

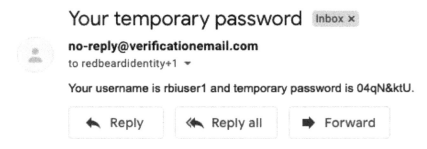

Figure 5.33 – New user account notification email

This process is all well and good for single accounts, but what if we need to bulk load a large population of users? Since creating accounts manually is tedious, let's examine the import function.

Bulk importing with CSV files

We can use CSV files to bootstrap our user pool with a large population of users. Let's take a look at this process:

1. Click the **Import users** button on the **Users and groups** page.

2. We can import users using CSV files and then create an import job to bulk load them into our user pool. The proper format for composing that CSV is provided to us upon clicking the **Download CSV** header button. We can populate as many values as we want per account as long as we populate our required attributes, which are email, phone, and `cognito:username`.

> **Tip**
>
> The `phone_verified`, `email_verified`, and `cognito:mfa_enabled` values are boolean true/false values and must be populated as well. If you require email or phone validation and are importing those users, the assumption is that those values are verified and thus must be set to `true`; otherwise, the import will fail. If those values are not required, you may either select `false` to force the users to go through the verification process or `true` to bypass this.

3. Next, click the **Create import job** button. This opens up a form where we give the batch job a name. This is where we also select or create the AWS IAM role that Amazon Cognito will use to write the logs from the batch load to AWS CloudWatch. Let's select our file and hit **Create job**:

Figure 5.34 – Creating our import batch job

4. This will start the job, and its status will appear in the list.

5. The sample file we used only had two additional records, so even though it is marked as **Failed**, we can confirm that the correct count of records was imported. However, Amazon Cognito reported that **Too many users have failed or been skipped during the import.**:

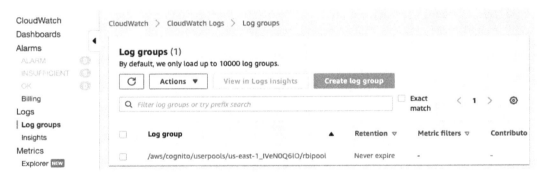

Figure 5.35 – Output of the import process

6. The results of the import are logged in Amazon CloudWatch. We can review the events of the import there to see why Amazon Cognito considered the batch a failure. From the Amazon CloudWatch service, we can click on **Log Groups** on the left-hand side of the window. This will show our Amazon Cognito user pool logs:

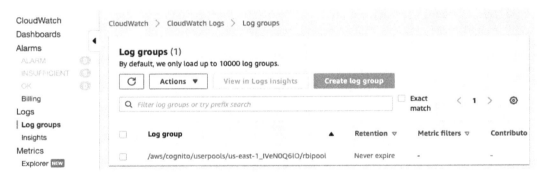

Figure 5.36 – Our Amazon Cognito CloudWatch log group

7. The **Log events** page shows two successful imports, and dozens of failures thereafter:

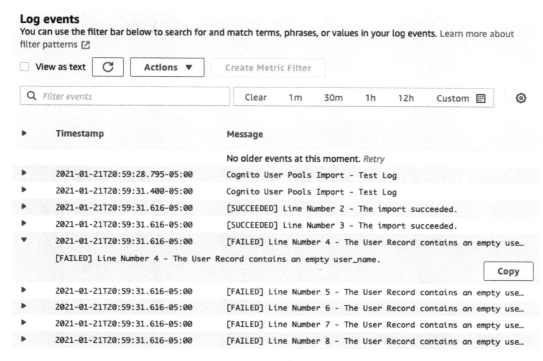

Figure 5.37 – Log events from the user import

Looking at this gives us a clue as to why Amazon Cognito labeled the import as a failure. We used Excel to edit the CSV file. Our two entries corresponded to the successful imports on lines 2 and 3, but it appears as though Amazon Cognito read every additional line on the Excel sheet as a malformed entry. Now that we understand why it was falsely labeled a failure, we can move on.

8. Upon returning to the **Users** and **Groups** page of our user pool, we can confirm that the two new imported entries are there:

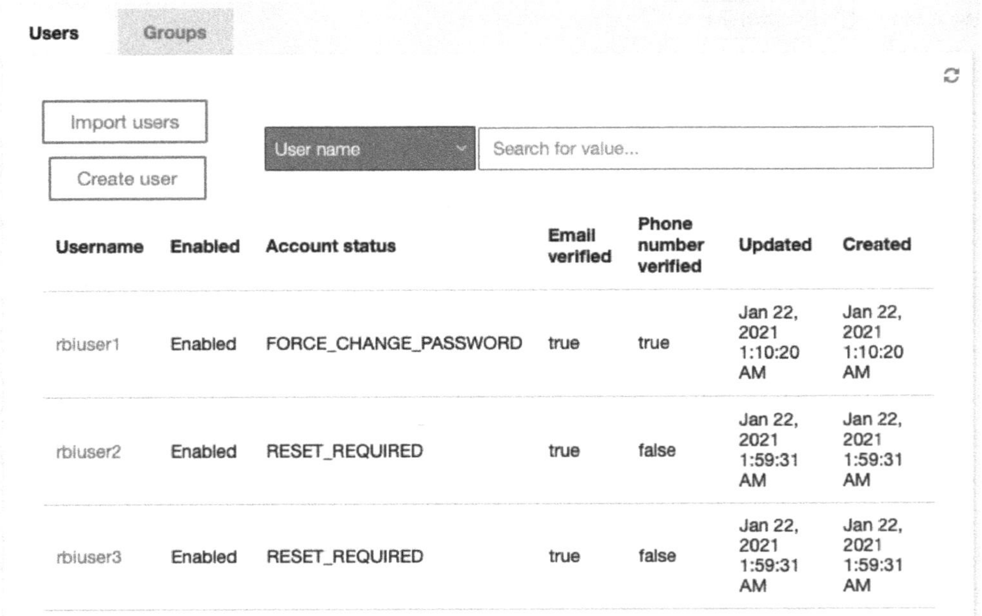
Users
Groups
Import users
Create user
User name
Search for value...
Username
Enabled
Account status
Email verified
Phone number verified
Updated
Created
rbiuser1
Enabled
FORCE_CHANGE_PASSWORD
true
true
Jan 22, 2021 1:10:20 AM
Jan 22, 2021 1:10:20 AM
rbiuser2
Enabled
RESET_REQUIRED
true
false
Jan 22, 2021 1:59:31 AM
Jan 22, 2021 1:59:31 AM
rbiuser3
Enabled
RESET_REQUIRED
true
false
Jan 22, 2021 1:59:31 AM
Jan 22, 2021 1:59:31 AM

Figure 5.38– Additional user accounts in our user pool

There were a lot of steps and options for creating our user pool. However, consider what a user pool is and what it can do. A user pool is a full-featured identity provider for applications. When the process is recontextualized through that lens, it makes sense that it is complicated. This is an opportunity to use the AWS CLI to simplify this complicated task.

Creating a user pool using the AWS CLI

Now, let's create another user pool, but this time using the AWS CLI. This will demonstrate the value of the CLI for simplifying complex tasks, assuming we understand the commands, parameters, and desired values we require for our environment. Fortunately, if we recall our lessons from *Chapter 2, An Introduction to the AWS CLI*, we will remember a few tricks that will simplify this task:

1. From the AWS CLI, we can generate a CLI skeleton, which will ensure we create the user pool accurately. We can do this with the following command:

```
$ aws cognito-idp create-user-pool --generate-cli-
skeleton yaml-input
```

This will produce the template for the user pool, which we can then copy and place into an IDE or text editor:

```
PoolName: ''  # [REQUIRED] A string used to name the user
pool.
Policies: # The policies associated with the new user
pool.
(Truncated for space, you can view the full template at
https://github.com/jonlehtinen/ImplementingAWSIdentity/
blob/main/create-user-pool_template.yml)
    Name: verified_phone_number # [REQUIRED] Specifies
the recovery method for a user. Valid values are:
verified_email, verified_phone_number, admin_only.
```

As we can see, all the steps that made creating the user pool so laborious with the Management Console are represented in this skeleton.

2. From our IDE or text editor, we can enter the values we want into the template and save it as a YAML file. If we are uncertain of certain values, we can refer to our existing user pool as an example by executing the following command:

```
$ aws cognito-idp describe-user-pool --user-pool-id
us-east-1_IVeN0Q61O -output yaml
```

The user pool ID can be found by opening the user pool from the Management Console, on the **General settings** page. This provides us with our existing pool's values, which we can use as a reference for our new CLI-made user pool as needed. The full output of the command can be seen at https://github.com/jonlehtinen/ImplementingAWSIdentity/blob/main/rbicli.yml.

It is important to note that while many of the values are good to reference, the output of the describe-user-pool command is ultimately a different format than the create-user-pool CLI skeleton template requires. As such, it is important that we are mindful when copying values from one to another. Additionally, any parameters and values that are found in the skeleton template, but we do not intend to use with our user pool, such as LambdaConfig and all of its sub-entries, will need to be removed from the template.

3. Once the template is ready, we can execute the following command to create the pool:

```
$ aws cognito-idp create-user-pool --cli-input-yaml
file://rbipoolcli.yml
```

If all goes well, the user pool will be created. If the parser finds issues with our template, it will tell us what is wrong and how to fix it, as shown in the following example:

```
> aws cognito-idp create-user-pool --cli-input-yaml
file://rbipoolcli.yml


Parameter validation failed:
Invalid length for parameter SmsVerificationMessage,
value: 0, valid range: 6-inf
Invalid length for parameter SmsAuthenticationMessage,
value: 0, valid range: 6-inf
```

In this case, we had left out these two parameters and values. We can copy and paste them from our reference example and update the `rbipoolcli.yml` file and try again. Once we have cleared all the parsing errors, the output of a successful command will be a description of our new user pool, similar to as if we had used `describe-user-pool`. You can see the `rbipoolcli.yml` file at `https://github.com/jonlehtinen/ImplementingAWSIdentity/blob/main/rbipoolcli.yml`.

4. If we refresh the Cognito service, we will be able to see our new user pool:

Figure 5.39 – The CLI-created user pool is now available

5. A couple steps remain for configuring this user pool to be fully functional, all of which were addressed using the Management Console wizard. These need to be handled with ad hoc CLI calls. The first of these is defining the domain for our authorization server. Let's follow the pattern we used for the first user pool we created and set the subdomain value equal to the name of the user pool name. From the CLI, do the following:

```
$ aws cognito-idp create-user-pool-domain --domain
rbipoolcli
```

If the command is successful, there will be no output. We can run a different command to validate that the domain was set:

```
$ aws cognito-idp describe-user-pool-domain --domain
rbipoolcli
DomainDescription:
  AWSAccountId: '451339973440'
  CloudFrontDistribution: d3oia8etllorh5.cloudfront.net
  CustomDomainConfig: {}
  Domain: rbipoolcli
  S3Bucket: aws-cognito-prod-iad-assets
  Status: ACTIVE
  UserPoolId: us-east-1_yGe1YAnTV
  Version: '20210123171729'
```

We will also see the domain listed under the **App Integration** menu in the user pool of the Management Console:

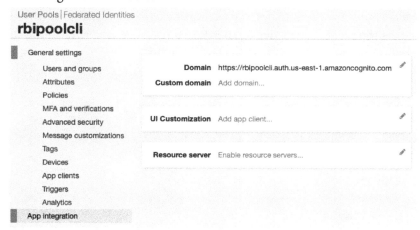

Figure 5.40 - The domain value on the rbipoolcli user pool

6. For the identity provider to be useful, we need to configure OIDC app clients that can interact with it on behalf of applications. Let's create an application client. As there are many options available to us when creating a client, this is a good opportunity to use another skeleton template:

```
$ aws cognito-idp create-user-pool-client --generate-cli-skeleton yaml-input
```

We can capture the template output and modify it in an IDE or text editor:

```
UserPoolId: ''  # [REQUIRED] The user pool ID for the
user pool where you want to create a user pool client.
ClientName: '' # [REQUIRED] The client name for the user
pool client you would like to create.
(This has been truncated for length. You can see
the full file at https://github.com/jonlehtinen/
ImplementingAWSIdentity/blob/main/rbipoolcliclient.yml.
PreventUserExistenceErrors: ENABLED # Use this setting to
choose which errors and responses are returned by Cognito
APIs during authentication, account confirmation, and
password recovery when the user does not exist in the
user pool. Valid values are: LEGACY, ENABLED.
```

This is another situation where referencing an existing object, such as a client in an existing pool that has the configurations we desired, may help us populate the values needed for the template. Similar to when we created the user pool with the CLI, anything that we do not wish to define right now, such as our LogoutURLs, should be removed or commented out of the template.

7. We intend to use this client for user authentication for an application with our user pool as the identity provider, so there are few values in particular we should highlight:

```
    AllowedOAuthFlows:
    - code
    AllowedOAuthFlowsUserPoolClient: true
    AllowedOAuthScopes:
    - aws.cognito.signin.user.admin
    - phone
    - openid
    - profile
    - email
```

```
CallbackURLs:
- https://openidconnect.net/callback
```

First, we need to enable the client to use the authorization code grant so that it can authenticate users. That is the `code` value beneath `AllowedOAuthFlows`. Next, we must set `AllowedOAuthFlowsUserPoolClient` to `True`, and set the `AllowedOAuthScopes` to include those scopes that we have access to in our user pool's authorization server. Finally, the `code` flow requires that we specify a callback URL, which is where the IDP redirects the user agent after authentication. Since we do not have any applications defined, I am specifying the callback URL of the OpenID Connect Playground, available at `https://openidconnect.net/#`. Alternatively, we could use any temporary value and update it once accurate callback URLs are known for a specific application.

8. Once our template has been configured to our liking, we can create our client:

```
$ aws cognito-idp create-user-pool-client --cli-input-
yaml file://rbipoolcliclient.yml
```

Assuming it was successful, we will see an output describing our client:

```
UserPoolClient:
    AccessTokenValidity: 1
    AllowedOAuthFlows:
    - code
    AllowedOAuthFlowsUserPoolClient: true
    AllowedOAuthScopes:
    - aws.cognito.signin.user.admin
    - phone
    - openid
    - profile
    - email
    CallbackURLs:
    - https://openidconnect.net/callback
    ClientId: 66c6kfb9rtvltnjkttgarrgak3
    ClientName: rbipoolcliclient
    ClientSecret:
1qgn7saucrfrfr4n51ad56nvit2vjcgjb0mka0abriohc318ult0
    CreationDate: '2021-01-23T13:09:32.317000-05:00'
    ExplicitAuthFlows:
```

```
  - ALLOW_CUSTOM_AUTH
  - ALLOW_USER_SRP_AUTH
  - ALLOW_REFRESH_TOKEN_AUTH
IdTokenValidity: 1
LastModifiedDate: '2021-01-23T13:09:32.317000-05:00'
PreventUserExistenceErrors: ENABLED
ReadAttributes:
- website
- zoneinfo
- address
- birthdate
- email_verified
- gender
- profile
- phone_number_verified
- preferred_username
- given_name
- locale
- middle_name
- picture
- updated_at
- custom:custom:costcenter
- name
- nickname
- phone_number
- family_name
- email
RefreshTokenValidity: 30
SupportedIdentityProviders:
- COGNITO
```

With the client created, we now have the application's entry point into the identity provider. However, we still need to create some users for our Cognito user pool. We'll do that next.

Importing users with the CLI

Now that we have a fully configured user pool, let's import the users from our CSV file into it. From the CLI, do the following:

1. The command to import the users is `create-user-import-job`. Just as we saw when we performed this function in the Management Console, this command requires us to provide values for `JobName`, `CloudWatchLogsRoleARN` – which is the role Amazon Cognito uses to write event logs to Amazon CloudWatch – and `UserPoolId` of the user pool where we will be importing the records. Our first task will be to aggregate those values from Amazon Cognito and AWS IAM so that we can build the command.

2. Once we have the values we need, we can assemble the command:

```
$ aws cognito-idp create-user-import-job --user-
pool-id us-east-1_yGe1YAnTV --cloud-watch-logs-role-arn
arn:aws:iam::451339973440:role/service-role/Cognito-
UserImport-Role --job-name importrbipoolcli
```

The output of this is as follows:

```
$ UserImportJob:
    CloudWatchLogsRoleArn: arn:aws:iam::451339973440:role/
service-role/Cognito-UserImport-Role
    CreationDate: '2021-01-22T13:23:52.520000-05:00'
    FailedUsers: 0
    ImportedUsers: 0
    JobId: import-iodmi00IBa
    JobName: importrbipoolcli
    PreSignedUrl: https://aws-cognito-idp-user-
import-iad.s3.amazonaws.com/451339973440/us-east-1_
yGe1YAnTV /import-iodmi00IBa?X-Amz-Security-Token=IQ
oJb3JpZ2luX2VjEOr%2F%2F%2F%2F%2F%2F%2F%2F%2F%2FwEaCX
VzLWVhc3QtMSJHMEUCIFaucD4oM6VPh9iM2%2FiW9UjiEISW2rkJ-
NYWMjuWe%2FHX%2FAiEA9xFMtDlqYo%2FZr96LzMkXfW1EOTe4Tot
%2BAEQOJsw0qLkqvQMIwv%2F%2F%2F%2F%2F%2F%2F%2F%2F%2FAR
ADGgw3NDU2MjM0Njc1NTUiDO4Mdkavkn7j292LiSqRA3bhbMFr3zm
%2FFcZeJ8hUQrK4xYRiD86MTmM07Blczl9jItrcXJKkyT1uUmHUf0
4WRHxMwksjZnNDGqztP8N1uCr6kwMAhNWpTjXtvVNOI6TqmMLjTyC
EGRwjBaPks2sHlmubrgWEV5YfOdQMmVYX9%2FhrSOwBSWAoSkLcBx
G5aYmhW4DGgxPcSBf6JQF%2BqeOj0IaGeOzoGeDMddBne%2BGURb%
2FzXkM0myTsq9pkNabn%2FsaptWoU11JOFK77DP%2FihIp%2FlaNR
NTni927uR3xGac0tvhU3pqE%2FkklWotHBZN6rZMxUtlLM8PuCv%2
```

F7VMjePbQqWjIFZjcklgfZb9Iv1blAwLJ7ZMlDhCIcEsYFaRg0HYEpWZ
a3Fw9I9FZcekySY3KIuDkWwl9My%2BbdEOTfOa0Mluxi9ekqJmOVT4uLv
VBOpDURe%2B7%2Fjy%2FRRT6KgqdLLVSPsNiZlXsczlkE%2F9OgavFfQj
fKpvvZ3kz5NWAm9I49nEvFMu%2BPA7NetUs0T%2BdJf5DujACXY9vW3Z0
p6CoP2gfuKN3I5MLaNrIAGOusBpk2rWpDWE02W4T0ZojuG4tWf%2BMsSQ
e7els7KrlTrW1%2BZ43LznZ17YMKx69KYSB1me97AhYEne8k6FaSSqVx%
2Ft3NIo%2BFdnD2cHntGSqLPdRmN8THYPE3pZy2jmQTiKHLg%2FC9s8ys
yc2YYVJFbKT6IUQmzohzVg8ZWWZogUaLHaqHbkBdZhGP51nsBD2UPVdz4
Ax%2BveCRWZDFcHTg4OAGk77wQJEcvq16qvzZXVBiBO6tSz910W1a%2B
C1M6Ieh33JSyBLmdwAjhfc67Ix%2BoAS9RN5isAtIVmTriP6ZkApGfvnC
qq1ZBnU1OFW3Z5g%3D%3D&X-Amz-Algorithm=AWS4-HMAC-SHA256&X-
Amz-Date=20210122T182352Z&X-Amz-SignedHeaders=host%3Bx-
amz-server-side-encryption&X-Amz-Expires=899&X-Amz-
Credential=ASIA23GU7GYR3RWBE2DS%2F20210122%2Fus-east-
1%2Fs3%2Faws4_request&X-Amz-Signature=299a07fb60b59aff2df
fc22476f7079a4bfb0e02b8e04f1a26575caf574200a3

```
SkippedUsers: 0
Status: Created
UserPoolId: us-east-1_yGe1YAnTV
```

This command creates a pre-signed URL where we can now `curl` our CSV file to trigger the import. We have a window of 15 minutes to upload our file to the pre-signed URL before we will need to generate a new one.

3. Next, we will use `curl` to send the CSV file to that pre-signed URL:

```
*     Trying 52.217.88.252...
* TCP_NODELAY set
* Connected to aws-cognito-idp-user-import-iad.
s3.amazonaws.com (52.217.88.252) port 443 (#0)
* ALPN, offering h2
* ALPN, offering http/1.1
* successfully set certificate verify locations:
*     CAfile: /etc/ssl/cert.pem
  CApath: none
* TLSv1.2 (OUT), TLS handshake, Client hello (1):
* TLSv1.2 (IN), TLS handshake, Server hello (2):
* TLSv1.2 (IN), TLS handshake, Certificate (11):
* TLSv1.2 (IN), TLS handshake, Server key exchange (12):
* TLSv1.2 (IN), TLS handshake, Server finished (14):
* TLSv1.2 (OUT), TLS handshake, Client key exchange (16):
```

```
* TLSv1.2 (OUT), TLS change cipher, Change cipher spec
(1):
* TLSv1.2 (OUT), TLS handshake, Finished (20):
* TLSv1.2 (IN), TLS change cipher, Change cipher spec
(1):
* TLSv1.2 (IN), TLS handshake, Finished (20):
* SSL connection using TLSv1.2 / ECDHE-RSA-AES128-GCM-
SHA256
<Truncating for length><Truncating for length>
< HTTP/1.1 100 Continue
* We are completely uploaded and fine
< HTTP/1.1 200 OK
< x-amz-id-2: vB/rs61GmOIjy09e+CMXMZXnQixvtUeDyNdcGdwQE/
Q2X0fDeoJG4yJd3Jn+TrSzJqdulq0u12Q=
< x-amz-request-id: E9283A75A43CE9C5
< Date: Fri, 22 Jan 2021 18:29:00 GMT
< x-amz-version-id: L4HgGM9HjY3kuU2efYDE4GSc869EMkXj
< x-amz-server-side-encryption: aws:kms
< x-amz-server-side-encryption-aws-kms-key-id:
arn:aws:kms:us-east-1:745623467555:key/5ac74669-371e-
437b-9070-d49608aa7ba5
< ETag: ''0eef29cd0fef7e5a0cd50ec5ca7fb2dc''
< Content-Length: 0
< Server: AmazonS3
<
* Connection #0 to host aws-cognito-idp-user-import-iad.
s3.amazonaws.com left intact
* Closing connection 0
```

4. Next, we can verify that the job has been created by running the following command:

```
$ aws cognito-idp list-user-import-jobs --user-pool-id
us-east-1_yGe1YAnTV --max-results 1
```

This will show us all the jobs in our user pool, up to the maximum number
of results that we set. Since we only care about the most recent, we set the
`--max-results` parameter to 1. This will confirm that the job has been created
and provide us with the imported `JobId`:

```
UserImportJobs:
- CloudWatchLogsRoleArn: arn:aws:iam::451339973440:role/
service-role/Cognito-UserImport-Role
  CreationDate: '2021-01-22T13:47:36.389000-05:00'
  FailedUsers: 0
  ImportedUsers: 0
  JobId: import-P7N4tCIdCj
  JobName: importrbipoolcli
  PreSignedUrl: https://aws-cognito-idp-user-
import-iad.s3.amazonaws.com/451339973440/us-east-1_
yGe1YAnTV /import-P7N4tCIdCj?X-Amz-Security-Token=IQ
oJb3JpZ21uX2VjEOr%2F%2F%2F%2F%2F%2F%2F%2F%2F%2FwEaCX
VzLWVhc3QtMSJHMEUCIG2Ek%2B%2FGJ7OpXYigudji8QAyFv71%2-
Bu%2FIzqyAusvY4qDZAiEArLFDaPYLXDZExxdBYeUvGbGwBj%2B
G2Qerid%2F%2FrFPPCJ4qvQMIwv%2F%2F%2F%2F%2F%2F%2F%2F
%2F%2FARADGgw3NDU2MjM0Njc1NTUiDFCjLmutSMzVtTvMXiqRA
0oHTrX26%2FnvyMOCOT1G%2FDPltiMy08SN3b4e5o1XzL5VbSDc
DDK%2FdmXjVTpzScw9r1rYxn55x9kfAmXjY9D8BuJ3rf%2F1hKd
%2FYMD3UScDO%2BLvEsnN5%2F5tX5wi2AIghzmZlgvIgpj4v6q%
2Bg13JSTvsRHyjHYPzkS3Rvh%2FSRw5bRr11fcrdcHgVNmSHJgU
B050sUKn6%2FdXB1NBXdIde3jaPEzN9IWm9tKCIkwei1A6oOfi8
ZBYAZH31qZyPi1muE%2B53z1awQOOUddI6TOFzXaru4ORYIXwoL
NSsb47gqsh6e%2Fc6HG1Ze2QwHQrTpUkAQXWQHzXlIdiyDP5t5L
1%2FGSS1ZfdWGKcXEhqQf0x5Tw0Sb2d2iUp%2BMeWrc0OAVCWQj
7b8q8SRfzDMD5pFs7N28xH%2F5J0UTMIzHpB8KPc5CSMZtMjhDi
Nk3mI2oYIV8I6orjLJiyxMLWWXhGbqJSGbsC2uHLkvOJXb2Blst
3CeiL447JD8QqHZVPP884rFuqGvYnBbEZZRy%2F%2B4G%2FrGvU
yG5XOqOGgcMIGNrIAGOusB637ZZ3HGtagu4gMbV4Nlch2Y7ruSI
GNuNe%2B4Jdc0mvKjnMtWPD%2BA1ysJQiRUbzFgJ3mMFEzXEBTp
drLPOCo1O1ildc7RG0W2ZVarFrYoIfkU7NrpigEBRyzrKnDNCJU
NBRm9iADIA0aUAAFBdRodBesSn31atQn0AJYgWCnfNmSatNS%2F
1DIz9qWnEKwj%2FcbmWFtGAN8q1rkEKIo33oS1wzjsRl094vE%2
BznA%2Bfr0dVE2LNEiyRGhPPdM3J5ZByeU01QH%2BXXjVoJqPTIEVdhl
j1tmDYEg%2FHW8V2hIBj15wi3zosAQZ7DlfI%2FttxA%3D%3D&X-Amz-
Algorithm=AWS4-HMAC-SHA256&X-Amz-Date=20210122T184736Z&X-
Amz-SignedHeaders=host%3Bx-amz-server-side-encryption&X-
Amz-Expires=899&X-Amz-Credential=ASIA23GU7GYRT5NAHQUZ%2F2
0210122%2Fus-east-1%2Fs3%2Faws4_request&X-Amz-Signature=0
```

```
eb9b127c76a62559784ac90abc537bdcb9e495ef40abf6814314104e8
15b5fc
    SkippedUsers: 0
    Status: Created
    UserPoolId: us-east-1_yGe1YAnTV
```

5. Next, we must start the job:

```
$ aws cognito-idp start-user-import-job --user-pool-id
''us-east-1_yGe1YAnTV'' --job-id ''import-0UnHP46OBe''
```

The output will be similar to the previous one, only now the status will have moved from Created to Pending, and eventually to Running.

6. We can rerun the list-user-import-jobs command from *step 4* to check the status of our import to verify that it succeeded:

```
UserImportJobs:
- CloudWatchLogsRoleArn: arn:aws:iam::451339973440:role/
service-role/Cognito-UserImport-Role
    CompletionDate: '2021-01-22T14:06:01.847000-05:00'
    CompletionMessage: Import Job Completed Successfully.
    CreationDate: '2021-01-22T14:04:59.342000-05:00'
    FailedUsers: 0
    ImportedUsers: 4
    JobId: import-0UnHP46OBe
    JobName: importrbipoolcli
...    PreSignedUrl: https://aws-cognito-idp-user-import-
iad.s3.amazonaws.com/...1f865bf0e2ba20c
    SkippedUsers: 0
    StartDate: '2021-01-22T14:06:01.062000-05:00'
    Status: Succeeded
    UserPoolId: us-east-1_yGe1YanTV
```

We can now return to the user pool in the Management Console and see the accounts that were created:

Figure 5.41 – The imported user pool accounts

We have spent a lot of time configuring the backend of our Amazon Cognito user pool, but we still haven't seen what it looks like when it is in front of an application. Fortunately, we do not need to integrate it with an app to test out its capabilities.

Exploring the hosted UI

Amazon Cognito offers a customizable hosted UI for user sign-in and sign-up. We can see the default UI by opening the link at the bottom of each app client, under the **App client settings** menu inside our user pool:

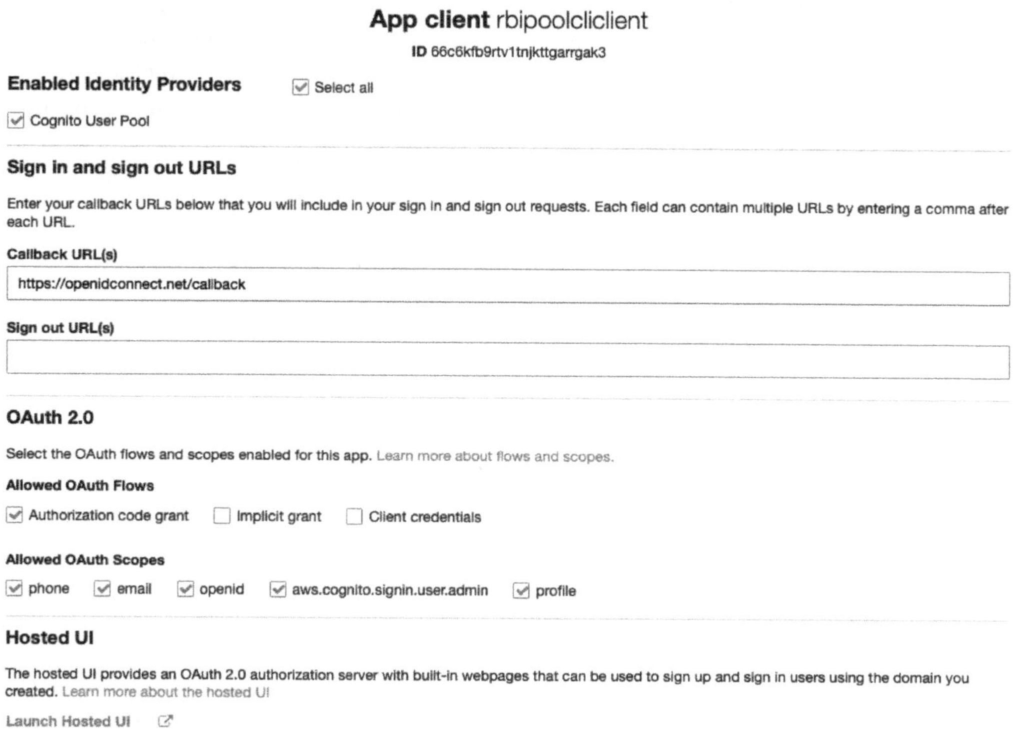

App client rbipoolcliclient

ID 66c6kfb9rtv1tnjkttgarrgak3

Enabled Identity Providers   ☑ Select all

☑ Cognito User Pool

Sign in and sign out URLs

Enter your callback URLs below that you will include in your sign in and sign out requests. Each field can contain multiple URLs by entering a comma after each URL.

Callback URL(s)

https://openidconnect.net/callback

Sign out URL(s)

OAuth 2.0

Select the OAuth flows and scopes enabled for this app. Learn more about flows and scopes.

Allowed OAuth Flows

☑ Authorization code grant   ☐ Implicit grant   ☐ Client credentials

Allowed OAuth Scopes

☑ phone   ☑ email   ☑ openid   ☑ aws.cognito.signin.user.admin   ☑ profile

Hosted UI

The hosted UI provides an OAuth 2.0 authorization server with built-in webpages that can be used to sign up and sign in users using the domain you created. Learn more about the hosted UI

Launch Hosted UI   ☐

Figure 5.42 – The hosted UI is available from the App Client details form

This is the default sign-in form:

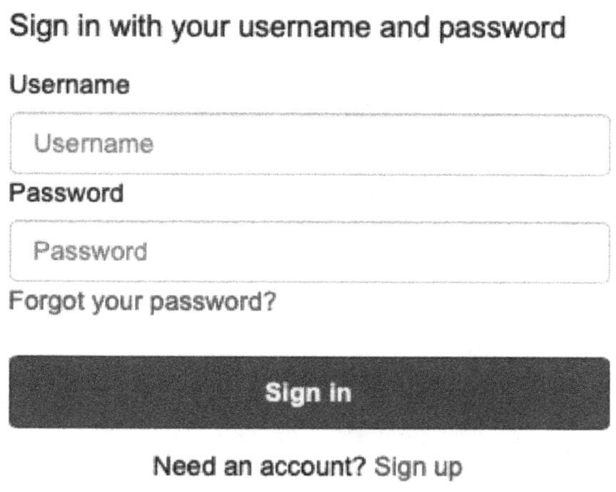

Sign in with your username and password

Username

Username

Password

Password

Forgot your password?

Sign in

Need an account? Sign up

Figure 5.43 – Amazon Cognito user pool's default form

If we wish to offer a branded experience, we can go to the UI customization menu in our pool and adjust the colors, border padding, and other CSS elements to adjust the look and feel of this hosted service so that it aligns with our own website. Amazon Cognito offers the option to run several different versions of the hosted UI, with distinct branding applied to a specific application client ID:

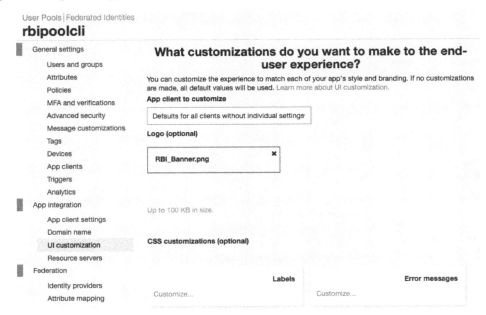

Figure 5.44 – Customization options for the hosted UI

Let's add a simple, and admittedly ugly, banner to our default form and relaunch the hosted UI:

Figure 5.45 – Applying branding to the hosted UI

Beyond branding, we can test out the user signup and sign-in experiences using this interface. Let's create an account by clicking the **Sign up** link:

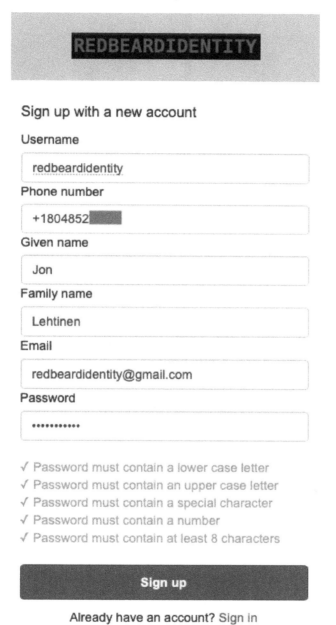

Figure 5.46 – New account creation form with the required attributes

This takes us to the new account creation form. It will ask for a few pieces of information, specifically the attributes that we defined, as required, when we set up our user pool. After populating that information, we can proceed:

Figure 5.47 – Email verification process

When we set up this user pool, we specified that users had to verify their email addresses as that was going to be our recovery mechanism. Before we can use this account, we must intercept the activation email that was sent to the address we used at signup:

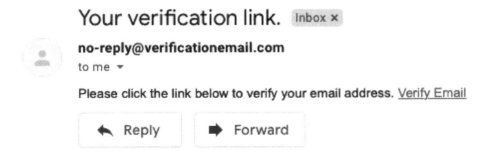

Figure 5.48 – The activation email with an activation link

The email will arrive in our inbox moments later. Upon clicking the link, we will get confirmation that our account is active:

Your registration has been confirmed!

Figure 5.49 – Confirmation of account activation

This user will now appear alongside the additional users we imported previously:

Username	Enabled	Account status	Email verified	Phone number verified	Updated	Created
rbiuser1	Enabled	RESET_REQUIRED	true	false	Jan 23, 2021 4:58:17 PM	Jan 23, 2021 4:58:17 PM
rbiuser2	Enabled	RESET_REQUIRED	true	false	Jan 23, 2021 4:58:17 PM	Jan 23, 2021 4:58:17 PM
rbiuser3	Enabled	RESET_REQUIRED	true	false	Jan 23, 2021 4:58:17 PM	Jan 23, 2021 4:58:17 PM
rbiuser4	Enabled	RESET_REQUIRED	true	false	Jan 23, 2021 4:58:17 PM	Jan 23, 2021 4:58:17 PM
redbeardidentity	Enabled	CONFIRMED	true	false	Jan 23, 2021 7:14:04 PM	Jan 23, 2021 7:12:45 PM

Figure 5.50 – New user in the pool, along with the confirmed account status

Now that we have created two user pools using both the Management Console and the CLI, we should have a greater appreciation for not just the feature richness of Amazon Cognito user pools, but also the complexity of their configuration. Amazon Cognito user pools are a customizable federated identity provider for applications, complete with user lifecycle management. We will explore practical applications of these capabilities for protecting resources in *Chapter 11*, *Bringing Your Users into AWS*. For now, these user pools provide us with what we need to look closely into Amazon Cognito's second major component, known as identity pools.

Creating an Amazon Cognito identity pool

Since we now have a user pool that can provide federated identities, we can create an identity pool. Doing so will allow the federated identities from that user pool to access AWS resources. To do this from the Management Console, follow these steps:

1. Go to the Amazon Cognito service and select **Manage Identity Pools**.

2. Since we have no existing identity pools, we are taken directly to the wizard to configure our first one. Let's call this one `rbiidentitypool`:

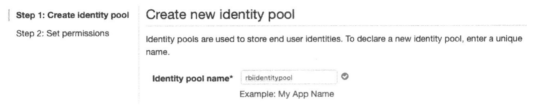

Figure 5.51 – Naming the new identity pool

3. An interesting capability of identity pools is that they allow unauthenticated users to obtain temporary credentials to access AWS resources. It may seem counterintuitive to permit this, but there may be use cases where access to a resource, such as placing a file into a bucket or adding an entry into an Amazon DynamoDB database, may be deemed so sufficiently low risk that identifying principals taking these actions may not be critical to the process. Whatever the justification may be, we have the option of applying a distinct trust role policy to unauthenticated users if we wish to use it. For our purposes, let's not allow unauthenticated identities:

▼ Unauthenticated identities ❶

Amazon Cognito can support unauthenticated identities by providing a unique identifier and AWS credentials for users who do not authenticate with an identity provider. If your application allows customers to use the application without logging in, you can enable access for unauthenticated identities. Learn more about unauthenticated identities.

☐   Enable access to unauthenticated identities

Enabling this option means that anyone with internet access can be granted AWS credentials. Unauthenticated identities are typically users who do not log in to your application. Typically, the permissions that you assign for unauthenticated identities should be more restrictive than those for authenticated identities.

Figure 5.52 – Identity pools support issuing temporary credentials to unauthenticated identities

4. We may optionally downgrade our authentication flow to a legacy pattern if required. Unless there is a good reason for us to do so, it is a best practice to leave the default enhanced authentication flow enabled:

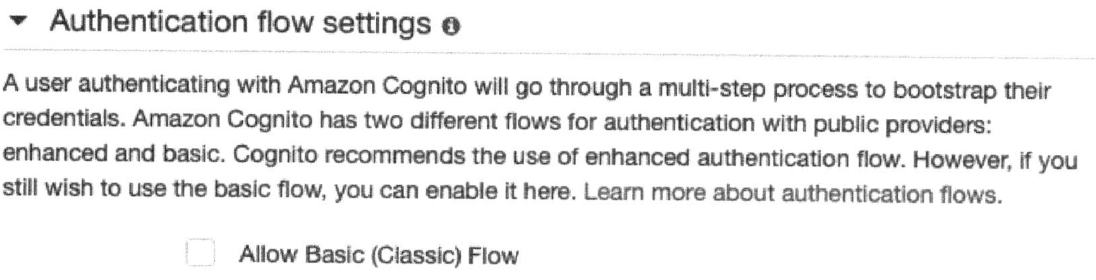

▼ Authentication flow settings ❶

A user authenticating with Amazon Cognito will go through a multi-step process to bootstrap their credentials. Amazon Cognito has two different flows for authentication with public providers: enhanced and basic. Cognito recommends the use of enhanced authentication flow. However, if you still wish to use the basic flow, you can enable it here. Learn more about authentication flows.

☐ Allow Basic (Classic) Flow

Figure 5.53 – Option to enable a legacy authentication flow

5. Finally, we must configure our authentication providers for the identity pool. This can be one or more Amazon Cognito user pools, any number of social providers, a SAML2 or OIDC identity provider, or any combination of them all. We will use the user pool and client we created earlier for this:

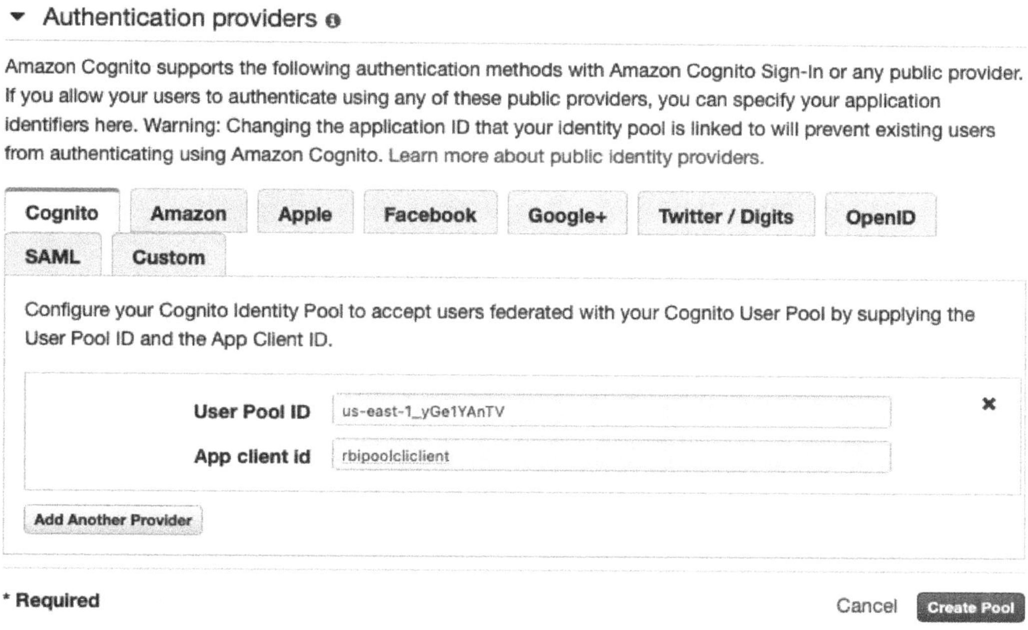

▼ Authentication providers ❶

Amazon Cognito supports the following authentication methods with Amazon Cognito Sign-In or any public provider. If you allow your users to authenticate using any of these public providers, you can specify your application identifiers here. Warning: Changing the application ID that your identity pool is linked to will prevent existing users from authenticating using Amazon Cognito. Learn more about public identity providers.

Cognito	Amazon	Apple	Facebook	Google+	Twitter / Digits	OpenID
SAML	Custom					

Configure your Cognito Identity Pool to accept users federated with your Cognito User Pool by supplying the User Pool ID and the App Client ID.

User Pool ID	us-east-1_yGe1YAnTV	✕
App client id	rbipoolcliclient	

Add Another Provider

* Required   Cancel   **Create Pool**

Figure 5.54 – Setting federated providers for the identity pool

6. Next, we will be prompted to create the two roles that the Amazon Cognito identity pool users will be able to assume. The first is the policy for authenticated identities. Let's make a policy that allows them to view and add items to a specific Amazon S3 bucket object:

```
{
    ''Version'': ''2012-10-17'',
    ''Statement'': [
        {
            ''Action'': [
''s3:ListBucket'',
''mobileanalytics:PutEvents'',
''cognito-sync:*'',
''cognito-identity:*''],
            ''Effect'': ''Allow'',
            ''Resource'': [''arn:aws:s3:::redbeardidentity-
bucket-1''],
            ''Condition'': {''StringLike'': {''s3:prefix'':
[''${cognito-identity.amazonaws.com:sub}/*'']}}
        },
        {
            ''Action'': [
                ''s3:GetObject'',
                ''s3:PutObject''
            ],
            ''Effect'': ''Allow'',
            ''Resource'': [''arn:aws:s3:::redbeardidentity-bucket-
1/${cognito-identity.amazonaws.com:sub}/*'']
        }
    ]
}
```

We will also define the role that unauthenticated identities get. We already precluded their access to this identity pool earlier, so we will leave this at the default unauthenticated policy provided by AWS.

7. After that, our identity pool is created, and we are presented with reference materials and SDKs for incorporating Amazon Cognito identity pools into our applications for a variety of programming languages:

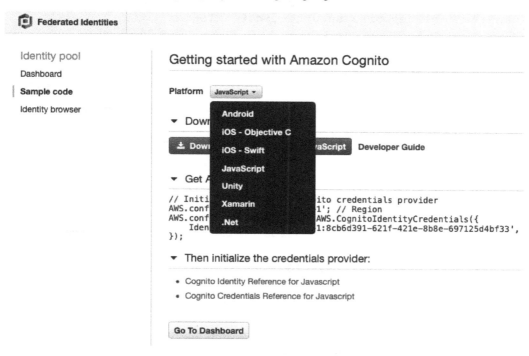

Figure 5.55 – SDK downloads for several languages

Unless we digress to building an application that leverages these SDKs, it is slightly more difficult to see identity pools in action. Unlike user pools, where everything can be immediately configured and consumed since the user pool is acting as the application's identity provider and user store, identity pools bridge application architecture and AWS resources. Identity pools facilitate the secure use of AWS services and resources by an application or the users of the application, but at a truly programmatic level. The decision to use identity pools and to leverage AWS resources directly from the application tier speaks to the PaaS identity focus of Amazon Cognito.

Creating an identity pool with the CLI

Creating an identity pool with the CLI follows the same pattern as other complex resources. First, we must generate the skeleton template to work from:

```
$ aws cognito-identity create-identity-pool --generate-cli-
skeleton yaml-input
```

This provides us with the following output:

```
IdentityPoolName: ''  # [REQUIRED] A string that you provide.
AllowUnauthenticatedIdentities: true # [REQUIRED] TRUE if the
identity pool supports unauthenticated logins.
AllowClassicFlow: true # Enables or disables the Basic
(Classic) authentication flow.
SupportedLoginProviders: # Optional key.
  KeyName: ''
DeveloperProviderName: '' # The ''domain'' by which Cognito
will refer to your users.
OpenIdConnectProviderARNs: # A list of OpendID Connect provider
ARNs.
- ''
CognitoIdentityProviders: # An array of Amazon Cognito user
pools and their client IDs.
- ProviderName: ''  # The provider name for an Amazon Cognito
user pool.
  ClientId: '' # The client ID for the Amazon Cognito user
pool.
  ServerSideTokenCheck: true # TRUE if server-side token
validation is enabled for the identity provider's token.
SamlProviderARNs: # An array of Amazon Resource Names (ARNs) of
the SAML provider for your identity pool.
- ''
IdentityPoolTags: # Tags to assign to the identity pool.
  KeyName: ''
```

Once we've made edits to omit variables that we will not use (such as the various provider names aside from Amazon Cognito), we can save it as a YAML file. Then, it is time to run the create command:

```
$ aws cognito-identity create-identity-pool --cli-input-yaml
file://rbiidentitypoolcli.yml
```

If this worked, the output will be a description of the new identity pool:

```
AllowClassicFlow: false
AllowUnauthenticatedIdentities: false
CognitoIdentityProviders:
```

```
- ClientId: rbiidentitypool
  ProviderName: cognito-idp.us-east-1.amazonaws.com/us-east-1_
yGe1YAnTV
  ServerSideTokenCheck: true
IdentityPoolId: us-east-1:538ab1fe-f633-468f-8904-f8c585d0fe7a
IdentityPoolName: rbiidentitypoolcli
IdentityPoolTags: {}
```

However, we are still not done. We still need to attach policies for authenticated and unauthenticated users to this identity pool. We can reuse existing roles or create net-new roles. Rather than reinvent the wheel, let's reuse the roles we created in our previous exercise. We can get their ARNs by running the following command against the first identity pool we created:

```
$ aws cognito-identity get-identity-pool-roles --identity-
pool-id us-east-1:8cb6d391-621f-421e-8b8e-697125d4bf33
```

This gets us the ARN values we need for both our authenticated and unauthenticated users:

```
IdentityPoolId: us-east-1:8cb6d391-621f-421e-8b8e-697125d4bf33
Roles:
  authenticated: arn:aws:iam::451339973440:role/Cognito_
rbiidentitypoolAuth_Role
  unauthenticated: arn:aws:iam::451339973440:role/Cognito_
rbiidentitypoolUnauth_Role
```

With that information, we can now set the new roles on our new identity pool instance:

```
$ aws cognito-identity set-identity-pool-roles --identity-
pool-id us-east-1:538ab1fe-f633-468f-8904-f8c585d0fe7a
--roles authenticated=arn:aws:iam::451339973440:role/
Cognito_rbiidentitypoolAuth_
Role,unauthenticated=arn:aws:iam::451339973440:role/Cognito_
rbiidentitypoolUnauth_Role
```

If successful, no output should appear in the CLI. We can validate that our roles have been set by running the `get-identity-pool-roles` command against the new identity pool, the output of which will show the two roles:

```
$ aws cognito-identity get-identity-pool-roles --identity-
pool-id us-east-1:538ab1fe-f633-468f-8904-f8c585d0fe7a


IdentityPoolId: us-east-1:538ab1fe-f633-468f-8904-f8c585d0fe7a
Roles:
   authenticated: arn:aws:iam::451339973440:role/Cognito_
rbiidentitypoolAuth_Role
   unauthenticated: arn:aws:iam::451339973440:role/Cognito_
rbiidentitypoolUnauth_Role
```

Once again, unless we start building applications that leverage AWS-native capabilities, it is difficult to demonstrate these identity pools in action.

Summary

We covered a lot of new ground in this chapter! In this chapter, we took our first steps into a very large identity service that is nearly completely separate from the identity service that we spent that last few chapters getting to know. However, now that we understand the capabilities of Amazon Cognito, as well as how it can be used to solve application identity in a PaaS context, we are prepared to incorporate it into a holistic cloud identity strategy. We can use services such as Amazon Cognito to facilitate and simplify the challenges that application teams have with user life cycle management and authentication, especially if they intend to fully enmesh their application architecture into the AWS ecosystem.

The next chapter will bring us back into managing access to AWS as an IaaS platform. However, it will do so via another fully featured identity provider service available on AWS that is totally different from Amazon Cognito. There, we will become familiar with AWS SSO and AWS organizations.

Questions

1. What are the two main components of Amazon Cognito?

2. What kind of tokens are issued by a Cognito user pool when it is acting as an authorization server and IDP?

3. How are identity pools different from user pools?

4. What role does Amazon Cognito play in an identity ecosystem compared to AWS IAM?

6
Introduction to AWS Organizations and AWS Single Sign-On

We've said several times that much of the confusion around applying identity to AWS stems from the various identity services available on the platform, and the ambiguity around their appropriate use. So far, we've split services into two groups: those that provide identity for AWS as an infrastructure-as-a-service platform, and those that offer identity capabilities in a **platform-as-a-service (PaaS)** context. AWS **Single Sign-On (SSO)** strains this motif. On the one hand, AWS SSO's primary function and capability focuses on facilitating access to AWS resources, specifically AWS accounts within an AWS organization. On the other, it is also capable of being an enterprise-grade identity provider for more than just AWS resources.

By the end of this chapter, we will understand AWS SSO's role in the AWS identity ecosystem, and how it operates as an identity service across AWS.

In this chapter, you will learn about the following:

- What is AWS SSO?
- AWS Organizations
- Configuring AWS SSO in the Management Console
- Configuring AWS SSO from the CLI

Technical requirements

To get the most out of this chapter, you will need the following:

- An AWS account
- A workstation running the AWS CLI
- A text editor or IDE to edit JSON/YAML files, such as Microsoft Visual Studio Code

What is AWS SSO?

AWS SSO is another IDaaS capability available from AWS, similar in some regards to Amazon Cognito. Whereas Amazon Cognito provides application identity capabilities, AWS SSO facilitates centralized administration for federated access to AWS accounts, as well as general identity provider capabilities. AWS SSO offers free account management, authentication services, and a strong authentication capability for AWS accounts and applications managed under an AWS organization. AWS SSO users can authenticate to one or more AWS accounts using either accounts and credentials managed by AWS SSO itself, or through accounts synched from an external authoritative source, such as an enterprise-managed directory and identity provider.

We mentioned that one of the primary use cases of AWS SSO was centralized user management and federated authentication to all AWS accounts within an AWS organization. We discussed AWS Organizations briefly at the end of *Chapter 3, IAM User Management*. Specifically, we said that as large organizations often have several AWS accounts to manage and maintain, ad hoc federated relationships between their existing IDP and each of those AWS IAM instances become an administrative burden over time:

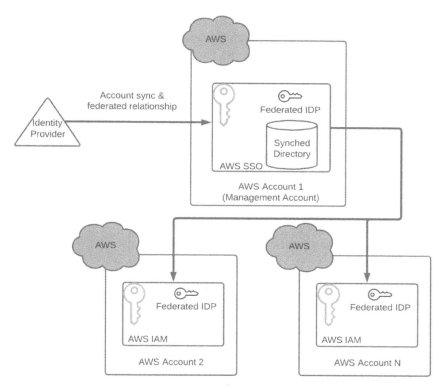

Figure 6.1 – AWS SSO managing SSO into managed AWS accounts in an AWS Organization

AWS SSO provides one connection for both directory synchronization from an on-premises user store, as well as a single service provider connection for all AWS accounts in use by the organization. The AWS SSO connection automatically registers the existing IDP as an IDP within the AWS IAM service of each downstream AWS account. This way, organizational administrators only need to tend the connection between their IDP and the main organizational AWS account, and every other AWS account included in their organization will be able to take advantage of delegated authentication without the per-account administrative overhead.

The deeply coupled integration between AWS Organizations and AWS SSO extends beyond just synching IDP objects into the member accounts' AWS IAM instances. For organizations that take advantage of **organizational units (OUs)** to organize their accounts, that same OU structure is replicated and presented within AWS SSO. This makes it easy to apply common permissions across accounts and manage access to specific accounts based upon user account attributes, group membership information, and other AWS IAM policy constructs, such as **service control policies (SCPs)**. Administrators can define permissions sets to enforce access control to specific services and resources when users sign into specific AWS accounts through AWS SSO.

In addition to managing access to AWS accounts, AWS SSO also provides application identity services for AWS apps (AWS' own offerings for business services such as instant messaging, email, and document management), third-party cloud SaaS providers, and any other SAML2-compliant service provider. Similar to Amazon Cognito, the AWS SSO user store can be the source of truth for identity information or those accounts can be synchronized from existing enterprise directories. Unlike Amazon Cognito, the service is less developer-focused in its consumption. There is no need to insert AWS SSO-specific code into an application for that application to consume the SAML tokens it issues for its users. AWS SSO behaves very similarly to any other commercial on-premises or IDaaS identity provider, complete with preconfigured integrations for major SaaS applications:

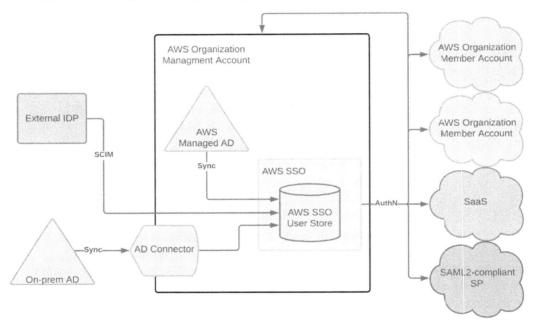

Figure 6.2 – Available identity flows through AWS SSO

AWS SSO's user store can be populated in several ways. In organizations with on-premises Active Directory as their source of identity, the recommendation is to populate the AWS SSO user store via AD Connector through the Amazon Cloud Directory service. As the name suggests, AD Connector bridges the two user stores. Alternatively, AD accounts can be synced directly to the AWS SSO user store when using the AWS Managed Active Directory also available in Amazon Cloud Directory. AWS SSO also supports **System for Cross-domain Identity Management**, or **SCIM**, which is a RESTful federated provisioning protocol supported by many identity providers. Accounts, groups, and credentials may optionally be managed directly within AWS SSO itself, free of entanglement from an external authoritative source.

Finally, AWS SSO offers audit and event logging capabilities. AWS SSO is meant to facilitate access to AWS accounts, and be the authoritative source of identity for not only those accounts, but also other federated applications. As such, every administrative action and user authentication event is captured in AWS CloudTrail, AWS' service for governance, operational auditing, risk auditing, and compliance monitoring.

Requirements to use AWS SSO

Unlike Amazon Cognito or AWS IAM, the AWS SSO service has preconditions before it can be activated on a given AWS account. These requirements are as follows:

- AWS Organizations must be enabled on the account.

- The account must be the management account of the organization.

- All AWS Organizations features must be enabled on the account, meaning the organizational relationship cannot only be merely "consolidated billing features."

- Decisions around how the user store will be populated must be settled if not using the built-in user store of AWS SSO. Whereas these points will not preclude activation of the service, failure to consider the following requirements may limit our deployment options if not considered before activation:

  i) The use of AWS Managed Microsoft Active Directory as the source of user identity will require activating AWS SSO in the same region where AWS SSO will be activated.

  ii) AWS Managed Microsoft Active Directory must be connected to the AWS Organizations management account.

  iii) An AD Connector to an on-premises Active Directory must also be in the same region where AWS SSO will be activated, also on the Organizations management account.

Once we have these preconditions addressed, we will be ready to enable AWS SSO on our AWS account. If we wanted to proceed directly with configuring AWS SSO, the act of enabling that service would also enable the AWS Organizations service, and set up an organization using our Redbeard Identity AWS account as the primary account. Let's instead examine AWS Organizations as a preamble to working with AWS SSO.

AWS Organizations

AWS Organizations is a service designed to facilitate the management and administration of AWS accounts. Whereas AWS Organizations is not necessarily an IAM service on its own, it can certainly be argued that it provides certain IAM-like functions. If AWS accounts are user accounts, and an AWS Organization-managed *organization* is a traditional organization, the AWS Organizations service is arguably the IAM system of that analogy. As AWS Organizations is deeply linked to AWS SSO, we will spend just enough time to ensure we understand the critical concepts about it that are necessary to ensure we understand that relationship.

Through AWS Organizations, enterprises or organizations with multiple AWS accounts can consolidate the management of every account down to a single, primary management account. This is great for simplifying some basic business processes, such as billing. A consolidated invoice for all the AWS accounts under a single organization is much less paperwork than needing to process one invoice per account. Similar can be said for other necessary, though tedious, processes such as software license reporting and audits. Additionally, AWS Organizations enables unified reporting on things such as resource utilization, service usage, and costs for the entire organization, with granularity available down to individual accounts.

As we mentioned both earlier in this chapter and in *Chapter 4, Access Management, Policies, and Permissions*, a key feature of AWS Organizations is access control offered by **SCPs**. SCPs allow us to define the maximum set permissions an account may have by selectively restricting the access to resources and services that the member accounts may grant their own users. These service control policies may be applied directly to member accounts, or to the **OU** where a number of member accounts reside. Should a specific account within an OU require a more restrictive authorization policy, supplemental policies may also be applied to that account, which will further restrict the access available to members of that account:

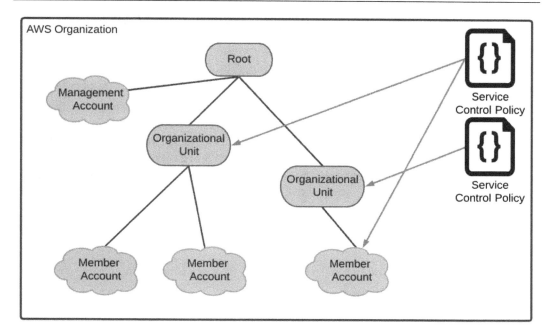

Figure 6.3 – Example of service control policies applied to an organizational structure

Each SCP is functionally a filter, and the most restrictive policy statements from each SCP will be applied to these authorization *filters* and stacked upon a member account. SCPs are never additive in what they enable users to do.

The management account can use AWS Organizations to enable policies and trusted services across all member accounts in the organization. Service control policies are an example of such a policy that, when enabled, will be applied to the member organizations. AWS SSO is an example of a trusted service that does the same. More specifically, a trusted service is allowed to create service-linked roles in the AWS IAM services of member organizations to enable cross-account capabilities. In the case of AWS SSO, that cross-account capability is single sign-on into those member accounts using AWS SSO as the identity provider.

There is more to AWS Organizations than we will address here. What is important is that we understand its function, and how it creates and manages the relationship between AWS accounts that allows AWS SSO to operate. Now, let's set up an organization in our account.

Configuring AWS Organizations using the Management Console

We've said it enough over the last couple of pages: the relationship between AWS Organizations and AWS SSO cannot be understated. As part of enabling AWS SSO, the wizard checks to see whether we have enabled AWS Organizations on our account. If it detects that we have not, it issues the prompt informing us of what AWS Organizations does and asks for permission to create an AWS organization using the current account as part of enabling AWS SSO. We can click the **Create AWS organization** button and proceed. Once we do, we see a new window explaining what the full process for enabling AWS SSO entails:

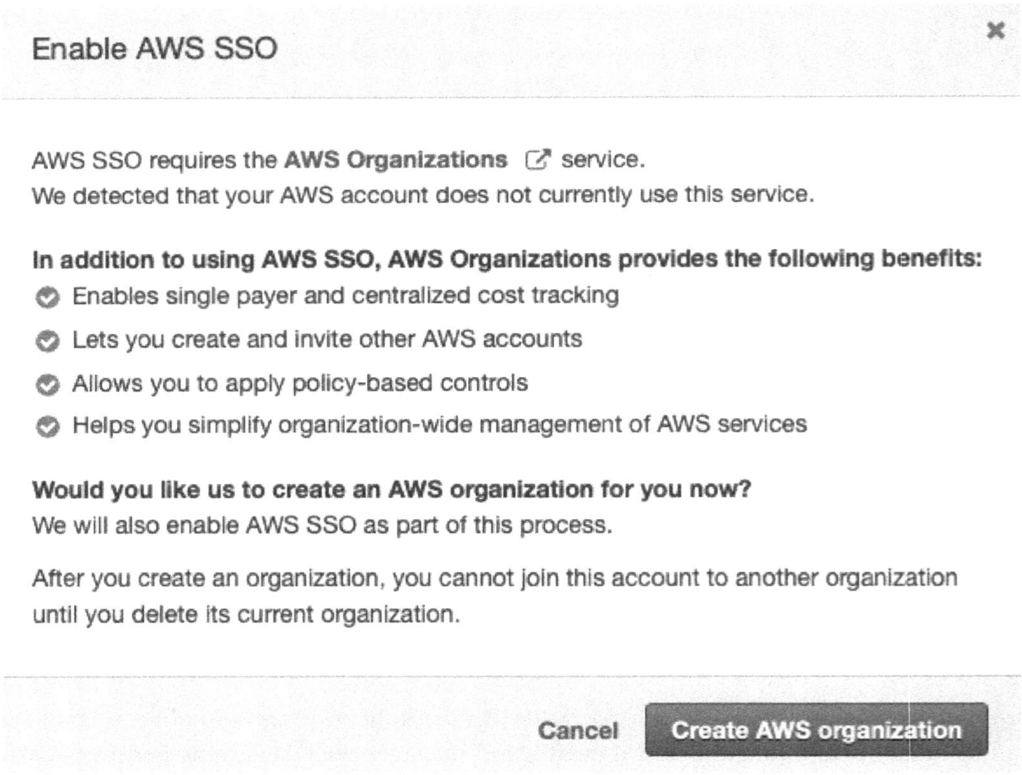

Enable AWS SSO ✕

AWS SSO requires the **AWS Organizations** ☑ service.
We detected that your AWS account does not currently use this service.

In addition to using AWS SSO, AWS Organizations provides the following benefits:

✅ Enables single payer and centralized cost tracking

✅ Lets you create and invite other AWS accounts

✅ Allows you to apply policy-based controls

✅ Helps you simplify organization-wide management of AWS services

Would you like us to create an AWS organization for you now?
We will also enable AWS SSO as part of this process.

After you create an organization, you cannot join this account to another organization until you delete its current organization.

Cancel   **Create AWS organization**

Figure 6.4 – AWS SSO has a dependency on AWS Organizations

We can continue, and an AWS organization will be created for us. Alternatively, we can simply enable an organization directly from the AWS Organizations service. The result will be the same. We will now be able to see our organization from the AWS Organizations service in the Management Console:

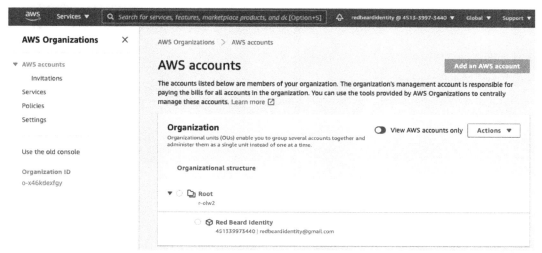

Figure 6.5 – The AWS Organizations dashboard

The first place we land is the dashboard. It shows us the at-a-glance view of everything that makes up our organization. This includes all of the accounts and OU we have configured so far, along with their hierarchy relative to the organizational root. In very complex organizations, the **View AWS accounts only** toggle will switch the organization view from the logical organizational structure to a bullet list of just the AWS accounts that comprise its membership. For now, the only account listed is our own. Let's correct that.

We created an additional AWS account, which we named RBI Sub Org 1, that we would like to add as a member account in our new AWS organization. To do this, we will start by clicking the **Add an AWS account** button on the dashboard of the AWS Organizations service from within our Redbeard Identity AWS account. Here we have two options, the first of which is creating an account directly from the AWS Organizations service:

AWS Organizations > AWS accounts > Add an AWS account

Add an AWS account

You can add an AWS account to your organization either by creating an account or by inviting an existing AWS account to join your organization.

● Create an AWS account	○ Invite an existing AWS account
Create an AWS account that is added to your organization.	Send an email request the owner of an account. If they accept, the account joins the organization.

Create an AWS account

AWS account name

> Sandbox

Email address of the account's owner

> account@domain.com

IAM role name
The management account can use this IAM role to access resources in the member account.

OrganizationAccountAccessRole

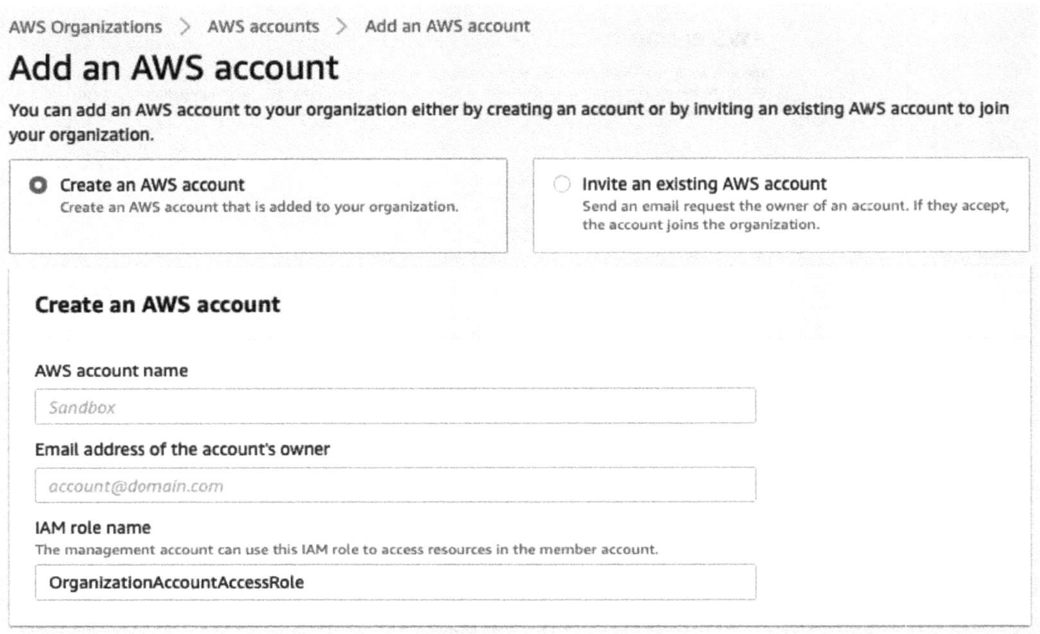

Figure 6.6 – We can create an AWS account from the AWS Organizations service

For anyone who has set up their own AWS account, it may seem odd that so little information would be required to set up a whole new account. Manually signing up for an account may not be an overly complicated process, but it does entail collecting significantly more pieces of contact and billing information than just an email address. However, recall that the management account handles billing for all membership accounts. As such, any net-new AWS accounts' charges will come back to the management account defined for the organization, tagged under that specific account's identifiers.

At any rate, in addition to the already existing account that we intended to add, let's create one more account directly from the AWS Organizations service. Once we populate the account's name and owner's email, we can optionally define tagging and create the AWS account:

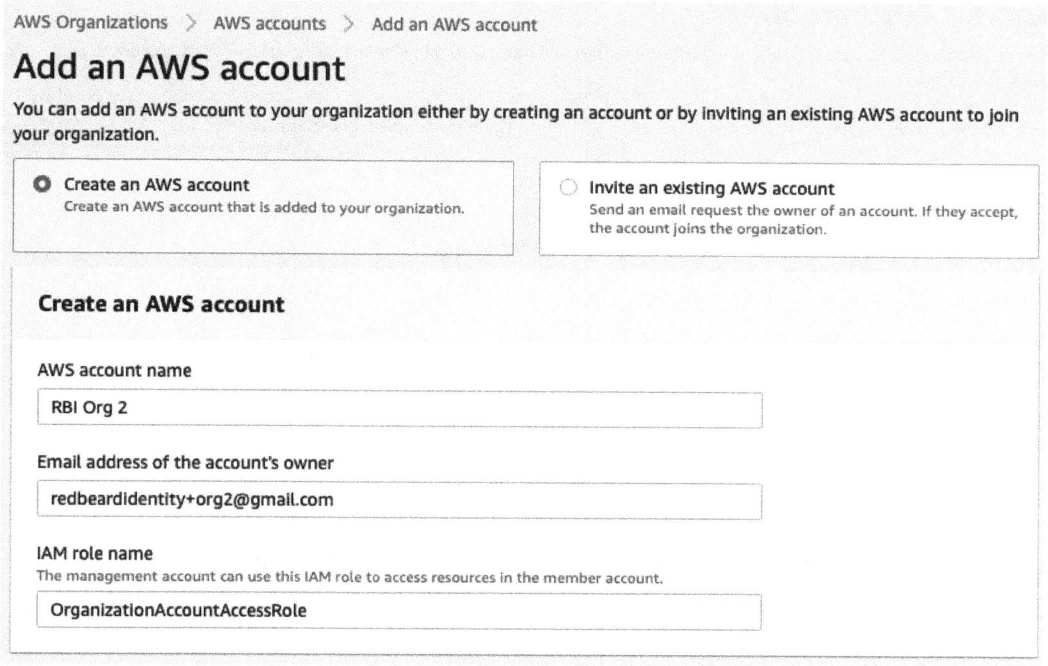

Figure 6.7 – Adding an AWS account to the organization

This kicks off the account provisioning process. The account owner will be notified by mail once it is complete, and the account will appear in the main organization hierarchy once provisioned.

Let's also invite the `RBI Sub Org 1` account into our organization. This time we will select **Invite an existing AWS account**:

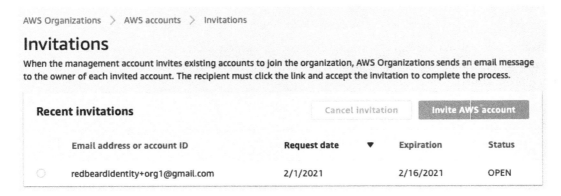

Figure 6.8 – Inviting an existing account to join the organization

We identify the account in question using either the account ID or the email address of the root user. Once sent, we can check the status of the invitation by going to the **Invitations** menu on the left-hand side of the AWS Organizations dashboard page:

AWS Organizations > AWS accounts > Invitations

Invitations

When the management account invites existing accounts to join the organization, AWS Organizations sends an email message to the owner of each invited account. The recipient must click the link and accept the invitation to complete the process.

Recent invitations			Cancel invitation	Invite AWS account
Email address or account ID		Request date ▼	Expiration	Status
○ redbeardidentity+org1@gmail.com		2/1/2021	2/16/2021	OPEN

Figure 6.9 – Invitations and their status

If we check the inbox of the root account user who owns the `RBI Sub Org 1` account, we can see the invitation:

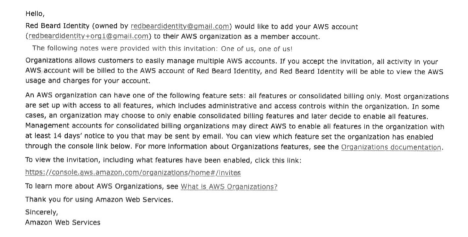

Figure 6.10 – The invitation to join the organization

If we follow the link, we are taken to the AWS Management Console sign-in page. By signing in as the root user, we will automatically be taken to the invitation we received in the RBI Sub Org 1 account's AWS Organizations service page. If we did not use the link from the email, we could simply sign in as the root user and go to the AWS Organizations service, where we will see an icon indicating there is a pending invitation. There we will find our invitation to join the Redbeard Identity management account's organization:

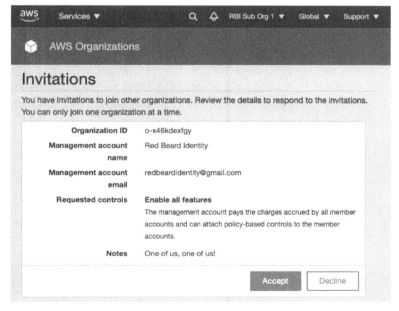

Figure 6.11 – The invitation as seen from the RBI Sub Org 1 AWS Organizations service

If we accept, we are prompted to confirm that we intend to join the organization identified by the unique ID displayed on the screen. It also warns us that this action will allow another administrator to be empowered to apply policies on our account, which could restrict the services and actions that we are able to take with resources within our account. Assuming we understand the implications of surrendering administrative control, we may click **Confirm** and continue:

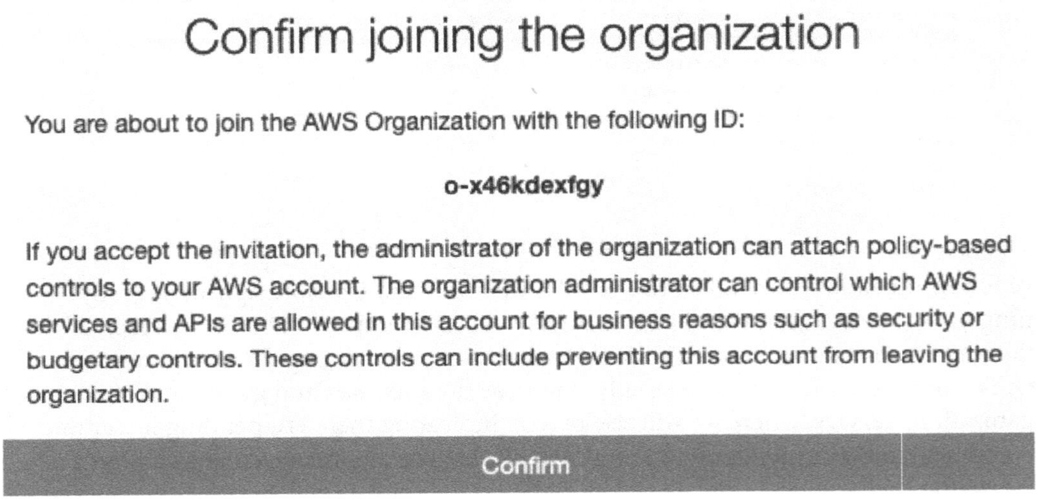

Confirm joining the organization

You are about to join the AWS Organization with the following ID:

o-x46kdexfgy

If you accept the invitation, the administrator of the organization can attach policy-based controls to your AWS account. The organization administrator can control which AWS services and APIs are allowed in this account for business reasons such as security or budgetary controls. These controls can include preventing this account from leaving the organization.

Confirm

Figure 6.12 – Confirmation prompt for joining the organization

After a brief bit of processing, we are dumped back into our AWS Organizations service inside of our account, and only now there is some new information on display. We can see whether our organization is limited to billing management or full policy management, as well as the organizational ID and email address of the root user for the management account. If we have joined the organization in error, we can leave; however, that option is not guaranteed to be available unless the management account allows it:

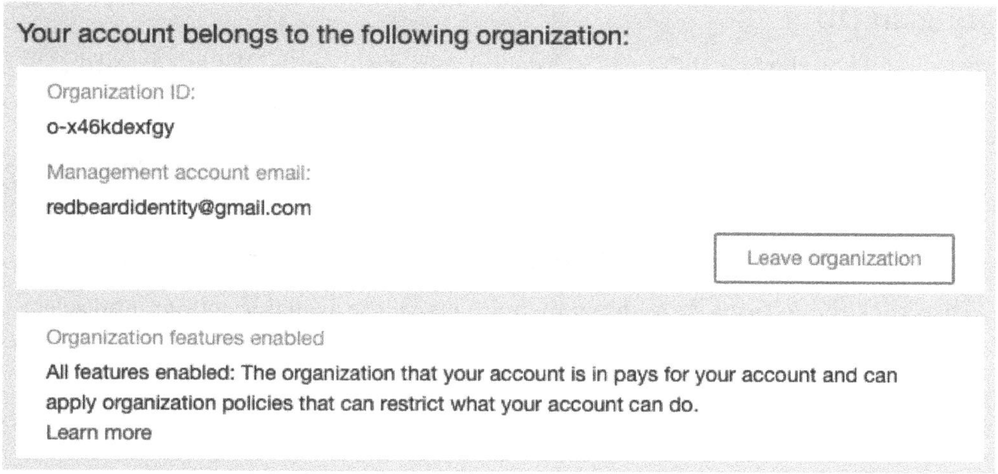

Your account belongs to the following organization:

Organization ID:

o-x46kdexfgy

Management account email:

redbeardidentity@gmail.com

Leave organization

Organization features enabled

All features enabled: The organization that your account is in pays for your account and can apply organization policies that can restrict what your account can do.
Learn more

Figure 6.13 – Organizational info from the member account perspective

Let's return to the Redbeard Identity account and look at our organizational structure now that we have more than our original account included in the hierarchy:

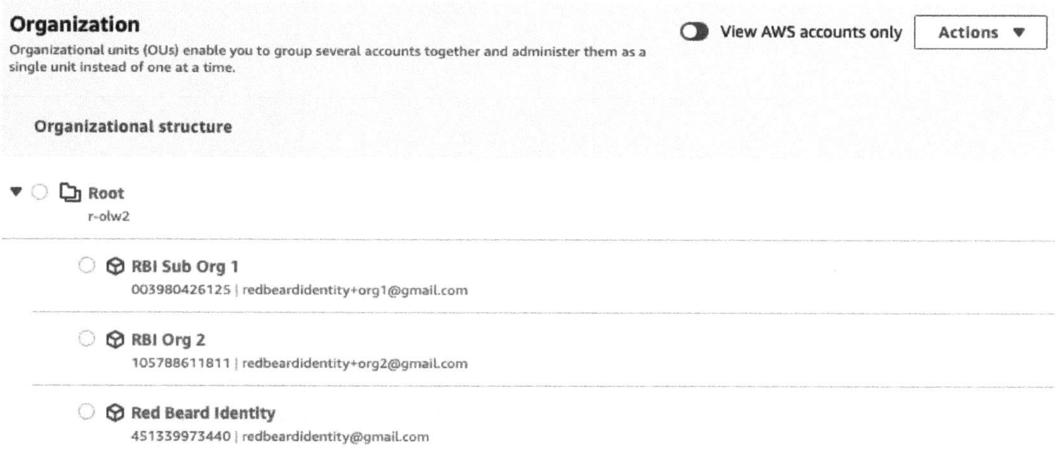

Organization

Organizational units (OUs) enable you to group several accounts together and administer them as a single unit instead of one at a time.

View AWS accounts only   Actions ▼

Organizational structure

▼ ○ ▢ **Root**
       r-olw2

       ○ ⊗ **RBI Sub Org 1**
            003980426125 | redbeardidentity+org1@gmail.com

       ○ ⊗ **RBI Org 2**
            105788611811 | redbeardidentity+org2@gmail.com

       ○ ⊗ **Red Beard Identity**
            451339973440 | redbeardidentity@gmail.com

Figure 6.14 – Organizational structure with three accounts

Presently, all three accounts are under the organizational root directory. If we enable any service control policies, tag policies, or other features available for organizational management, we will need to apply them individually to each account to maintain a distinction between the member accounts and the management account. This may not be much of a burden in such a small organization, but it would quickly become an administrative burden in a large one. We can create an organizational unit to facilitate administration. With an OU, we can apply policies to the OU directly, and those policies will apply to all accounts within that OU.

Creating an OU

We create an OU by starting at the AWS Organizations dashboard. OUs are defined relative to their encapsulating OU, which is represented by a directory on the AWS Organizations dashboard. As we do not have any depth to our organizational hierarchy yet, the only place where we can define an additional OU would be from the **Root** OU. To create the new OU, do the following:

1. Select the radio button next to the **Root** OU.

2. Then click the **Actions** drop-down menu to reveal the available actions, and select **Create new** underneath **Organizational unit**:

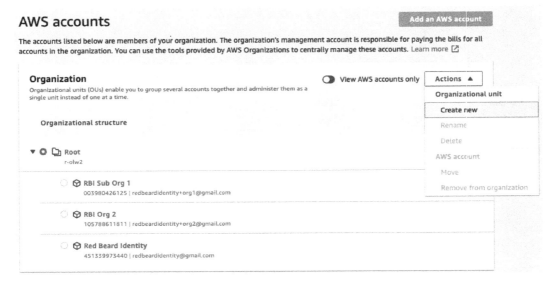

Figure 6.15 – Creating a new OU beneath Root

3. This opens a form where we get to name the new OU, and optionally apply some tagging to it. Since we created these new accounts using the Management Console, we will name this OU MgmtConsoleSubOrgs, and hit the **Create organizational unit** button to proceed:

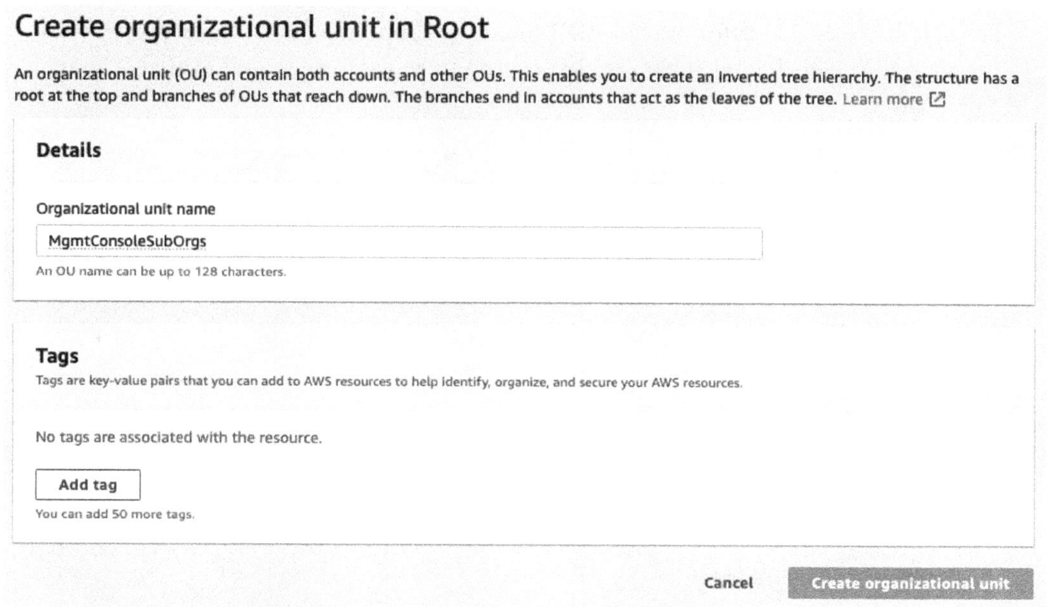

Create organizational unit in Root

An organizational unit (OU) can contain both accounts and other OUs. This enables you to create an inverted tree hierarchy. The structure has a root at the top and branches of OUs that reach down. The branches end in accounts that act as the leaves of the tree. Learn more

Details

Organizational unit name

MgmtConsoleSubOrgs

An OU name can be up to 128 characters.

Tags

Tags are key-value pairs that you can add to AWS resources to help identify, organize, and secure your AWS resources.

No tags are associated with the resource.

Add tag

You can add 50 more tags.

Cancel   Create organizational unit

Figure 6.16 – Naming and tagging the OU

4. Now we have our new **RBIMgmtConsoleSubAccts** OU available to us in our organization on the same level as our three accounts, though we still need to move the accounts into the OU:

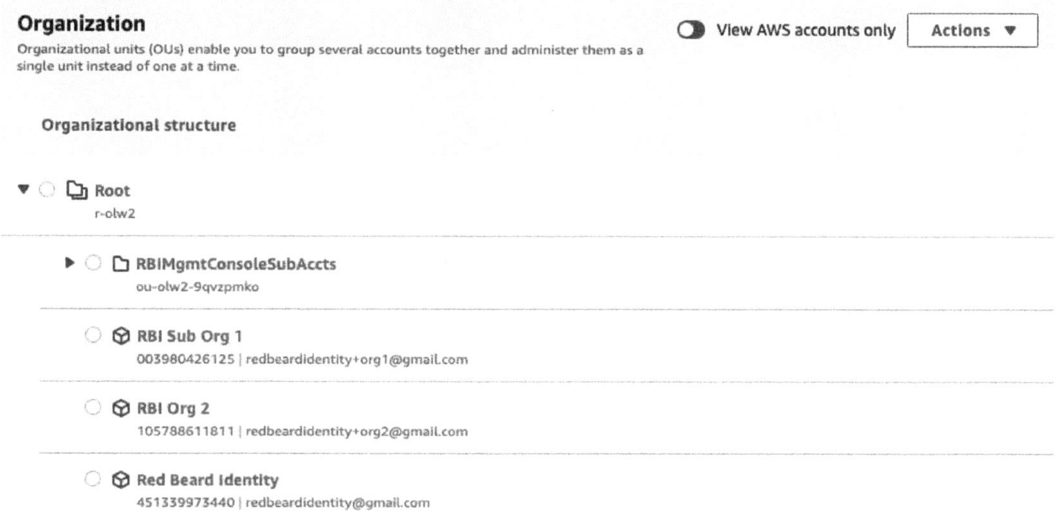

Organization

Organizational units (OUs) enable you to group several accounts together and administer them as a single unit instead of one at a time.

View AWS accounts only   Actions ▼

Organizational structure

▼ ○ ⬚ Root
       r-olw2

   ▶ ○ ⬚ RBIMgmtConsoleSubAccts
          ou-olw2-9qvzpmko

      ○ ⬚ RBI Sub Org 1
             003980426125 | redbeardidentity+org1@gmail.com

      ○ ⬚ RBI Org 2
             105788611811 | redbeardidentity+org2@gmail.com

      ○ ⬚ Red Beard Identity
             451339973440 | redbeardidentity@gmail.com

Figure 6.17 – The new OU is available in the organization

Let's click the radio button next to **RBI Sub Org 1**, and hit the **Actions** drop-down menu to discover that we now have the option to move the account available to us. Click **Move**:

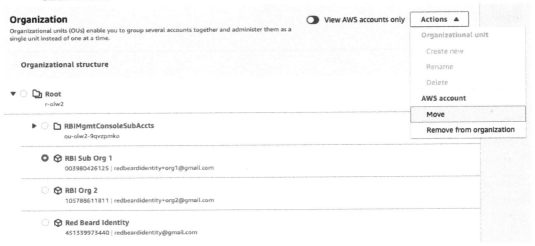

Figure 6.18 – Moving RBI Sub Org 1 to the new OU

5. This reveals a screen where we can select the OU where we would like to move the account. We only have two of them to choose from in this example. Let's click the radio button next to the **RBIMgmtConsoleSubAccts** OU and hit the **Move AWS account** button to continue:

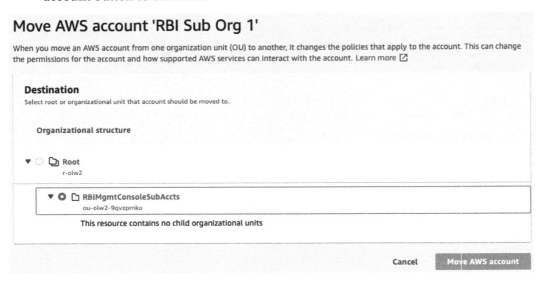

Figure 6.19 – Selecting the destination OU for the account

6. Once we are returned to the AWS Organizations dashboard, we will now see **RBI Sub Org 1** appearing inside of the **RBIMgmtConsoleSubAccts** OU:

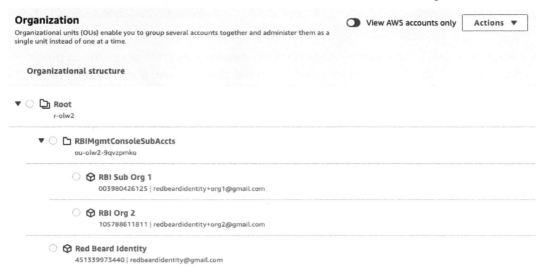

Figure 6.20 – The new organizational hierarchy

7. Let's repeat the process again with the **RBI Org 2** account.

8. Once we are done, we will have all of the member accounts in a single OU:

Figure 6.21 – The updated organizational hierarchy

With that work done, we can easily apply policies broadly to all the accounts within that OU without having to repeat the effort for each account within the OU.

Instead of starting to lock down these member accounts with an SCP, let's instead examine our organization and onboard some additional accounts using the AWS CLI.

AWS organizations in the AWS CLI

From the terminal, let's first take a look at our organization. The highest-level view comes with using the `describe-organization` command. From the terminal, we enter the following:

```
$ aws organizations describe-organization
```

This will provide details of the organization, including information about the master account, and what features and policies are enabled as part of the organization configuration. Here is how our organization appears:

```
[jonlehtinen@ ~ % aws organizations describe-organization
Organization:
  Arn: arn:aws:organizations::451339973440:organization/o-x46kdexfgy
  AvailablePolicyTypes:
  - Status: ENABLED
    Type: SERVICE_CONTROL_POLICY
  FeatureSet: ALL
  Id: o-x46kdexfgy
  MasterAccountArn: arn:aws:organizations::451339973440:account/o-x46kdexfgy/451339973440
  MasterAccountEmail: redbeardidentity@gmail.com
  MasterAccountId: '451339973440'
jonlehtinen@ ~ %
```

Figure 6.22 – The basic description of our organization

This may tell us about the organization as an AWS construct, but it is not giving us much information about the constituent membership of our organization. If we want to see the accounts that make up the organizations, we will need to use the `list-accounts` command. When executed using credentials belonging to the organization's management account, this command will show us all of the accounts inside of our organization, along with additional details about each one. From the terminal, we enter the following:

```
$ aws organizations list-accounts
```

And we see all the accounts in the current organization managed by the Redbeard Identity management account:

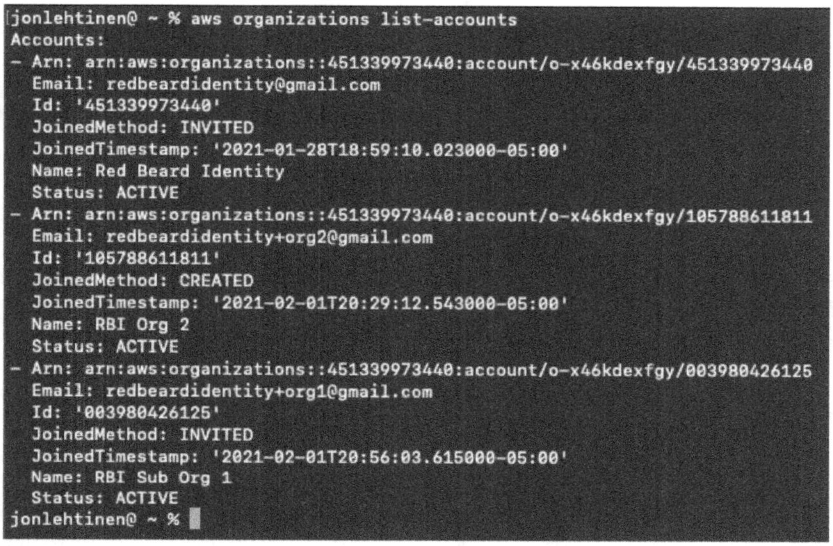

```
[jonlehtinen@ ~ % aws organizations list-accounts
Accounts:
- Arn: arn:aws:organizations::451339973440:account/o-x46kdexfgy/451339973440
  Email: redbeardidentity@gmail.com
  Id: '451339973440'
  JoinedMethod: INVITED
  JoinedTimestamp: '2021-01-28T18:59:10.023000-05:00'
  Name: Red Beard Identity
  Status: ACTIVE
- Arn: arn:aws:organizations::451339973440:account/o-x46kdexfgy/105788611811
  Email: redbeardidentity+org2@gmail.com
  Id: '105788611811'
  JoinedMethod: CREATED
  JoinedTimestamp: '2021-02-01T20:29:12.543000-05:00'
  Name: RBI Org 2
  Status: ACTIVE
- Arn: arn:aws:organizations::451339973440:account/o-x46kdexfgy/003980426125
  Email: redbeardidentity+org1@gmail.com
  Id: '003980426125'
  JoinedMethod: INVITED
  JoinedTimestamp: '2021-02-01T20:56:03.615000-05:00'
  Name: RBI Sub Org 1
  Status: ACTIVE
jonlehtinen@ ~ %
```

Figure 6.23 – All the accounts in our organization, as seen from the management account

But what does this look like if we try to see all the accounts using the programmatic credentials of a superuser account in one of the member organizations? We can run the same command under a profile created for use with one of the managed organizations to find out:

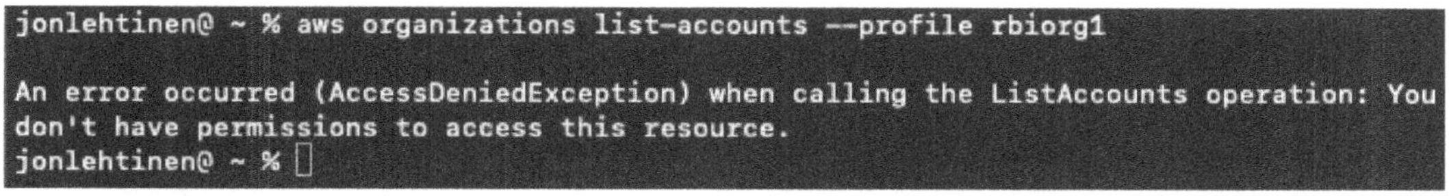

```
jonlehtinen@ ~ % aws organizations list-accounts --profile rbiorg1

An error occurred (AccessDeniedException) when calling the ListAccounts operation: You
don't have permissions to access this resource.
jonlehtinen@ ~ %
```

Figure 6.24 – Managed organizations cannot use the list-accounts command

Though this user account has the AdministratorAccess managed policy attached to it within the RBI Sub Org 1 account, those permissions are insufficient to execute the command against the Redbeard Identity management account. However, since we never applied any SCPs that restricted what the member accounts could do, the RBI Sub Org 1 account could leave the organization, and even create its own. Let's do that using the CLI:

```
$ aws organizations leave-organization --profile rbiorg1
```

If this is successful, there will be no response. We can confirm the account has left the organization by executing the list-accounts command once more as the management account:

```
$ aws organizations list-accounts
```

The output of which confirms `RBI Sub Org 1` is once again an independent account:

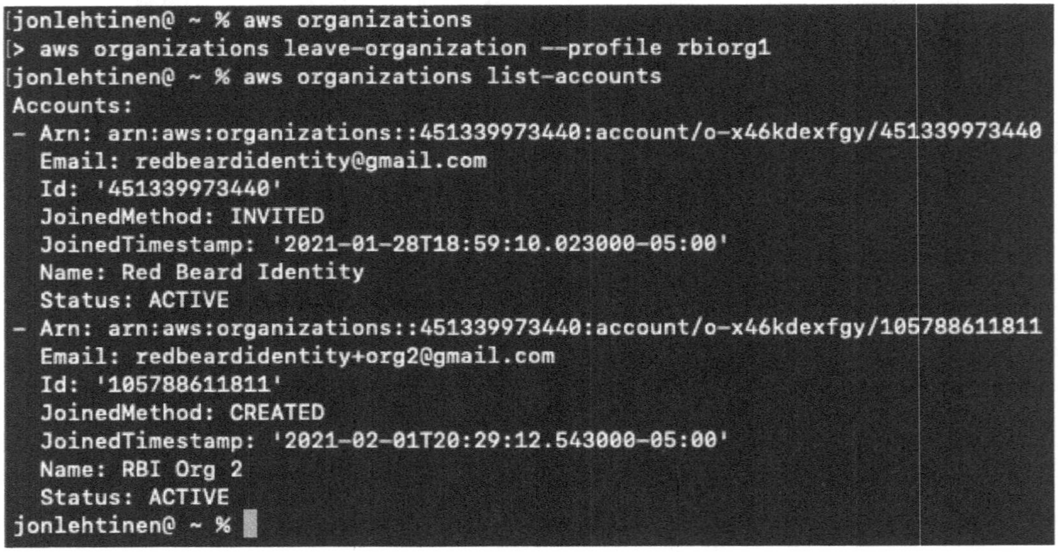

```
[jonlehtinen@ ~ % aws organizations
[> aws organizations leave-organization --profile rbiorg1
[jonlehtinen@ ~ % aws organizations list-accounts
Accounts:
- Arn: arn:aws:organizations::451339973440:account/o-x46kdexfgy/451339973440
  Email: redbeardidentity@gmail.com
  Id: '451339973440'
  JoinedMethod: INVITED
  JoinedTimestamp: '2021-01-28T18:59:10.023000-05:00'
  Name: Red Beard Identity
  Status: ACTIVE
- Arn: arn:aws:organizations::451339973440:account/o-x46kdexfgy/105788611811
  Email: redbeardidentity+org2@gmail.com
  Id: '105788611811'
  JoinedMethod: CREATED
  JoinedTimestamp: '2021-02-01T20:29:12.543000-05:00'
  Name: RBI Org 2
  Status: ACTIVE
jonlehtinen@ ~ %
```

Figure 6.25 – RBI Sub Org 1 has left the organization

With `RBI Sub Org 1` having become independent, let's use it as the base to create a new organization from the AWS CLI.

> **Tip**
> It is easy to get confused when administrating several different accounts under a single organization using the CLI. Be sure to validate that each command is under the correct profile before executing.

Creating an organization in the AWS CLI

We can verify that the `RBI Sub Org 1` account has left the organization by checking the Organizations service in the Management Console, or by running the `describe-account` command in the CLI. Since we don't want to bother signing into the `RBI Sub Org 1` Management Console, let's enter the following into the terminal:

```
$ aws organizations describe-account --account-id 003980426125
--profile rbiorg1
```

We should note the parameters in use with that command to ensure we used the correct credentials, and also evaluated the correct account. --account-id uses the unique account number to specify which account will be evaluated with the command. The number referenced previously corresponds to the RBI Sub Org 1 account. The --profile parameter ensures that the correct AWS IAM user account is used to execute the call, specifically one that actually exists within the RBI Sub Org 1 account. The output of this command confirms that RBI Sub Org 1 is not a member of any organization:

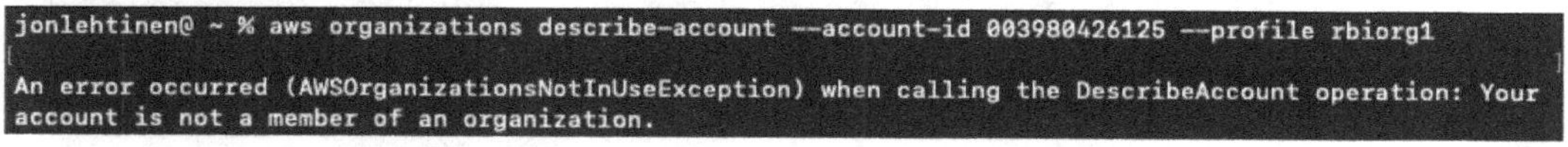

```
jonlehtinen@ ~ % aws organizations describe-account --account-id 003980426125 --profile rbiorg1
[
An error occurred (AWSOrganizationsNotInUseException) when calling the DescribeAccount operation: Your
account is not a member of an organization.
```

Figure 6.26 – Confirming the account is not a member of an organization

What would happen if we used the Redbeard Identity management account's credentials to examine that account? If we clear the --profile parameter and use the default credentials that belong to the Redbeard Identity account to execute the command, we get the following:

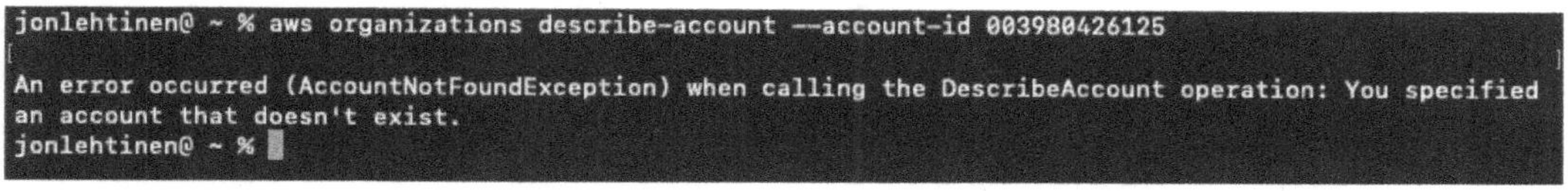

```
jonlehtinen@ ~ % aws organizations describe-account --account-id 003980426125
[
An error occurred (AccountNotFoundException) when calling the DescribeAccount operation: You specified
an account that doesn't exist.
jonlehtinen@ ~ %
```

Figure 6.27 – The RBI Sub Org 1 account does not exist

From the perspective of the Redbeard Identity account, there is no account by that account number since it is not found within the organization. Having established that our RBI Sub Org 1 account is independent, let's proceed with creating an organization with it as the management account.

Creating the organization is a simple command, with only one parameter. We must decide to create the organization with either all features enabled, or only consolidated billing enabled. As AWS SSO is just one of many features available in a complete organization, we will set that parameter to ALL:

```
$ aws organizations create-organization --feature-set ALL
--profile rbiorg1
```

The output of this command when successful looks very similar to what we saw when we used the `describe-organization` command on the organization owned by our Redbeard Identity account. However, now we can see the information from RBI Sub Org 1 populating the attributes for the management account:

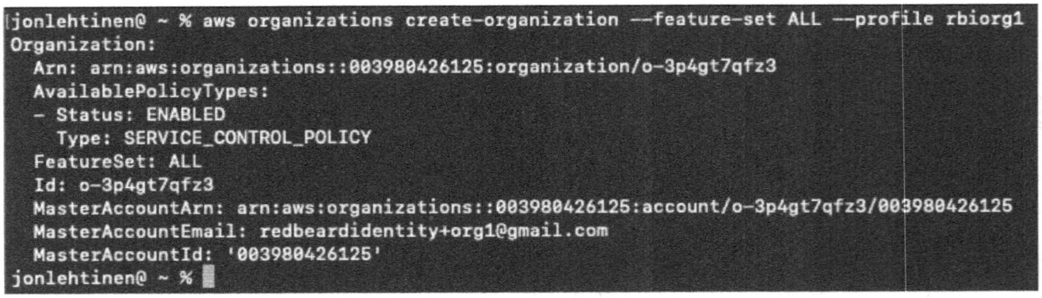

```
[jonlehtinen@ ~ % aws organizations create-organization --feature-set ALL --profile rbiorg1
Organization:
  Arn: arn:aws:organizations::003980426125:organization/o-3p4gt7qfz3
  AvailablePolicyTypes:
  - Status: ENABLED
    Type: SERVICE_CONTROL_POLICY
  FeatureSet: ALL
  Id: o-3p4gt7qfz3
  MasterAccountArn: arn:aws:organizations::003980426125:account/o-3p4gt7qfz3/003980426125
  MasterAccountEmail: redbeardidentity+org1@gmail.com
  MasterAccountId: '003980426125'
jonlehtinen@ ~ %
```

Figure 6.28 – Creation of a new organization

Now that we have an organization, it's time to invite some members. Similar to how we added a pair of accounts to our organization using the Management Console, we will first invite and join this organization using an existing account. After that, we will create a new account from our organization directly.

Adding member organizations

We created yet another AWS account, this time called RBI Sub Org 3. We also created a profile with programmatic credentials for it in our AWS CLI config. Now we can send an invitation from the RBI Sub Org 2 management account to this new account to join our organization. This is done with the intuitively named `invite-account-to-organization` command but requires some rather unintuitive parameters if we attempt to assemble the command directly from the CLI. Similar to what we saw when doing this with the Management Console, we need to identify the account we want to invite using either the account ID number or the email address of the root account user. We could either run this:

```
$ aws organizations invite-account-to-organization --target
Id=281142516251,Type=ACCOUNT --profile rbiorg1
```

Or we could run this:

```
$ aws organizations invite-account-to-organization --target
Id=redbeardidentity+org3@gmail.com,Type=EMAIL --profile rbiorg1
```

Let's run the second. The output creates something called a handshake:

```
[jonlehtinen@ ~ % aws organizations invite-account-to-organization --target Id=redbeardidentity+org3@gmail.com,Type=EMAIL
profile rbiorg1
Handshake:
  Action: INVITE
  Arn: arn:aws:organizations::003980426125:handshake/o-3p4gt7qfz3/invite/h-5991a5ffc26b442fb53ca9b878866b48
  ExpirationTimestamp: '2021-02-22T15:02:59.967000-05:00'
  Id: h-5991a5ffc26b442fb53ca9b878866b48
  Parties:
  - Id: 3p4gt7qfz3
    Type: ORGANIZATION
  - Id: redbeardidentity+org3@gmail.com
    Type: EMAIL
  RequestedTimestamp: '2021-02-07T15:02:59.967000-05:00'
  Resources:
  - Resources:
    - Type: MASTER_EMAIL
      Value: redbeardidentity+org1@gmail.com
    - Type: MASTER_NAME
      Value: RBI Sub Org 1
    - Type: ORGANIZATION_FEATURE_SET
      Value: ALL
    Type: ORGANIZATION
    Value: o-3p4gt7qfz3
  - Type: EMAIL
    Value: redbeardidentity+org3@gmail.com
  State: OPEN
jonlehtinen@ ~ %
```

Figure 6.29 – The invitation handshake

The handshake is what occurs behind the scenes when an existing AWS account is invited by a management account to join its organization. We did not see these handshakes when inviting accounts to our organizations through our Management Console, but they were there. In fact, we will now see the same artifacts from that invitation process, such as the invitation email that was sent to the RBI Sub Org 3 account owner:

Hello,

RBI Sub Org 1 (owned by redbeardidentity+org1@gmail.com) would like to add your AWS account (redbeardidentity+org3@gmail.com) to their AWS organization as a member account.

Organizations allows customers to easily manage multiple AWS accounts. If you accept the invitation, all activity in your AWS account will be billed to the AWS account of RBI Sub Org 1, and RBI Sub Org 1 will be able to view the AWS usage and charges for your account.

An AWS organization can have one of the following feature sets: all features or consolidated billing only. Most organizations are set up with access to all features, which includes administrative and access controls within the organization. In some cases, an organization may choose to only enable consolidated billing features and later decide to enable all features. Management accounts for consolidated billing organizations may direct AWS to enable all features in the organization with at least 14 days' notice to you that may be sent by email. You can view which feature set the organization has enabled through the console link below. For more information about Organizations features, see the Organizations documentation.

To view the invitation, including what features have been enabled, click this link:

https://console.aws.amazon.com/organizations/home#/invites

To learn more about AWS Organizations, see What is AWS Organizations?

Thank you for using Amazon Web Services.

Sincerely,
Amazon Web Services

Figure 6.30 – Invitation to join the organization

But rather than complete the handshake through email, let's examine the acceptance process through the CLI. Using the `rbiorg3` profile, we can check whether we have any outstanding invitations using the `list-handshakes-for-account` command:

```
$ aws organizations list-handshakes-for-account --profile
rbiorg3
```

This gives us the details about the invitation:

```
[jonlehtinen@ ~ % aws organizations list-handshakes-for-account --profile rbiorg3
Handshakes:
- Action: INVITE
  Arn: arn:aws:organizations::003980426125:handshake/o-3p4gt7qfz3/invite/h-5991a5ffc26b442fb53ca9b878866b48
  ExpirationTimestamp: '2021-02-22T15:02:59.967000-05:00'
  Id: h-5991a5ffc26b442fb53ca9b878866b48
  Parties:
  - Id: 3p4gt7qfz3
    Type: ORGANIZATION
  - Id: redbeardidentity+org3@gmail.com
    Type: EMAIL
  RequestedTimestamp: '2021-02-07T15:02:59.967000-05:00'
  Resources:
  - Resources:
    - Type: MASTER_EMAIL
      Value: redbeardidentity+org1@gmail.com
    - Type: MASTER_NAME
      Value: RBI Sub Org 1
    - Type: ORGANIZATION_FEATURE_SET
      Value: ALL
    Type: ORGANIZATION
    Value: o-3p4gt7qfz3
  - Type: EMAIL
    Value: redbeardidentity+org3@gmail.com
  State: OPEN
```

Figure 6.31 – RBI Sub Group 3's handshake to join the RBI Sub Org 1 organization

Rather than accept this handshake now, let's instead ignore it. We used RBI Sub Org 1 as an excuse to demonstrate how to create AWS organizations in the CLI, but at the end of the day, we want to aggregate all of our accounts under the original Redbeard Identity account. So, let's use the AWS CLI to do the following:

1. Decline the handshake to RBI Sub Org 3 to join the RBI Sub Org 1 organization.

2. Delete the RBI Sub Org 1 organization.

3. Invite RBI Sub Org 1 and RBI Sub Org 3 into the existing Redbeard Identity organization.

4. Place RBI Sub Org 1 into the existing RBIMgmtConsoleSubAccounts OU.

5. Create a new OU for RBI Sub Org 3 at the same level in the org and put it there.

We can decline the handshake using the `decline-handshake` command under the correct profile. The handshake has a unique ID that we can use to either accept or reject its invitation to join the organization, which we saw when we listed the handshakes currently extended to the RBI Sub Org 3 account. From the terminal, run the following:

```
$ aws organizations decline-handshake --handshake-id
h-5991a5ffc26b442fb53ca9b878866b48 --profile rbiorg3
```

Once run, the output shows the handshake details once more, only this time the `State` attribute is no longer showing as OPEN:

```
[jonlehtinen@ ~ % aws organziations decline-ha
> aws organizations decline-handshake --handshake-id h-5991a5ffc26b442fb53ca9b878866b48 --profile rbiorg3
Handshake:
  Action: INVITE
  Arn: arn:aws:organizations::003980426125:handshake/o-3p4gt7qfz3/invite/h-5991a5ffc26b442fb53ca9b878866b48
  ExpirationTimestamp: '2021-02-22T15:02:59.967000-05:00'
  Id: h-5991a5ffc26b442fb53ca9b878866b48
  Parties:
  - Id: redbeardidentity+org3@gmail.com
    Type: EMAIL
  - Id: 3p4gt7qfz3
    Type: ORGANIZATION
  RequestedTimestamp: '2021-02-07T15:02:59.967000-05:00'
  Resources:
  - Resources:
    - Type: MASTER_EMAIL
      Value: redbeardidentity+org1@gmail.com
    - Type: MASTER_NAME
      Value: RBI Sub Org 1
    - Type: ORGANIZATION_FEATURE_SET
      Value: ALL
    Type: ORGANIZATION
    Value: o-3p4gt7qfz3
  - Type: EMAIL
    Value: redbeardidentity+org3@gmail.com
  State: DECLINED
jonlehtinen@ ~ %
```

Figure 6.32 – Declining the handshake

`State` is now `Declined`, which terminates this invitation. Next, we will delete the unnecessary RBI Sub Org 1 organization:

```
$ aws organizations delete-organization --profile rbiorg1
```

There is no output if the command is successful, but we can verify that the account is no longer part of an organization by running the `describe-organization` command once more:

```
$ aws organizations describe-organization --profile rbiorg1
```

If we get an error indicating that the account is not a member of an organization, then we are good to move onward. As the Redbeard Identity account, we will issue invitations to both `RBI Sub Org 1` and `RBI Sub Org 3` to join the organization we created earlier in the Management Console:

```
$ aws organizations invite-account-to-organization --target
Id=redbeardidentity+org1@gmail.com,Type=EMAIL
Handshake:
  Action: INVITE
  Arn: arn:aws:organizations::451339973440:handshake/o-
x46kdexfgy/invite/h-2dd0a3050c13464ba0560a57748c4737
  ExpirationTimestamp: '2021-02-22T15:51:41.959000-05:00'
  Id: h-2dd0a3050c13464ba0560a57748c4737
  Parties:
  - Id: redbeardidentity+org1@gmail.com
    Type: EMAIL
  - Id: x46kdexfgy
    Type: ORGANIZATION
  RequestedTimestamp: '2021-02-07T15:51:41.959000-05:00'
  Resources:
  - Resources:
    - Type: MASTER_EMAIL
      Value: redbeardidentity@gmail.com
    - Type: MASTER_NAME
      Value: Redbeard Identity
    - Type: ORGANIZATION_FEATURE_SET
      Value: ALL
    Type: ORGANIZATION
    Value: o-x46kdexfgy
  - Type: EMAIL
    Value: redbeardidentity+org1@gmail.com
  State: OPEN

$ aws organizations invite-account-to-organization --target
Id=redbeardidentity+org3@gmail.com,Type=EMAIL
Handshake:
  Action: INVITE
```

```
   Arn: arn:aws:organizations::451339973440:handshake/o-
x46kdexfgy/invite/h-2fbff6438566499999f21eb40a1d57d4
   ExpirationTimestamp: '2021-02-22T15:52:11.087000-05:00'
   Id: h-2fbff6438566499999f21eb40a1d57d4
   Parties:
   - Id: redbeardidentity+org3@gmail.com
     Type: EMAIL
   - Id: x46kdexfgy
     Type: ORGANIZATION
   RequestedTimestamp: '2021-02-07T15:52:11.087000-05:00'
   Resources:
   - Resources:
     - Type: MASTER_EMAIL
       Value: redbeardidentity@gmail.com
     - Type: MASTER_NAME
       Value: Redbeard Identity
     - Type: ORGANIZATION_FEATURE_SET
       Value: ALL
     Type: ORGANIZATION
     Value: o-x46kdexfgy
   - Type: EMAIL
     Value: redbeardidentity+org3@gmail.com
   State: OPEN
```

Now, we will accept each of those handshakes under their corresponding profiles. First,
RBI Sub Org 1:

```
$ aws organizations accept-handshake --handshake-id h-2dd0a3050
c13464ba0560a57748c4737 --profile rbiorg1
```

The output is similar to when we declined the previous offer under the `RBI Sub Org 3` profile, only this time the `State` attribute on the handshake changed to `ACCEPTED`. We can validate that `RBI Sub Org 1` is now part of the Redbeard Identity organization by running the `describe` command under its profile:

```
[jonlehtinen@ ~ % aws organizations describe-organization --profile rbiorg1
Organization:
  Arn: arn:aws:organizations::451339973440:organization/o-x46kdexfgy
  AvailablePolicyTypes:
  - Status: ENABLED
    Type: SERVICE_CONTROL_POLICY
  FeatureSet: ALL
  Id: o-x46kdexfgy
  MasterAccountArn: arn:aws:organizations::451339973440:account/o-x46kdexfgy/451339973440
  MasterAccountEmail: redbeardidentity@gmail.com
  MasterAccountId: '451339973440'
jonlehtinen@ ~ %
```

Figure 6.33 – RBI Sub Org 1 is back in the Redbeard Identity organization

Now we repeat the process and accept the handshake meant for `RBI Sub Org 3`:

```
$ aws organizations accept-handshake --handshake-id
h-2fbff6438566499999f21eb40a1d57d4 --profile rbiorg3
```

We get the same output if successful and can validate it the same way:

```
[jonlehtinen@ ~ % aws organizations accept-handshake --handshake-id h-2fbff6438566499999f21eb40a1d57d4 --profile rbiorg3
Handshake:
  Action: INVITE
  Arn: arn:aws:organizations::451339973440:handshake/o-x46kdexfgy/invite/h-2fbff6438566499999f21eb40a1d57d4
  ExpirationTimestamp: '2021-02-22T15:52:11.087000-05:00'
  Id: h-2fbff6438566499999f21eb40a1d57d4
  Parties:
  - Id: '281142516251'
    Type: ACCOUNT
  - Id: x46kdexfgy
    Type: ORGANIZATION
  RequestedTimestamp: '2021-02-07T15:52:11.087000-05:00'
  Resources:
  - Resources:
    - Type: MASTER_EMAIL
      Value: redbeardidentity@gmail.com
    - Type: MASTER_NAME
      Value: Red Beard Identity
    - Type: ORGANIZATION_FEATURE_SET
      Value: ALL
    Type: ORGANIZATION
    Value: o-x46kdexfgy
  - Type: EMAIL
    Value: redbeardidentity+org3@gmail.com
  State: ACCEPTED
jonlehtinen@ ~ % aws organizations describe-organization --profile rbiorg3
Organization:
  Arn: arn:aws:organizations::451339973440:organization/o-x46kdexfgy
  AvailablePolicyTypes:
  - Status: ENABLED
    Type: SERVICE_CONTROL_POLICY
  FeatureSet: ALL
  Id: o-x46kdexfgy
  MasterAccountArn: arn:aws:organizations::451339973440:account/o-x46kdexfgy/451339973440
  MasterAccountEmail: redbeardidentity@gmail.com
  MasterAccountId: '451339973440'
jonlehtinen@ ~ %
```

Figure 6.34 – RBI Sub Org 3 is now part of the Redbeard Identity organization

Now that every account is inside of a single organization, let's move them into specific OUs. This will be done with the `move-account` command; however, before we can use that command, we need to know the unique identifiers for each of the OUs in the organization, including the root OU, as `move-account` requires the source and destination IDs of the OUs where accounts will be moving. Let's get the ID for our root OU first:

```
$ aws organizations list-roots
```

This provides details on the top-level, or root, OU in our organization's hierarchy. This is where the management account resides, as well as the accounts that just joined the organization:

```
[jonlehtinen@ ~ % aws organizations list-roots
Roots:
- Arn: arn:aws:organizations::451339973440:root/o-x46kdexfgy/r-olw2
  Id: r-olw2
  Name: Root
  PolicyTypes:
  - Status: ENABLED
    Type: SERVICE_CONTROL_POLICY
jonlehtinen@ ~ %
```

Figure 6.35 – Root OU information

With the root OU's ID, we can drill down and discover additional OUs using another command:

```
$ aws organizations list-organizational-units-for-parent
--parent-id r-olw2
```

This command lets us discover nested OUs starting at the specific "parent" OU's ID. In our use case, the parent ID is the root OU. This command reveals the following:

```
[jonlehtinen@ ~ % aws organizations list-organizational-units-for-parent --parent-id r-olw2
OrganizationalUnits:
- Arn: arn:aws:organizations::451339973440:ou/o-x46kdexfgy/ou-olw2-9qvzpmko
  Id: ou-olw2-9qvzpmko
  Name: RBIMgmtConsoleSubAccts
jonlehtinen@ ~ %
```

Figure 6.36 – The OUs beneath the root OU

This gives us the ID for the other OU within our organization. As such, let's now move `RBI Sub Org 1` back into the `RBIMgmtConsoleSubAccts` OU:

```
$ aws organizations move-account --account-id 003980426125
--source-parent-id r-olw2 --destination-parent-id ou-olw2-
9qvzpmko
```

There will be no output if the command is successful. We can verify the account moved to the new OU by running the `list-account-for-parent` command. This command will list all the accounts within a specific OU, though it will not show OUs, nor accounts within OUs found within the specified parent OU. Let's run it against the OU where we moved the `RBI Sub Org 1` account to demonstrate the nuance:

```
$ aws organizations list-accounts-for-parent --parent-id
ou-olw2-9qvzpmko
```

This outputs information for the accounts found under that parent OU's ID. The ID in that command corresponds to the `RBIMgmtConsoleSubAccts` OU and does show that the account was successfully moved:

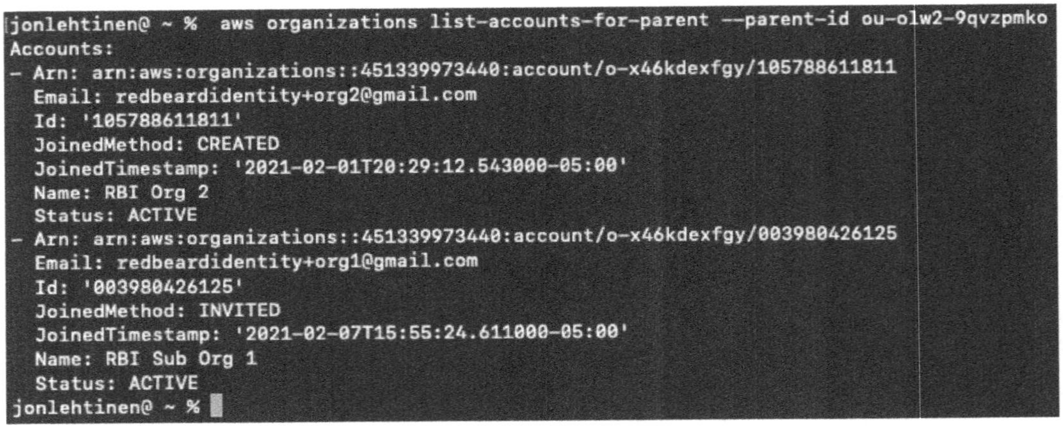

```
jonlehtinen@ ~ %  aws organizations list-accounts-for-parent --parent-id ou-olw2-9qvzpmko
Accounts:
- Arn: arn:aws:organizations::451339973440:account/o-x46kdexfgy/105788611811
  Email: redbeardidentity+org2@gmail.com
  Id: '105788611811'
  JoinedMethod: CREATED
  JoinedTimestamp: '2021-02-01T20:29:12.543000-05:00'
  Name: RBI Org 2
  Status: ACTIVE
- Arn: arn:aws:organizations::451339973440:account/o-x46kdexfgy/003980426125
  Email: redbeardidentity+org1@gmail.com
  Id: '003980426125'
  JoinedMethod: INVITED
  JoinedTimestamp: '2021-02-07T15:55:24.611000-05:00'
  Name: RBI Sub Org 1
  Status: ACTIVE
jonlehtinen@ ~ %
```

Figure 6.37 – The RBI Org 1 account is now in the RBIMgmtConsoleSubAccts OU

Let's also run that command against the root OU:

```
$ aws organizations list-accounts-for-parent --parent-id r-olw2
```

We see the management account and the new `RBI Sub Org 3` account listed under the root OU in the output:

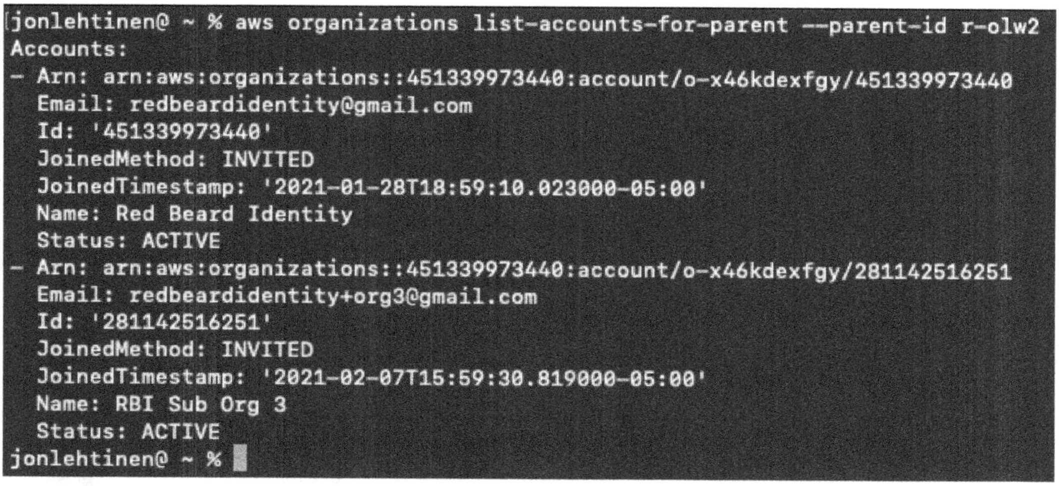

```
[jonlehtinen@ ~ % aws organizations list-accounts-for-parent --parent-id r-olw2
Accounts:
- Arn: arn:aws:organizations::451339973440:account/o-x46kdexfgy/451339973440
  Email: redbeardidentity@gmail.com
  Id: '451339973440'
  JoinedMethod: INVITED
  JoinedTimestamp: '2021-01-28T18:59:10.023000-05:00'
  Name: Red Beard Identity
  Status: ACTIVE
- Arn: arn:aws:organizations::451339973440:account/o-x46kdexfgy/281142516251
  Email: redbeardidentity+org3@gmail.com
  Id: '281142516251'
  JoinedMethod: INVITED
  JoinedTimestamp: '2021-02-07T15:59:30.819000-05:00'
  Name: RBI Sub Org 3
  Status: ACTIVE
jonlehtinen@ ~ %
```

Figure 6.38 – The accounts within the root OU

However, we do not see the `RBIMgmtConsoleSubAcct` OU, nor the accounts found within it. This command will only reveal the accounts within the OU, not the full list of objects within the hierarchy.

Finally, let's create a new OU and move the `RBI Sub Org 3` account into it. We do this with the `create-organizational-unit` command. All we need for parameters are the ID of the OU where we would like this OU to live and a name for the OU. Let's make this OU one level deeper in the `RBIMgmtConsoleSubAcct` OU:

```
$ aws organizations create-organizational-unit --parent-id
ou-olw2-9qvzpmko --name RBICliSubAcct
```

The output will provide the ARN, ID, and name of the new OU:

```
[jonlehtinen@ ~ % aws organizations create-organizational-un
> aws organizations create-organizational-unit --parent-id ou-olw2-9qvzpmko --name RBICliSubAcct
OrganizationalUnit:
  Arn: arn:aws:organizations::451339973440:ou/o-x46kdexfgy/ou-olw2-tww7ves0
  Id: ou-olw2-tww7ves0
  Name: RBICliSubAcct
jonlehtinen@ ~ %
```

Figure 6.39 – The new OU is created

We can capture the new OU's ID in order to move the `RBI Sub Org 3` account into it:

```
$ aws organizations move-account --account-id 281142516251
--source-parent-id r-olw2 --destination-parent-id ou-olw2-
tww7ves0
```

If successful, there will be no output. However, we can list the accounts for the OU to verify that it was moved:

```
$ aws organizations list-accounts-for-parent --parent-id
ou-olw2-tww7ves0
```

And the output will confirm it was successful:

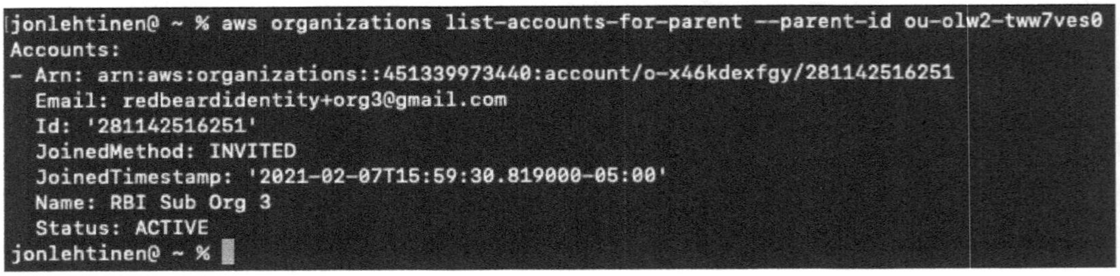

```
[jonlehtinen@ ~ % aws organizations list-accounts-for-parent --parent-id ou-olw2-tww7ves0
Accounts:
- Arn: arn:aws:organizations::451339973440:account/o-x46kdexfgy/281142516251
  Email: redbeardidentity+org3@gmail.com
  Id: '281142516251'
  JoinedMethod: INVITED
  JoinedTimestamp: '2021-02-07T15:59:30.819000-05:00'
  Name: RBI Sub Org 3
  Status: ACTIVE
jonlehtinen@ ~ %
```

Figure 6.40 – The account is in the new OU

We have just finished making several radical changes to the Redbeard Identity organization using the CLI. Though we can logically track how those changes impacted the organization, if we look back at it in the Management Console, we can see the changes represented visually:

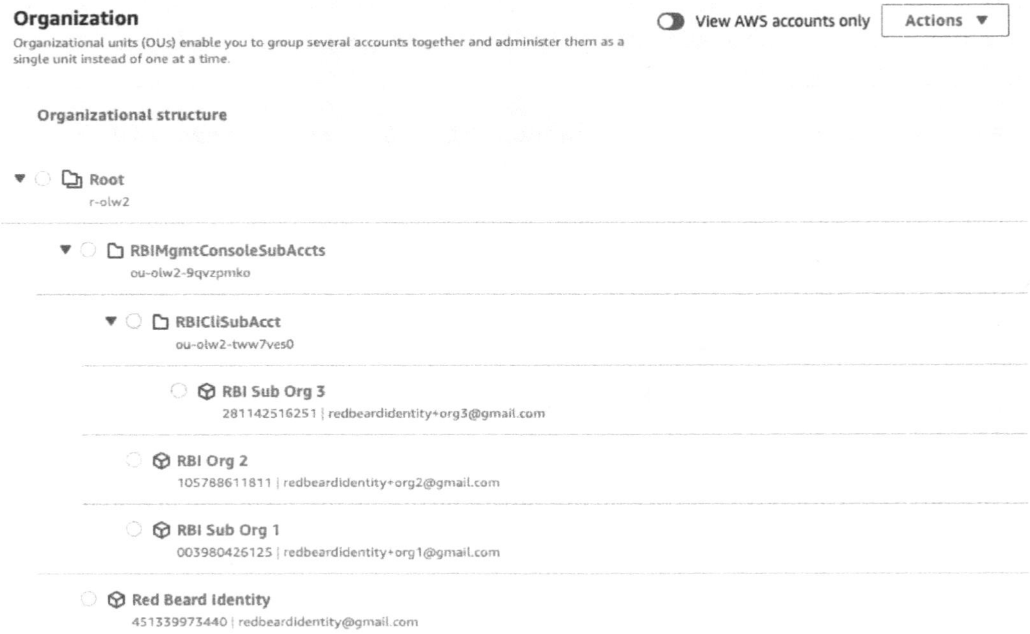

Figure 6.41 – The exploded view of the organization in the Management Console

> **Tip**
>
> We are not following AWS' best practice guidance for administrating organizations with the Redbeard Identity organization example. Ideally, the management account will only be used for administrative functions for managing the organization. This includes setting and modifying permission sets and policies for OUs and accounts within the organization, configuring the available services available across all accounts within the organization such as AWS SSO, resource utilization reporting, and consolidated billing. Whereas we are using the Redbeard Identity account and organization to explore these services, please consider following AWS' guidance when designing your own organization and implementing services between accounts within it.

There is much more to the AWS Organizations service than we have covered in this chapter. However, we would quickly lose focus if we were to drill down into each potential capability of the service. By now, we should know enough about how AWS Organizations works to ensure that we will be able to understand its influence and interactions with AWS SSO.

Configuring AWS SSO in the Management Console

In this section, we will configure AWS SSO to be an identity provider using the Management Console. For this exercise, we will set up AWS SSO as the identity store and identity provider without connecting it to any pre-existing installations, as those scenarios will be explored more fully in Chapters 10 and 12. Our objective here is to become familiar with the service and its basic administration before we leap into those other deployment patterns.

As usual, we start by signing into our Management Console. If we have not configured the AWS SSO service with this account, we are greeted with a screen that invites us to enable the service:

Getting Started Guide │ AWS SSO Prerequisites

Figure 6.42 – The AWS SSO activation banner

The banner informs us that when we enable AWS SSO, we will allow it to create AWS IAM roles for each of the AWS accounts within our AWS organization. It also warns us that those organization member accounts will be able to assign applications to our AWS SSO users. Since this is part of the value proposition for AWS SSO from our perspective, we should be inclined to proceed. If we had not created an organization on this account yet, we would also be told that this process will create one for us and assign this account as the management account:

Figure 6.43 – AWS SSO will create an organization if one does not exist

After a few moments, all the organization and IAM configuration tasks complete and we are taken to the AWS SSO dashboard. This page gives us a list of recommended setup tasks that are necessary to address as part of an overall deployment strategy if we are using this in a production environment. We will work through these recommended setup steps momentarily, but not necessarily in that order:

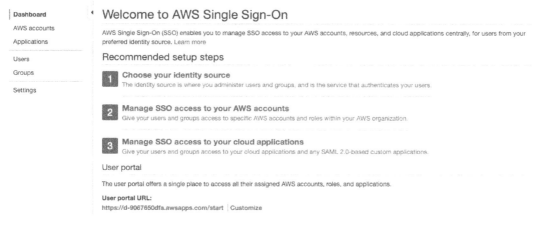

Figure 6.44 – The AWS SSO dashboard

We can review and make adjustments to the user management and authentication settings for our AWS SSO service by clicking on the **Settings** menu on the left side of the screen. Here we can designate the identity store where our users and groups will be managed, as well as the authentication service provider for our user accounts. By default, AWS SSO is set to handle all of those services.

AWS SSO settings

As we mentioned earlier, we have alternative options for populating the local identity store in AWS SSO. Some of these options include synching the identities from Active Directory, such as a forest connected to our AWS account through an AD connector, or a connection to AWS' own managed Active Directory service:

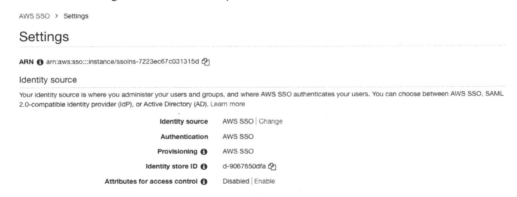

Figure 6.45 – Identity source configuration settings in AWS SSO

Alternatively, we could use just-in-time provisioning with an external identity provider and create the users upon authentication into AWS SSO from that external IDP. These options are spelled out if we select the **Change** option next to **Identity source**.

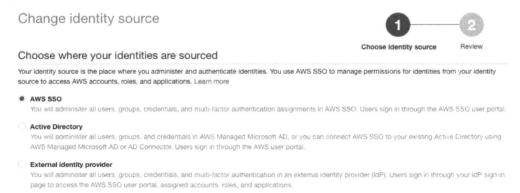

Figure 6.46 – Options for changing the source of identity for AWS SSO

Returning to the **Settings** page, we see additional configuration items under the **Identity source** area. Currently, the **Provisioning** option is set to use AWS SSO as its authoritative source for account creation. As we mentioned when looking at options for alternative identity sources for our AWS SSO service, we have the option to use an external IDP to populate AWS SSO's local user store with our users' identity information. In addition to just-in-time provisioning using SAML2, we can also set up synchronization between an external IDP and the AWS SSO user store using SCIM2. This will allow the identity provider to automatically push new user accounts into AWS SSO, as well as to update or remove existing ones.

Finally, the **Attributes for access control** option allows us to define which attributes to use in policies to determine control to AWS accounts and resources. This is an example of **Attribute-Based Access Control (ABAC)**. ABAC is a fine-grained authorization method based upon the attributes of a user. Rather than using groups, roles, or other entitlement management strategies to make access control decisions, ABAC looks at a user's attributes and allows or disallows access to a resource or action based upon the value of those attributes. When implemented correctly, ABAC alleviates a lot of administrative overhead for determining birthright entitlements, requesting additional access, and enforcing an authorization policy as access control is now innate to the user object and its attributes. We can see an example of how ABAC works in the following example:

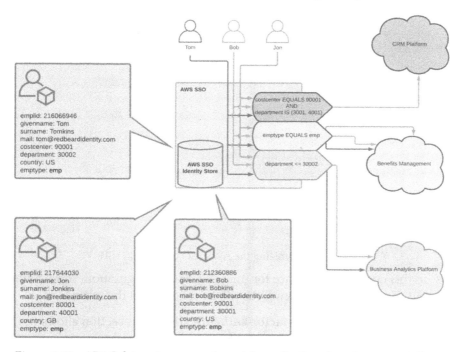

Figure 6.47 – ABAC determines access to certain applications based on user attributes

We can enable ABAC within our AWS SSO instance and define the attributes we will use to gate access to AWS accounts. We could go further with ABAC, and even include user attributes as part of an AWS IAM policy to determine access to specific services and resources in the connected AWS accounts. We will dive into the practicalities of managing administrative access through this and other mechanisms in *Chapter 10, Administrative Single Sign-On* to the AWS backplane.

The next item we will configure on the settings screen is the user portal. The user portal is like the home page available with other IDaaS providers, in that it provides a single place for users to view and launch all of the applications, AWS accounts, and roles they have been assigned through AWS SSO:

User portal

The user portal is a central place where your users can see and access their assigned AWS accounts, roles, and applications. Share this URL with your users to get them started with AWS SSO.

User portal URL   https://d-9067650dfa.awsapps.com/start

Customize

Figure 6.48 – User portal URL configuration

By default, the user portal comes assigned with a URL that is prefixed by our AWS SSO instance's identity store ID. Since that makes for a poor user experience, we can customize it to something slightly more human-readable:

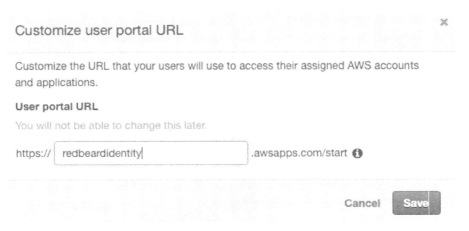

Customize user portal URL

Customize the URL that your users will use to access their assigned AWS accounts and applications.

User portal URL

You will not be able to change this later.

https:// redbeardidentity .awsapps.com/start ⓘ

Cancel   **Save**

Figure 6.49 – Customizing the AWS SSO user portal URL

We can only do this customization once for this AWS SSO configuration, so we better keep it on-brand. Upon saving, the new URL will be reflected on the settings page.

The next configuration item is **Multi-factor authentication**. This section allows us to define the MFA policy we want to apply to AWS SSO transactions:

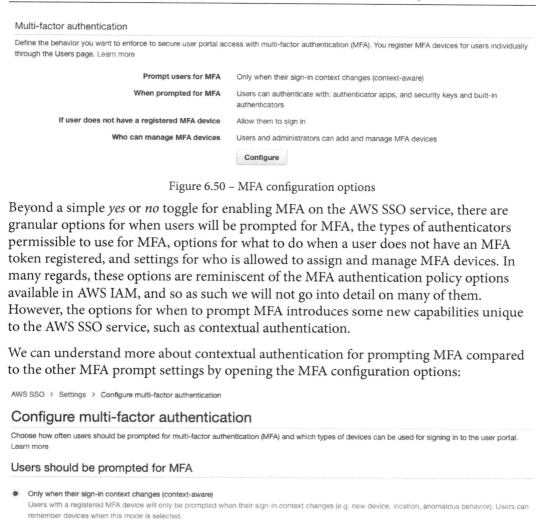

Figure 6.50 – MFA configuration options

Beyond a simple *yes* or *no* toggle for enabling MFA on the AWS SSO service, there are granular options for when users will be prompted for MFA, the types of authenticators permissible to use for MFA, options for what to do when a user does not have an MFA token registered, and settings for who is allowed to assign and manage MFA devices. In many regards, these options are reminiscent of the MFA authentication policy options available in AWS IAM, and so as such we will not go into detail on many of them. However, the options for when to prompt MFA introduces some new capabilities unique to the AWS SSO service, such as contextual authentication.

We can understand more about contextual authentication for prompting MFA compared to the other MFA prompt settings by opening the MFA configuration options:

Figure 6.51 – Options for MFA prompts in AWS SSO

Here we have the usual options for disabling MFA and requiring MFA upon every authentication attempt. Best practice and good security hygiene suggest that we should enable MFA, but we should not discount the tedious user experience that forcing users to respond to a prompt or enter a code on every application authentication event would entail. Context-aware MFA walks a balance between security and user experience by only prompting registered MFA users to use MFA at authentication time if there is something abnormal about the context of their authentication event compared to a baseline. The variables that determine this context include the device used, the location that the request is coming from, requests coming from IP addresses that are known to be malicious or suspicious, or other similar anomalies. This approach offers the security of MFA without burdening users.

Some of the additional MFA configuration options available to us in this settings menu include the types of authenticators available for our users to register, as well as options for how we handle user authentication when a user who does not currently have a token attempts to sign in. AWS SSO supports FIDO2 - and U2F-compatible security keys, TOTP authenticator applications, and, when using a compatible browser, built-in FIDO2 authenticators such as Touch ID and Windows Hello. For ease of administration, we will opt to enforce MFA device registration at user authentication time. This will work for demonstration purposes; however, it is important to consider the provenance of MFA device registration for production deployments. Some use cases may require the additional security that comes with having only an administrator able to issue and revoke MFA tokens.

Now that we have walked through the essential settings in our AWS SSO service, let's start creating some users within our user store.

Creating and managing users

As we are not synchronizing an existing identity source into our AWS SSO identity store, we will need to manually create a couple of users in order to continue exploring what the service can do. We can do this from the Management Console, from the **Users** menu within AWS SSO:

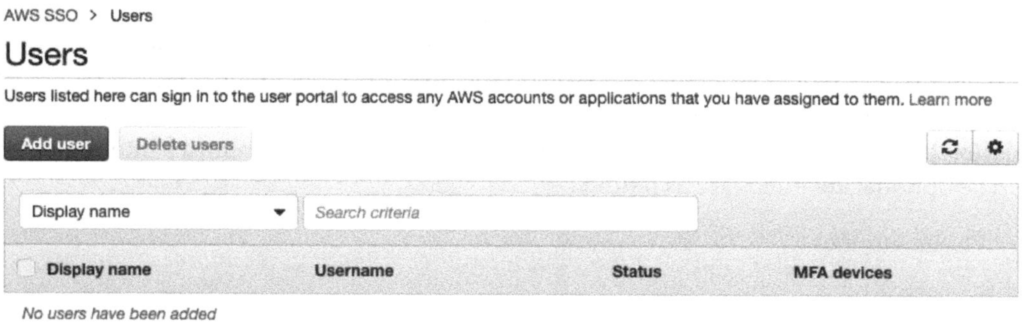

AWS SSO > Users

Users

Users listed here can sign in to the user portal to access any AWS accounts or applications that you have assigned to them. Learn more

Add user	Delete users				C	⚙

Display name	▼	Search criteria

	Display name	Username	Status	MFA devices

No users have been added

Figure 6.52 – The empty user store of our AWS SSO service

AWS SSO is a directory. We can create a user within that directory by providing only the attributes that are required to disambiguate and issue credentials to each user within that directory. We are required to provide a username, email address, first name, last name, and display name for each user record we create. Credentials are either randomly generated and require change upon the first logon or will be created by the account owner as a part of the account activation process:

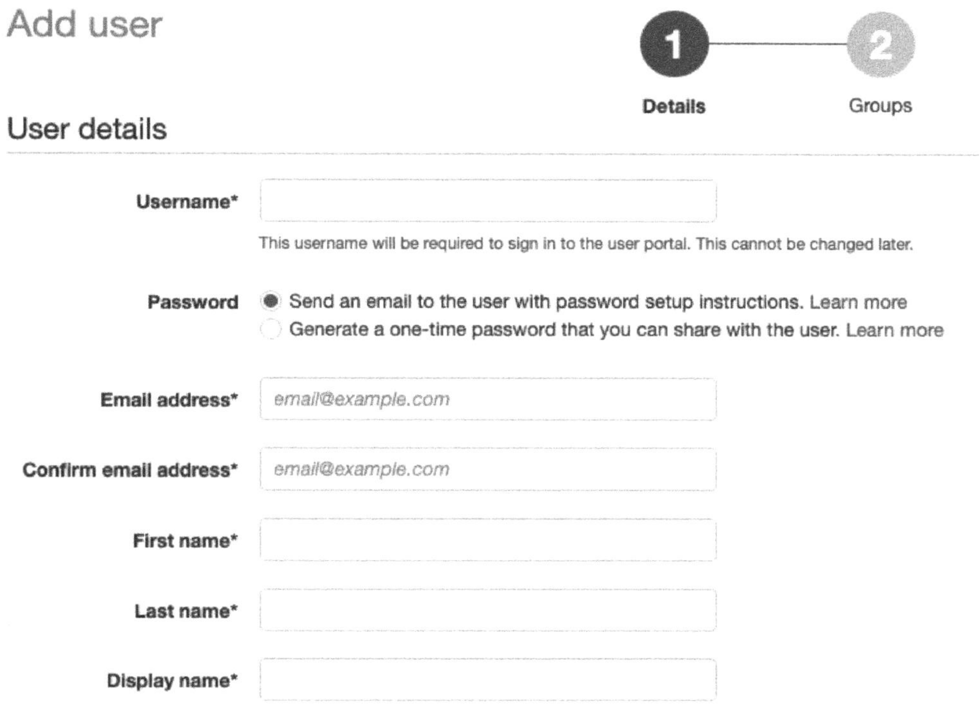

Add user

① —— ②

Details   Groups

User details

Username*

This username will be required to sign in to the user portal. This cannot be changed later.

Password   ● Send an email to the user with password setup instructions. Learn more
○ Generate a one-time password that you can share with the user. Learn more

Email address*   email@example.com

Confirm email address*   email@example.com

First name*

Last name*

Display name*

Figure 6.53 – Add user form

Aside from the required attributes, there are several additional sub-attribute categories with optional attributes that may be used to enrich the user record. We cannot customize the attributes available within this directory. In organizations with unique directory schemas, administrators must arrange attribute mappings into these attributes from their local directories as part of synchronizing user information from on-premises into this one. The optional attributes are as follows:

- Phone number
- User type
- Title
- Employee number
- Cost center
- Organization
- Division
- Department
- Manager
- Street address
- Locality
- Region
- Postal code
- Country
- Formatted (referring to the formatted street address)
- Nickname
- Preferred language
- Locale
- Time zone

Let's create a few users using the required attributes along with a few of the optional ones. We will be creating three records in total, with the following attributes:

Attribute	User 1	User 2	User 3
Username	rbiuser1	rbiuser2	rbiuser3
Email address	redbeardidentity+rbiuser1@gmail.com	redbeardidentity+rbiuser2@gmail.com	redbeardidentity+rbiuser3@gmail.com
First name	Tom	Jon	Bob
Last name	Tomkins	Jonkins	Bobkins
User type	emp	emp	emp
Employee number	216066946	217644030	212360886
Cost center	90001	80001	90001
Department	30002	40001	30001
Country	US	GB	US

Table 6.1 – The attributes for our test users

As we create each one, we will have an option to place it into a group. We will skip this for now. We also could choose to have an email sent inviting the user to set up their account or have a temporary password be created that will need to be changed at the first logon. We will examine both of those flows.

We now have our user store populated with some users. The default view gives us an at-a-glance security view by showing the account enablement status and MFA device association for the accounts within our store:

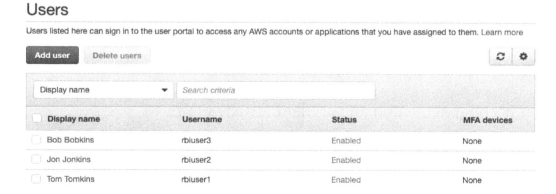

Figure 6.54 – Our populated user store

We can check on each record in our store by selecting it. Once inside a specific user record, we get options that are reminiscent of some features we had already seen across both AWS IAM and Amazon Cognito in both general user account information, last logon information, MFA device enrollment information, and details on the verification status of the email address used to register the account:

Figure 6.55 – User account management in AWS SSO

If we wanted to enrich the record and populate any of the attributes that we initially skipped during account creation, we have the option to do so now. In addition to those attributes, a new attribute has appeared for the user record called User Id. The User Id is the immutable and unique identifier for the user account.

Now that we have accounts in our user store, let's take a look at account activation from the user perspective. In the next section, we will look at registration, password management, and MFA device registration.

User account activation and credential activation

Let's activate this account for Tom Tomkins. When we created it, we opted to have an invitation email sent to his email address:

Hello Tom Tomkins,

Your AWS Organization (AWS Account #451339973440) uses AWS Single Sign-On (SSO) to provide access to AWS accounts and business applications.

Your administrator has invited you to access the AWS Single Sign-On (SSO) user portal. Accepting this invitation activates your AWS SSO user account so that you can access assigned AWS accounts and applications. Click on the link below to accept this invitation.

Accept invitation

This invitation will expire in 7 days.

Accessing your AWS SSO User Portal
After you've accepted the invitation, you can access your AWS SSO user portal by using the information below.

Your User portal URL:
https://redbeardidentity.awsapps.com/start

Your Username:
rbiuser1

Figure 6.56 – Invitation to join AWS SSO

When we click the acceptance link, we are taken to the AWS SSO logon form. It is pre-configured to accommodate a password reset for our user. As we set a new password, we are prompted to comply with the service's password policy. Once we've entered and confirmed our password, we can proceed onward:

New user sign up

Enter your user information

Username: rbiuser1

New password

·············

Use: ×
⊘ 8-64 characters
⊘ Uppercase & lowercase letters
⊘ Numbers
⊘ Non-alphanumeric characters

Confirm password

⚠ Passwords must match
☐ Show password

Set new password

Figure 6.57 – Setting a compliant password

We are told that our password has been reset, and we are immediately redirected to an MFA device registration screen to enroll. We will use the Touch ID built into our workstation for `rbiuser1`:

Register MFA device

Username: rbiuser1 (not you?)

Your organization requires multi-factor authentication (MFA) for added security during sign-in. Each time you sign in, you'll be prompted for your password and an MFA device. Learn more ⤴

Select one of the options below to get started:

Authenticator app
Authenticate using a code generated by an app installed on your mobile device or computer.

Security key
Authenticate by touching a hardware security key such as YubiKey, Feitian, etc.

Built-in authenticator
Authenticate using a fingerprint scanner or camera built-in to your computer such as Apple TouchID, Windows Hello, etc.

Next

Figure 6.58 – AWS SSO MFA registration form

Upon selecting the built-in authenticator option, we see the prompt indicating that our browser is attempting to access the built-in authentication hardware:

Figure 6.59 – OS prompt to allow the browser to access Touch ID

If we allow the browser to proceed, we are then prompted by the browser to allow the specific website to access the built-in authenticator. If we allow it, the built-in authenticator will be successfully enrolled as the `rbiuser1` MFA device:

Built-in authenticator registered

⊘ Your built-in authenticator has been successfully registered. You can now use it when prompted for additional verification at sign in.

rbiuser1's MFA 1 Rename

Type and description: Security key or built-in authenticator

Done

Figure 6.60 – Successful registration of Touch ID as rbiuser1's MFA device

Having registered all of our credentials, we are now ready to access the user application portal that acts as the launch point for AWS SSO users to access the applications and accounts they are authorized to through AWS SSO. As we have not connected any applications or AWS accounts to AWS SSO, we don't have anything to use with our account:

Single Sign-On   MFA devices | Sign

You do not have any applications.

Figure 6.61 – rbiuser1's empty application portal

We will now remediate that as we assign Management Console access to our AWS SSO users to our AWS Organizations accounts.

Connecting AWS accounts to AWS SSO

We finally have everything we need to use AWS SSO to control access to our AWS accounts. We will investigate the basics of connecting the accounts to users, as well as using permissions sets to define the permissions for those users within those accounts in this chapter. However, the interplay between permissions sets, OU, individual accounts, and AWS SSO groups in determining access to AWS accounts and the services within those accounts on a per-user level provides ample opportunity for administrators to either implement a sustainable and scalable access management pattern or find themselves in an overly complex administrative nightmare. The patterns that will work for your specific use cases may vary from another organization's implementation. This is why we will be spending time in *Chapter 8, An Ounce of Prevention – Planning Your Administrative Model*, thinking through our use cases, architecture, and implementation prior to working through the practical implementation through the remainder of the book. For now, let's just get our users into their own AWS account using SSO.

We can see all of the AWS accounts in our organization in the AWS accounts section of the AWS SSO service. We have two views available to us. Similar to how the organization was presented in AWS Organizations, we can either see all the accounts in a list, or we can expand the root OU and view the accounts at each level of the OU hierarchy:

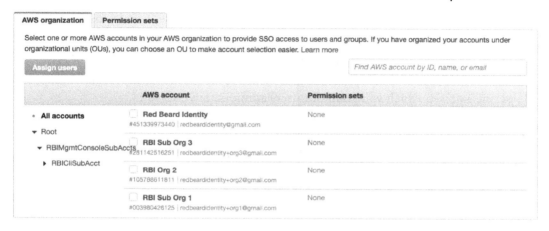

Figure 6.62 – The AWS accounts in the organization as seen in AWS SSO

In addition to the account view, we can also view the permission sets that are defined for use within our organization. We will need to create at least one permission set before we will be able to assign a user to an account. Permission sets can be thought of as the roles that the AWS SSO users will be assuming as they federate into member AWS accounts.

We should recall that part of the value of AWS SSO is how it automatically creates the AWS IAM federated IDP provider in the member organizations that it federates users into. Furthermore, we should also recall that when we use a federated identity provider in AWS IAM, the federated principal has no corresponding user account object in the downstream AWS account or that account's AWS IAM service. Rather, the federated principal exchanges its IDP's token for an AWS IAM role to assume. By defining a permission set, we are setting up the access limits for the assumed roles that will be available for the federated users to assume in the member AWS organizations.

We can define several different permissions sets on a specific account, and in turn, assign one or more of those permission sets to the user or group for them to use when accessing an account. Let's start by creating a permission set that gives our users administrative access to our AWS accounts:

1. We start by clicking the **Create permission set** button on the **Permission set** tab.

2. Similar to when we create roles, we can either use existing managed policy objects as a baseline for our permission set or create a custom permission set. Let's keep it simple and use an AWS-provided job function policy as the foundation for the permission set, which will grant our users administrative access:

Figure 6.63 – Selecting existing or custom policies for permission set creation

3. AWS has several policies aligned to "job functions." We've created roles and group entitlements based on the AdministratorAccess job function policy before and know that it gives full administrator access to an account, so we will select that and proceed:

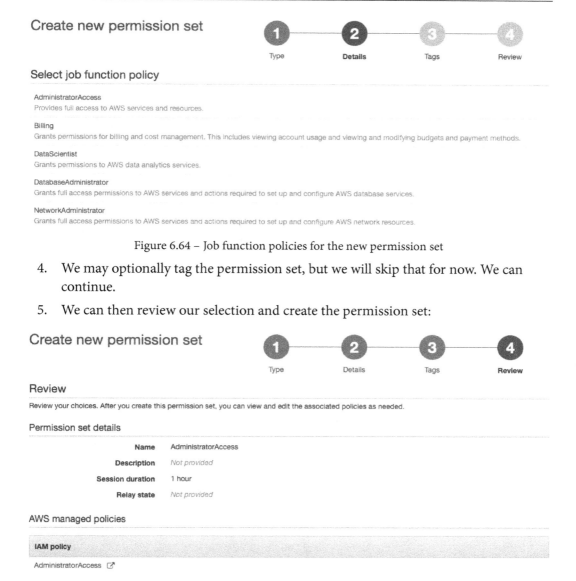

Create new permission set

Type   **Details**   Tags   Review

Select job function policy

AdministratorAccess
Provides full access to AWS services and resources.

Billing
Grants permissions for billing and cost management. This includes viewing account usage and viewing and modifying budgets and payment methods.

DataScientist
Grants permissions to AWS data analytics services.

DatabaseAdministrator
Grants full access permissions to AWS services and actions required to set up and configure AWS database services.

NetworkAdministrator
Grants full access permissions to AWS services and actions required to set up and configure AWS network resources.

Figure 6.64 – Job function policies for the new permission set

4. We may optionally tag the permission set, but we will skip that for now. We can continue.

5. We can then review our selection and create the permission set:

Create new permission set

Type   Details   Tags   **Review**

Review

Review your choices. After you create this permission set, you can view and edit the associated policies as needed.

Permission set details

Name	AdministratorAccess
Description	*Not provided*
Session duration	1 hour
Relay state	*Not provided*

AWS managed policies

IAM policy

AdministratorAccess ☑

Figure 6.65 – The AdministratorAccess permission set

Now that we have our permission set, let's add a user to an account. Let's start by adding `rbiuser1` to `RBI Sub Org 1`:

1. First, we click RBI Sub Org 1 on the **AWS organization** tab within the AWS accounts section of AWS SSO.

2. This opens the **Details** page for `RBI Sub Org 1`. Here we can see the users that are currently assigned to this account, any defined permissions sets for the account, and the IAM identity providers currently configured within that account's AWS IAM service. There isn't much here at present:

RBI Sub Org 1

Details

Account name	RBI Sub Org 1
Account ID	003980426125
Email	redbeardidentity+org1@gmail.com

Assigned users and groups

The following users or groups can access this AWS account from their user portal. Learn more

Assign users

User/group	Permission sets
You have not yet assigned any users or groups to this account.	

▾ Permission sets

Permission sets define the level of access that assigned users and groups have to this AWS account. The sets are stored in AWS SSO and appear in this account as IAM roles. You can update any of the permission sets associated with this AWS account to reapply or reset your permissions policies in IAM. Learn more

Update

Permission sets	Description
You have not yet created any permission sets for this account.	

▾ IAM identity provider

AWS SSO creates an IAM identity provider in each AWS account. Identity Provider enables the AWS account to trust AWS SSO for allowing SSO access. If the identity provider was deleted or modified, you can repair it.

You have not yet created a permission set. Once you create a permission set, an IAM identity provider will be created automatically for you.

Figure 6.66 – RBI Sub Org 1's details

3. Let's add `rbiuser1` to this account by clicking the **Assign users** button.

4. We see all of our AWS SSO users listed here and available for assignment. Similar to the AWS IAM access policy, we could also choose to assign a group to have access to the account, and then assign the users we want to have access to the account to that group. Although that is a better pattern, and we will be building out those patterns in a future implementation in later chapters; for now, let's simply select `rbiuser1` and move on:

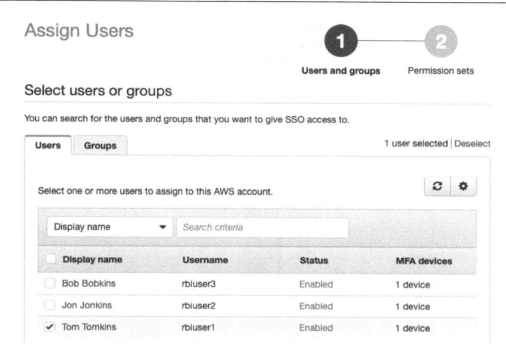

Figure 6.67 – Assigning rbiuser1 to the RBI Sub Org 1 account

5.   Next, we select the permission sets we wish to attach to this user for the account. We only have one permission set at present, but we could choose to attach multiple different permission sets if we wanted to give this user an option of roles to use with the `RBI Sub Org 1` account. Let's select the **AdministratorAccess** permission set and continue:

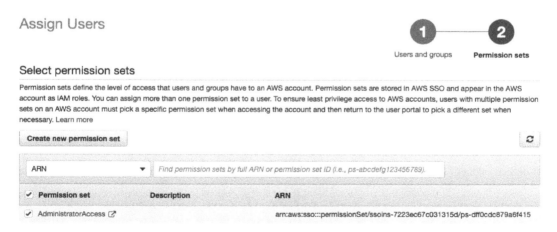

Figure 6.68 – Assigning the rbiuser1 account the permission set it will use with the AWS account

6. After a brief amount of processing, the process is complete. We can see what this process did by expanding the details section. There we see the entire workflow that AWS SSO undertook to enable `rbiuser1` to federate into the `RBI Sub Org 1` account:

Figure 6.69 – AWS SSO completes the configuration of SSO into the member account

If we return to the `RBI Sub Org 1` account, we can now see updated information for the users, permission sets, and identity providers configured there, including the `rbiuser1` account. The next time we sign into the application portal as that user, we should see that account available for us to use:

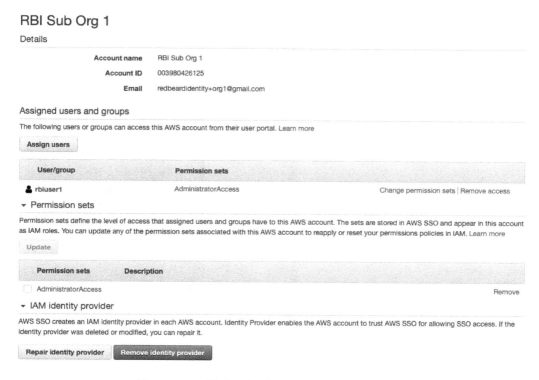

Figure 6.70 – Updated information in RBI Sub Org 1

Let's sign in to our application portal as rbiuser1. Once we do, we see an account has been assigned to us:

Figure 6.71 – The RBI Sub Org 1 AWS account is now available for rbiuser1

We have options for both the Management Console and for the command line. Let's start with the Management Console. Sure enough, when we click on the link, we are taken to the Management Console. We can see the assumed role under which we are operating within this account:

Figure 6.72 – Federated logon to RBI Sub Org 1 under an assumed role

The AWS IAM service within this account shows several roles and an IDP listed on the dashboard. The only AWS IAM object that we created in this account was a generic administrator account that we configured upon first configuring the account. All the other objects were placed within the AWS IAM service automatically by AWS Organizations and AWS SSO from the management account:

IAM dashboard

Sign-in URL for IAM users in this account

https://003980426125.signin.aws.amazon.com/console 🗗 | Customize

IAM resources

Users: 1   Roles: 5

Groups: 0   Identity providers: 1

Customer managed policies: 0

Figure 6.73 – Roles and IDP created automatically through AWS Organizations and AWS SSO

Now that we are satisfied that we can access the Management Console, let's check out how programmatic access works. Returning to the user application portal, we click on the link that says Command line or programmatic access. We get a window with some information:

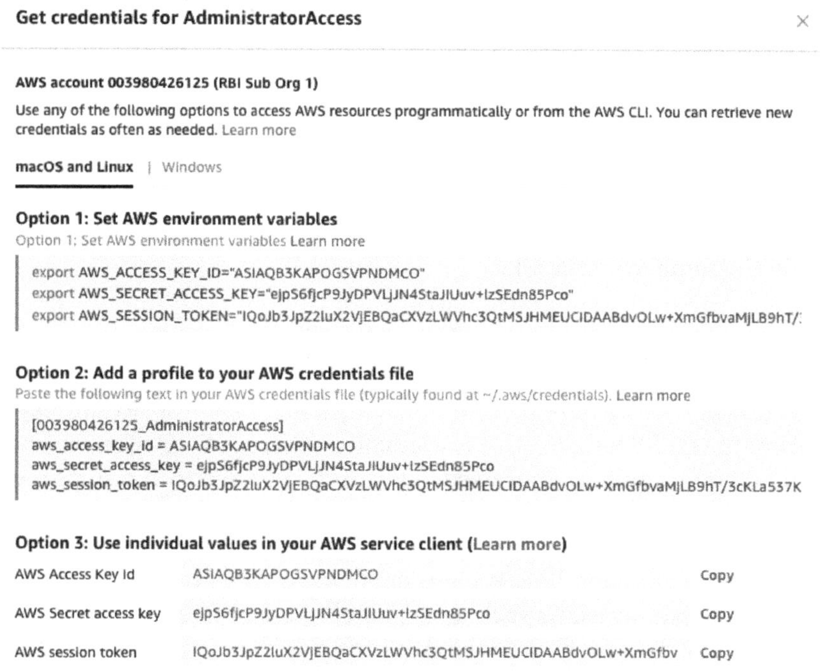

Get credentials for AdministratorAccess   ✕

AWS account 003980426125 (RBI Sub Org 1)

Use any of the following options to access AWS resources programmatically or from the AWS CLI. You can retrieve new credentials as often as needed. Learn more

macOS and Linux | Windows

Option 1: Set AWS environment variables
Option 1: Set AWS environment variables Learn more

```
export AWS_ACCESS_KEY_ID="ASIAQB3KAPOGSVPNDMCO"
export AWS_SECRET_ACCESS_KEY="ejpS6fjcP9JyDPVLjJN4StaJIUuv+lzSEdn85Pco"
export AWS_SESSION_TOKEN="IQoJb3JpZ2luX2VjEBQaCXVzLWVhc3QtMSJHMEUCIDAABdvOLw+XmGfbvaMjLB9hT/.
```

Option 2: Add a profile to your AWS credentials file
Paste the following text in your AWS credentials file (typically found at ~/.aws/credentials). Learn more

```
[003980426125_AdministratorAccess]
aws_access_key_id = ASIAQB3KAPOGSVPNDMCO
aws_secret_access_key = ejpS6fjcP9JyDPVLjJN4StaJIUuv+lzSEdn85Pco
aws_session_token = IQoJb3JpZ2luX2VjEBQaCXVzLWVhc3QtMSJHMEUCIDAABdvOLw+XmGfbvaMjLB9hT/3cKLa537K
```

Option 3: Use individual values in your AWS service client (Learn more)

AWS Access Key Id	ASIAQB3KAPOGSVPNDMCO	Copy
AWS Secret access key	ejpS6fjcP9JyDPVLjJN4StaJIUuv+lzSEdn85Pco	Copy
AWS session token	IQoJb3JpZ2luX2VjEBQaCXVzLWVhc3QtMSJHMEUCIDAABdvOLw+XmGfbv	Copy

Figure 6.74 – Programmatic credentials for RBI Sub Org 1 through AWS SSO

We are given a set of temporary programmatic credentials that are generated upon our successful sign-in through AWS SSO. We can use them in several ways, such as were described in *Chapter 2*, *An Introduction to the AWS CLI*. However, as these are only good for between 1 to 12 hours (depending upon the AWS SSO configuration), the easiest way to use them without cluttering a config file would be exporting the value in a terminal session. We can copy all of the export commands by clicking on the box with those commands, and then simply paste and execute inside of our terminal window:

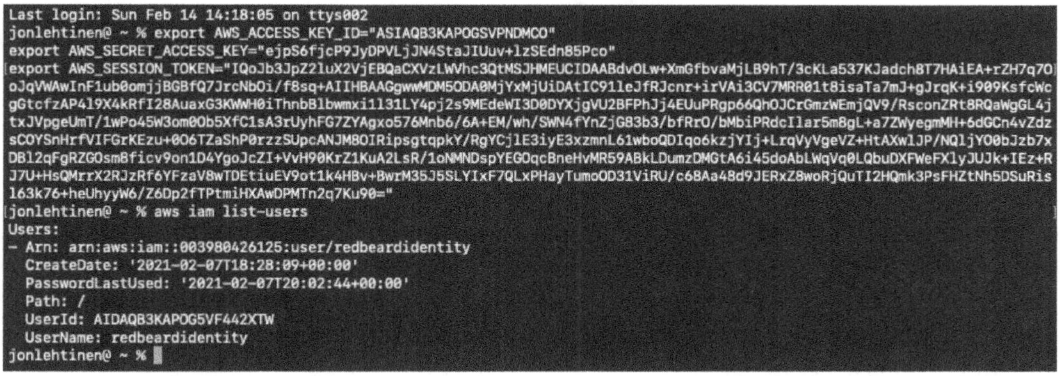

```
Last login: Sun Feb 14 14:18:05 on ttys002
jonlehtinen@ ~ % export AWS_ACCESS_KEY_ID="ASIAQB3KAPOGSVPNDMCO"
export AWS_SECRET_ACCESS_KEY="ejpS6fjcP9JyDPVLjJN4StaJIUuv+lzSEdn85Pco"
export AWS_SESSION_TOKEN="IQoJb3JpZ2luX2VjEBQaCXVzLWVhc3QtMSJHMEUCIDAABdvOLw+XmGfbvaMjLB9hT/3cKLa537KJadch8T7HAiEA+rZH7q7O
oJqVWAwInF1ub0omjjBGBfQ7JrcNbOi/f8sq+AIIHBAAGgwwMDM5ODA0MjYxMjUiDAtIC911eJfRJcnr+irVAi3CV7MRR01t8isaTa7mJ+gJrqK+i909KsfcWc
gGtcfzAP4l9X4kRfI28AuaxG3KWWH0iThnbB1bwmxi1l31LY4pj2s9MEdeWI3D0DYXjgVU2BFPhJj4EUuPRgp66QhOJCrGmzWEmjQV9/RsconZRt8RQaWgGL4j
txJVpgeUmT/1wPo45W3om0Ob5XfC1sA3rUyhFG7ZYAgxo576Mnb6/6A+EM/wh/SWN4fYnZjG83b3/bfRrO/bMbiPRdcI1ar5m8gL+a7ZWyegmMH+6dGCn4vZdz
sCOYSnHrfVIFGrKEzu+0O6TZaShP0rzzSUpcANJM8OIRipsgtqpkY/RgYCj1E3iyE3xzmnL61wboQDIqo6kzjYIj+LrqVyVgeVZ+HtAXwlJP/NQljYO0bJzb7x
DB12qFgRZGOsmB8ficv9on1D4YgoJcZI+VvH90KrZ1KuA2LsR/1oNMNDspYEGOqcBneHvMR59ABkLDumzDMGtA6i45doAbLWqVq0LQbuDXFWeFX1yJUJk+IEz+R
J7U+HsQMrrX2RJzRf6YFzaV8wTDEtiuEV9ot1k4HBv+BwrM35J5SLYIxF7QLxPHayTumoOD31ViRU/c68Aa48d9JERxZ8woRjQuTI2HQmk3PsFHZtNh5DSuRis
163k76+heUhyyW6/Z6Dp2fTPtmiHXAwDPMTn2q7Ku90="
jonlehtinen@ ~ % aws iam list-users
Users:
- Arn: arn:aws:iam::003980426125:user/redbeardidentity
  CreateDate: '2021-02-07T18:28:09+00:00'
  PasswordLastUsed: '2021-02-07T20:02:44+00:00'
  Path: /
  UserId: AIDAQB3KAPOG5VF442XTW
  UserName: redbeardidentity
jonlehtinen@ ~ %
```

Figure 6.75 – Executing CLI commands using temporary credentials through AWS SSO

Immediately afterward, we are able to execute a command. The ARN on the returned AWS IAM account aligns to the account number of `RBI Sub Org 1`. Normally, the commands we issue without a `--profile` parameter execute against our default profile and AWS account, which is the Redbeard Identity account. By using the `export` command, this terminal session will override the defaults defined in the `.aws/config` file.

Now that we've set up single sign-on between our accounts using the Management Console, let's do the same from the CLI.

Configuring AWS SSO from the CLI

We usually tear down what we have built in the Management Console to address the steps to recreate it from the CLI. However, this is a situation where AWS Organizations and AWS SSO's tight coupling already addressed many of the initial creation steps required to begin the tasks that we would perform from the AWS SSO service. We functionally already created a new AWS SSO service and identity store when we created an AWS organization using the command line. What is left to us to do with the CLI involves user assignment to member accounts. As such, to get a full picture of how to create an AWS SSO instance from scratch, refer back to the *AWS Organization in the AWS CLI* section earlier in this chapter.

The CLI does not have many options for account management for the identity store. Unlike Amazon Cognito, which is a full-featured platform for application identity use cases, AWS SSO uses the identity store primarily as its attribute and credential store for identities that were synched from an external authoritative source. As such, and though it will feel like a large gap in this chapter, there are no options to create users in the identity store from the CLI. We may only list/describe users and groups with the `identitystore` command:

```
$ aws identitystore list-users --identity-store-id d-9067650dfa
--filters AttributePath=UserName,AttributeValue=rbiuser1
Users:
- UserId: 9067650dfa-bdaf7f4d-50db-44ff-9dd0-9aa5ca24f64b
  UserName: rbiuser1
```

Recognizing that limitation, let's focus on what we can do. We'll create a permission set that enables view-only access to an account, and then assign that to `rbiuser2` for `RBI Org 2`. The first thing we will do is create that new permission set, and we can either look at the CLI skeleton or the documentation to see what values we will need to make the command execute successfully.

> **Reminder**
>
> AWS CLI documentation can be found at `https://awscli.amazonaws.com/v2/documentation/api/latest/index.html`.

From the CLI, we will run the `create-permission-set` command:

```
$ aws sso-admin create-permission-set --name ReadOnly
--instance-arn arn:aws:sso:::instance/ssoins-7223ec67c031315d
--session-duration PT12H
```

If successful, the output will show the details of the new permission set:

```
PermissionSet:
  CreatedDate: '2021-02-14T15:29:54.560000-05:00'
  Name: ReadOnly
  PermissionSetArn: arn:aws:sso:::permissionSet/ssoins-
7223ec67c031315d/ps-4fdc1dcd0d2cdd66
  SessionDuration: PT12H
```

This command creates the permission set within our AWS SSO instance, with a session duration of 12 hours. Now that we have the permission set, we need to populate it with some policy objects to make it useful, otherwise it is merely an empty permission set that grants no access. We do this with the `attach-managed-policy-to-permission-set` command:

```
$ aws sso-admin attach-managed-policy-to-permission-set
--instance-arn arn:aws:sso:::instance/ssoins-7223ec67c031315d
--permission-set-arn arn:aws:sso:::permissionSet/ssoins-
7223ec67c031315d/ps-4fdc1dcd0d2cdd66 --managed-policy-arn
arn:aws:iam::aws:policy/job-function/ViewOnlyAccess
```

If successful, there will be no output. However, we can verify that there is now an attached managed policy by running the `list-managed-policies-in-permission-set` command:

```
$ aws sso-admin list-managed-policies-in-permission-set
--instance-arn arn:aws:sso:::instance/ssoins-7223ec67c031315d
--permission-set-arn arn:aws:sso:::permissionSet/ssoins-
7223ec67c031315d/ps-4fdc1dcd0d2cdd66
```

This will show the attached policies as the output:

```
AttachedManagedPolicies:
- Arn: arn:aws:iam::aws:policy/job-function/ViewOnlyAccess
  Name: ViewOnlyAccess
```

With all of that prework addressed, we are now ready to assign the `rbiuser2` user account to the `RBI Org 2` AWS account for AWS SSO. We do this with `create-account-assignment`, along with several parameters:

```
$ aws sso-admin create-account-assignment --instance-arn
arn:aws:sso:::instance/ssoins-7223ec67c031315d --target-id
105788611811 --target-type AWS_ACCOUNT --permission-set-
arn arn:aws:sso:::permissionSet/ssoins-7223ec67c031315d/
ps-4fdc1dcd0d2cdd66 --principal-type USER --principal-id
9067650dfa-f5605b5a-79c6-4834-adc0-ef2d6d563dd2
```

Let's walk through the parameters so we understand what is going on:

- `instance-arn` refers to the AWS SSO instance in the management account of the AWS organization.
- `target-id` is the managed AWS account's unique account ID number.
- `target-type` specifies whether the target is an application or an AWS account.
- `permission-set-arn` is the ARN of the permission set from the management AWS account, which will be provisioned into the target member AWS account, and will determine the access available to the principal with the assumed role in that target account.
- `principal-type` can either refer to a specific user or a group.
- `principal-id` is the unique GUID for either the user or group selected as the principal.

Assuming we didn't make a mistake with all of those parameters, we will get this as our output:

```
AccountAssignmentCreationStatus:
  PermissionSetArn: arn:aws:sso:::permissionSet/ssoins-
7223ec67c031315d/ps-4fdc1dcd0d2cdd66
  PrincipalId: 9067650dfa-f5605b5a-79c6-4834-adc0-ef2d6d563dd2
  PrincipalType: USER
  RequestId: c28f7b77-5c0d-4878-9187-861ae690b083
```

```
Status: IN_PROGRESS
TargetId: '105788611811'
TargetType: AWS_ACCOUNT
```

The status indication suggests that it is a process that could take some time. We could validate that the process has been completed by checking the request ID using the `describe-account-assignment-creation-status` command:

```
$ aws sso-admin describe-account-assignment-creation-status
--instance-arn arn:aws:sso:::instance/ssoins-7223ec67c031315d
--account-assignment-creation-request-id c28f7b77-5c0d-4878-
9187-861ae690b083
```

The output provides the new status:

```
AccountAssignmentCreationStatus:
    CreatedDate: '2021-02-14T15:45:39.977000-05:00'
    PermissionSetArn: arn:aws:sso:::permissionSet/ssoins-
7223ec67c031315d/ps-4fdc1dcd0d2cdd66
    PrincipalId: 9067650dfa-f5605b5a-79c6-4834-adc0-ef2d6d563dd2
    PrincipalType: USER
    RequestId: c28f7b77-5c0d-4878-9187-861ae690b083
    Status: SUCCEEDED
    TargetId: '105788611811'
    TargetType: AWS_ACCOUNT
```

As we can see our request succeeded, we should theoretically be able to open up our SSO start page as `rbiuser2` and see the account. After sign-on, we see the account and role:

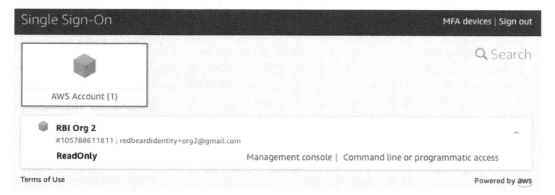

Figure 6.76 – rbiuser2 with ReadOnly access to RBI Org 2

Sure enough, we are now able to sign into that account, though we cannot create anything there:

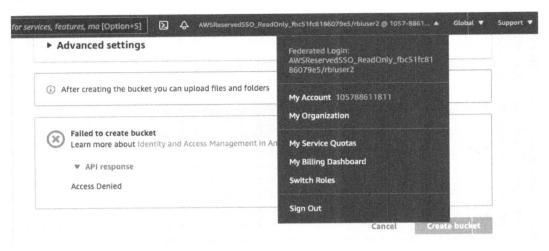

Figure 6.77 – The assumed read-only role in the RBI Org 2 account

By now, deleting the account assignments, permission sets, and managed role assignments to those permission sets should be fairly intuitive based on the patterns we have seen when using the CLI so far. As such, let's instead recap the massive amount of information that we picked up over the course of this chapter.

Summary

This was another tremendously long chapter filled with a ton of information. That said, this truly was only an introduction. AWS SSO has only recently become the strategic cornerstone for multi-account AWS account management in conjunction with AWS Organizations, and new best practices and patterns are still being established. That said, we learned how AWS Organizations is used to both bring existing accounts under centralized management as well as to provision net-new accounts within an organization. AWS SSO provides authentication and authorization for those accounts, as well as to third-party SaaS providers and AWS applications. Access to AWS accounts is governed by permission sets, which provide the template for the local AWS IAM roles that the users will assume in the target AWS accounts through identity federation.

The next chapter will provide a high-level overview of the remaining identity and identity-adjacent services that we need to be familiar with when implementing identity on AWS. We will take a look at services that address directory services, encryption and secrets management, and logging and auditing.

Questions

1. True/False: AWS Organizations can function without AWS SSO.

2. True/False: AWS SSO can function without AWS Organizations.

3. True/False: AWS IAM accounts are provisioned into the managed AWS accounts from the management AWS account through AWS SSO.

4. Which of these is not a valid place where AWS SSO can source and synchronize user information into its identity store?

   a. AWS Managed Active Directory Service

   b. External federated identity provider

   c. On-premises Active Directory using AD Connector

   d. Amazon Cloud Directory

5. What is the role of the management account in an AWS organization?

6. How do permission sets work to limit access to connected AWS account services and resources in managed AWS accounts?

Further reading

- AWS Organizations User Guide – `https://docs.aws.amazon.com/organizations/latest/userguide/orgs_introduction.html?org_product_rc_usergude`

- AWS Single Sign-On User Guide – `https://docs.aws.amazon.com/singlesignon/latest/userguide/what-is.html`

7
Other AWS Identity Services

We are coming to the end of this section, where we have introduced and explored the identity services that are available on AWS. The two previous chapters deep dived into the customer and enterprise identity services, but in this chapter, we will be taking a slightly different approach. This chapter will provide a brief overview of several additional identity and identity-adjacent services. While familiarity with these services and their use cases is an important part of a well-rounded education for implementing identity on AWS, these services don't merit as deep a dive, nor as much of a practical exploration for our purposes, before we move on to the next section of this book.

The first service we will look at is **AWS Directory Service**. This service primarily deals with supporting **Active Directory (AD)** workloads on AWS and extending an organization's (organization meaning enterprise, not organization as in AWS Organization) **AD footprint** into AWS. The next two services deal with secrets management. **AWS Secrets Manager** provides secure storage for application passwords, automated secrets rotation, and programmatic secret retrieval for AWS services and applications through API calls that replace the plaintext secret value in configs and code. AWS Secrets Manager works in conjunction with **AWS Key Management Service**, which provides users control over the encryption master keys used to secure their data at rest. The final two services address monitoring, logging, and audit. **Amazon CloudWatch** provides operational metrics and monitoring, as well as application and service logging capabilities. **AWS CloudTrail** is also a logging service, but it focuses on auditing actions that have been made by services, principals, and entities within the AWS account itself.

In this chapter, we will cover the following topics:

- Understanding AWS Directory Service

- Encryption and secrets management

- Logging and auditing

Technical requirements

To get the most out of this chapter, you will need an AWS account.

Understanding AWS Directory Service

Microsoft AD is a complex and feature-rich enterprise directory service. Beyond basic LDAP capabilities for user management and authentication, it can also be used for machine management, including device authentication and authorization, DNS, certificate authority services, endpoint policy management and enforcement, and federation services. Over the years, it has been positioned and marketed as a one-stop-shop for enterprise workloads. Unfortunately, the feature-richness that made AD an enterprise mainstay for over 20 years is also why it can become insecure or misconfigured. This is why AD implementations are at the heart of so many security incidents. Its monolithic nature, broad set of services, and wide network port utilization also make it a tempting target for bad actors and limit its capability to securely operate outside of an established network perimeter.

Though traditional on-premises AD may not be naturally suited for an internet-as-the-backplane, cloud-first world, there is still a huge ecosystem of enterprise software designed for Active Directory. Organizations are looking to move themselves into the cloud, but many of those organizations are not willing to abandon functional software and business processes to get there. AWS Directory Service aims to help organizations bridge their existing on-premises AD into the cloud, make their existing user and computer objects available there, and facilitate the management of AWS resources through AD.

Another major feature of AWS Directory Service is its potential to serve as the user store for **AWS SSO**. As we covered in *Chapter 6, Introduction to AWS Organizations and AWS Single Sign-On*, some of the Active Directory services available through AWS Directory Service can be defined as identity stores for AWS SSO. This lets us use our AD credentials to access AWS accounts and resources. Doing so has certain prerequisites, such as making sure the AD service is deployed within the same region as the AWS SSO service on the AWS Organizations' management account.

Amazon and AWS offer several enterprise productivity applications, such as **Amazon WorkDocs**, **Amazon WorkMail**, **Amazon WorkSpaces**, and **Amazon Chime**. These are more like traditional SaaS apps than AWS services. AWS Directory Service can be the authenticating user store for them in the same way that AWS SSO can.

There are three major flavors of AWS Directory Service. Next, we'll quickly examine each of them and their intended use cases.

AWS Managed Microsoft AD

The first solution is **AWS Managed Microsoft AD**. Aside from having a name that rolls off the tongue, this is the most feature-rich option available within AWS Directory Service. AWS Managed Microsoft AD provides turnkey instances of Microsoft Active Directory that are automatically deployed on AWS. AWS handles the infrastructure support and operations for running the AD domain, including backups, patching, replication, and monitoring.

Besides the value proposition of delegating infrastructure support, this version of the service can also satisfy certain regulatory requirements. AWS Managed Microsoft AD can be made compliant with the US **Health Insurance Portability and Accountability Act (HIPAA)** and the **Payment Card Industry Data Security Standard (PCI-DSS)** regulations.

AWS Managed Microsoft AD comes in two varieties: Standard Edition and Enterprise Edition. Though both versions are fully functional Active Directory deployments, there are important capacity and feature differences between them. Both editions start with a pair of domain controllers spread across availability zones that have been deployed within a single region, and the same automatic configuration process that deploys the required network interfaces and security groups to use the service within the VPC and subnets within the account where we activate it:

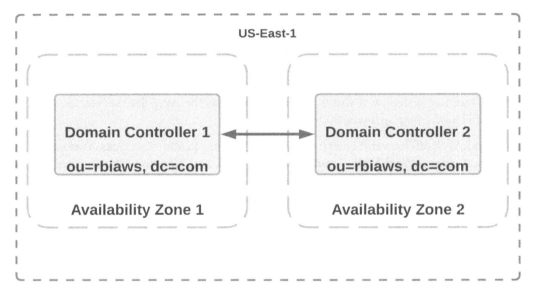

Figure 7.1 – Default Multi-Availability Zone deployment

The domain controllers available within the Standard Edition are sized more modestly than those that come with the Enterprise Edition. The Standard Edition is best for small or mid-sized organizations that won't go beyond 30,000 directory objects. Enterprise Edition is capable of handling enterprise workloads and up to 500,000 objects.

Both versions support adding additional domain controllers. However, the Standard Edition is constrained to keeping those DCs within a single region, though those DCs may be deployed across all the AZs within that region. In contrast, the Enterprise Edition is capable of handling global workloads and multi-region replication:

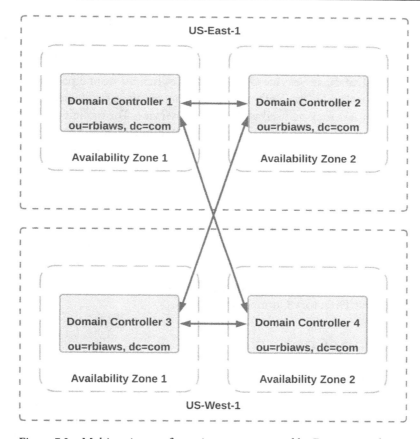

Figure 7.2 – Multi-region configurations are supported by Enterprise Edition

Since many organizations already have an on-premises AD infrastructure and are not willing to migrate their entire AD topology and workload to the cloud, AWS Managed Microsoft AD also supports trust between the domain that's created within the service. A trust allows users and objects from one domain to access the resources in another domain:

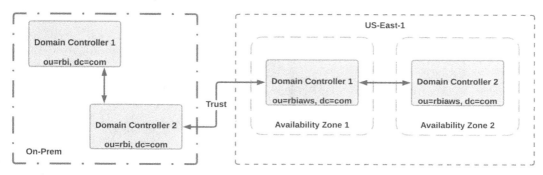

Figure 7.3 – Example of a trust between two AD forests

There is plenty of nuance around what types of trusts and their directionality are appropriate for a given use case. We won't go into that here, but it is important to keep it in mind when implementing a secure AD architecture for a given organization's use case.

Amazon Relational Database Service (RDS) instances benefit from a pairing with AWS Managed Microsoft AD. The Kerberos tokens that are used for device and user authentication within AD may be used with the RDS service. The benefit of this arrangement is that no passwords need to be transmitted during logon time, and administrators and their access can be governed using existing business logic managed through AD.

Finally, the AWS Managed Microsoft AD service facilitates the management of certain AWS resources. EC2 Windows and Linux instances can be joined and managed within the domain using the same tools and processes as they would on-premises. Existing instances may be manually joined to the domain, and new instances can be instantly joined by adding a specific IAM policy to that resource.

Now that we've gone over the capabilities of the AWS Managed Microsoft AD service, let's look at what the Active Directory Connector can do.

Active Directory Connector

The AWS Managed Microsoft AD service offers a full AD instance within an AWS account, which is valuable for organizations looking to realize the benefits of running AD as a managed service. However, implicit to that service is replicating on-premises identity information into the cloud. Much of the often nebulous concern around putting identity data inside a public cloud has evaporated as cloud platforms such as AWS have proven secure. That said, some organizations may either prefer not to accept that risk or are precluded from doing so due to regulatory requirements. For these use cases, the **Active Directory Connector** allows AWS resources to redirect directory requests to an on-premises AD environment without importing any of that directory information into the AWS account:

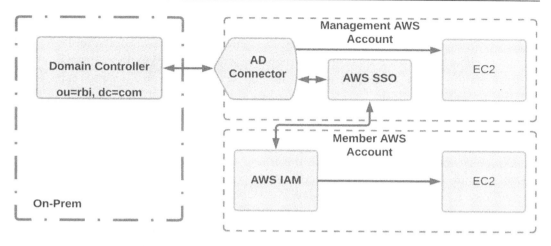

Figure 7.4 – Authentication and management flows through the AD Connector

The AD Connector can perform many of the same functions as AWS Managed Microsoft AD, including authentication to AWS resources such as Amazon EC2 instances, authentication to AWS accounts within an AWS Organization with the on-premises AD as the identity store, and automatically joining and managing AWS resources into the on-premises AD domain. However, there are some additional limitations of using the AD Connector compared to the Managed Microsoft AD Service. Each unique domain must be paired with its own connector. AD Connectors do not support multi-region configurations. Perhaps most significantly, each AWS region is limited to a default quota of 10 AD connectors per account. This could be a constraint on orgs with complicated, on-premises AD forests with multiple domains and trusts.

If we don't want to use the full capabilities of AD, we can save ourselves some complexity by looking into the third variant of AWS Directory Service called **Simple Active Directory**.

Simple Active Directory

Simple Active Directory is a managed AD service for lightweight AD workloads. Compared to the previous two Directory Service variants we looked at, this one is the most spartan in terms of features and service capacity. Strictly speaking, it is not truly AD; it runs a **Samba 4** Active Directory Compatible Server.

Simple Active Directory is a managed service, similar to AWS Managed Microsoft AD. Upon initialization, two domain controllers are deployed in a single region across two different availability zones, and the directory itself can be managed using the typical tools used to manage user and computer objects within AD. It supports user binds, groups, Kerberos, and machine joins for EC2 instances, but that is the end of its similarities with its more robust sister service.

Simple Active Directory comes in two sizing options, small or large, supporting 2,000 directory objects and 20,000 directory objects, respectively. Its limited capacity and capabilities make it useful for application authentication and LDAPS services in small organizations. Similar to AD Connector, this service is also limited on a per-region basis to 10 directories.

Amazon Cognito

AWS Directory Service is geared toward enterprise use cases. Although we won't relitigate Amazon Cognito here, it merits mention that Amazon Cognito is specifically highlighted as the directory solution for application developers looking for directory services for customer use cases. For additional information on Amazon Cognito, please revisit *Chapter 5, Introducing Amazon Cognito*.

Next, we will turn our attention to how AWS offers secrets and encryption key management in relation to IAM use cases.

Encryption and secrets management

Confidentiality is one of the three pillars of information security. Encryption preserves the confidentiality of data both in transit and at rest. To decrypt encrypted data, we need the appropriate keys.

AWS offers services for both managing the cryptographic keys that encrypt the data used within an AWS account, as well as a service for preserving secrets used for accessing AWS resources. We will go over these services briefly.

AWS Key Management Service

Several services within AWS offer encryption for the data at rest and in transit. S3 buckets, RDS instances, EBS volumes, and other resources leverage encryption to secure the data they store. By default, each AWS service capable of leveraging AWS KMS can generate their own instance of a default, AWS-managed encryption key that is used to encipher that data for that AWS account. However, some organizations would prefer to retain control of their encryption keys. In either case, AWS Key Management Service manages those keys.

AWS Key Management Service lets us create and control something called **customer master keys** (**CMKs**). CMKs are what we can use with AWS KMS to generate the encryption keys to encipher our data within AWS. By default, each AWS account offers an AWS-managed CMK for encryption services. If you prefer not to use the default AWS-managed key or create additional keys to use per resource or service, you can create and manage them within the AWS Key Management Service. You can then select the specific CMK you wish to use with a given resource when creating that resource:

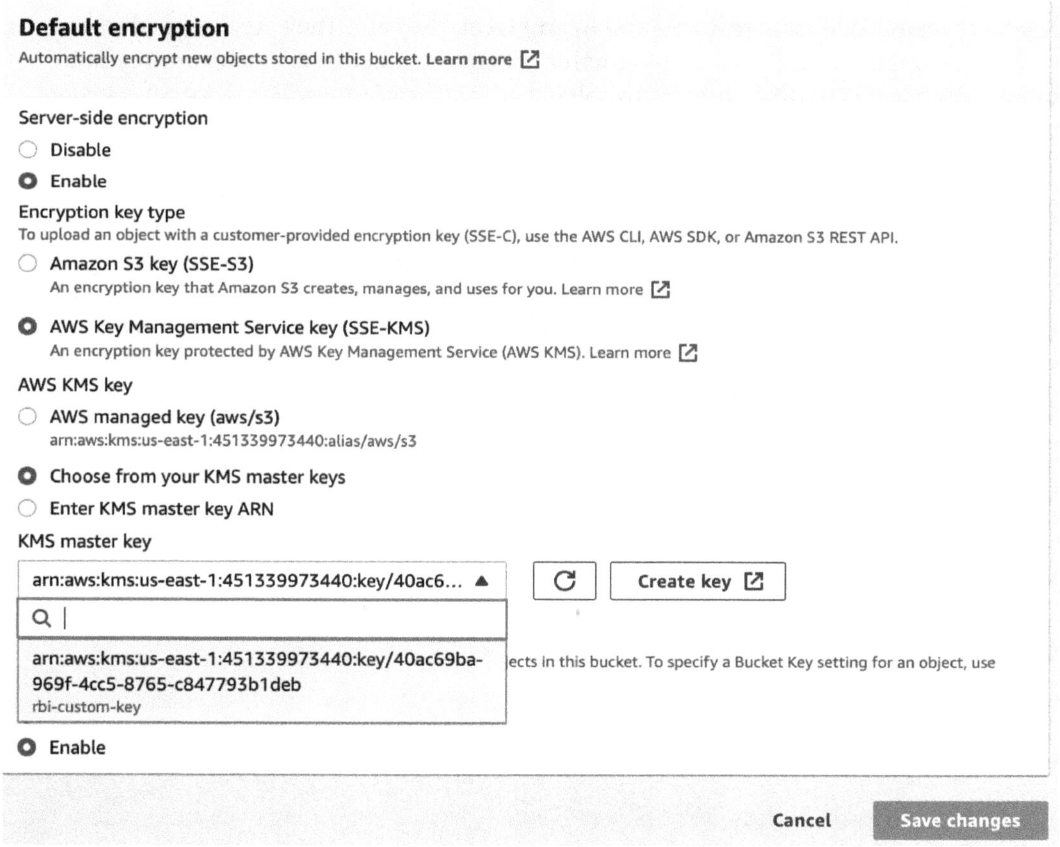

Default encryption

Automatically encrypt new objects stored in this bucket. **Learn more** [↗]

Server-side encryption

○ Disable

● Enable

Encryption key type

To upload an object with a customer-provided encryption key (SSE-C), use the AWS CLI, AWS SDK, or Amazon S3 REST API.

○ Amazon S3 key (SSE-S3)

   An encryption key that Amazon S3 creates, manages, and uses for you. Learn more [↗]

● AWS Key Management Service key (SSE-KMS)

   An encryption key protected by AWS Key Management Service (AWS KMS). Learn more [↗]

AWS KMS key

○ AWS managed key (aws/s3)

   arn:aws:kms:us-east-1:451339973440:alias/aws/s3

● Choose from your KMS master keys

○ Enter KMS master key ARN

KMS master key

| arn:aws:kms:us-east-1:451339973440:key/40ac6... ▲ | C | Create key [↗] |

Q |

arn:aws:kms:us-east-1:451339973440:key/40ac69ba-   jects in this bucket. To specify a Bucket Key setting for an object, use
969f-4cc5-8765-c847793b1deb
rbi-custom-key

● Enable

Cancel   Save changes

Figure 7.5 – Selecting a custom KMS key to use with S3 bucket encryption

In addition to creating or bringing along our own keys, AWS Key Management Service provides services for cryptographic functions and key administration. It provides the hardware security modules, interfaces, and services to create, edit, view, and delete all the cryptographic keys that we would wish to use within our AWS account. We can perform programmatic encryption/decryption functions, sign messages, verify the integrity of messages, export key pairs, and generate random numbers. It also allows us to automate how cryptographic material is rotated within our own customer-managed customer master key.

These may seem like obtuse features, and to an extent, they are. They are designed to be leveraged programmatically for cryptographic functions rather than by an administrator. There is another service that aligns with solving administrator use cases – AWS Secrets Manager.

AWS Secrets Manager

Simply put, AWS Secrets Manager stores secrets. These secrets can be for AWS-specific resources, such as Amazon RDS instances, or they could be for other things such as service account passwords, API keys, or OAuth tokens. The secrets that are stored within AWS Secrets Manager are encrypted using keys managed by AWS Key Management Service:

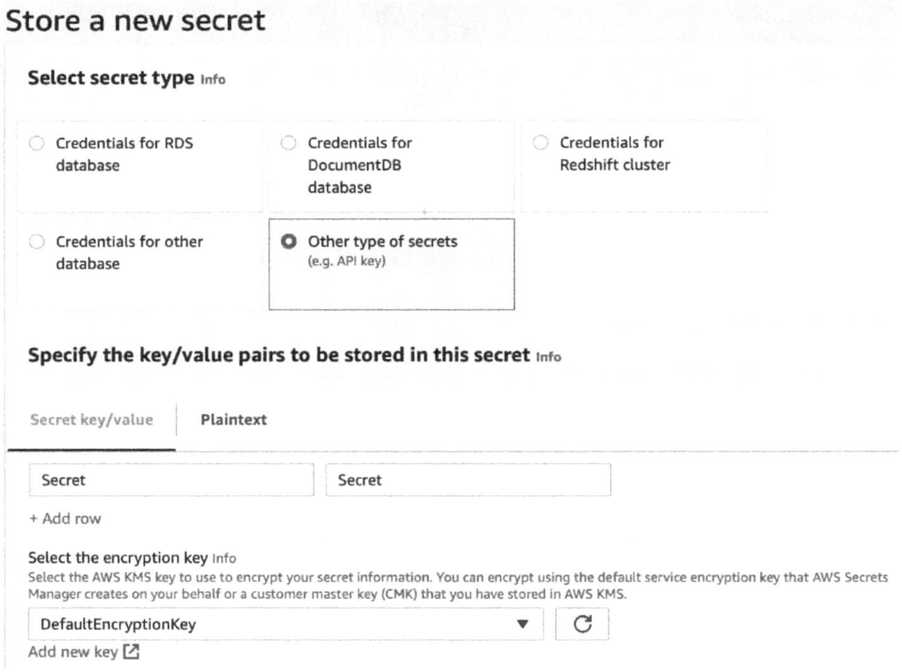

Figure 7.6 – Available secrets in Secrets Manager

Aside from secure secret storage, there are two huge advantages to using AWS Secrets Manager. The first is support for automatic secret rotation. When AWS Secrets Manager is used with a supported AWS service such as Amazon RDS, Amazon Redshift, or Amazon DocumentDB, AWS Secrets Manager can be configured to automatically rotate the credential both within its secret store and within the connected service at a specified interval. AWS Secrets Manager can do this for other secrets as well, though an AWS Lambda function to cycle the **Other type of secrets** secret, stored at the connected service, will need to be written and connected to the rotation schedule:

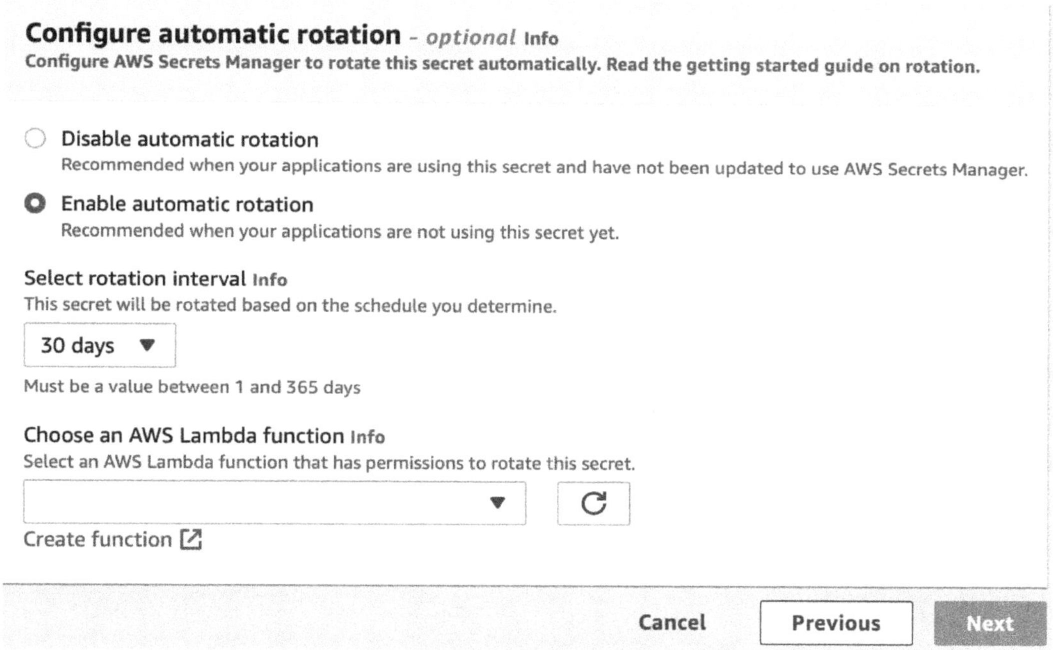

Configure automatic rotation - *optional* Info
Configure AWS Secrets Manager to rotate this secret automatically. Read the getting started guide on rotation.

◯ Disable automatic rotation
    Recommended when your applications are using this secret and have not been updated to use AWS Secrets Manager.

◉ Enable automatic rotation
    Recommended when your applications are not using this secret yet.

Select rotation interval Info
This secret will be rotated based on the schedule you determine.

| 30 days   ▼ |

Must be a value between 1 and 365 days

Choose an AWS Lambda function Info
Select an AWS Lambda function that has permissions to rotate this secret.

| ▼ | | C |

Create function ↗

Cancel   Previous   Next

Figure 7.7 – Secrets Manager supports automatic rotation

The second advantage that AWS Secrets Manager provides is code snippets for secrets retrieval. This replaces a variable for the secret in a config file within the code, which is an insecure programming practice. Once a secret has been stored in AWS Secrets Manager, we can copy a specific code snippet that references our secret and insert it into our application to take advantage of AWS Secrets Manager:

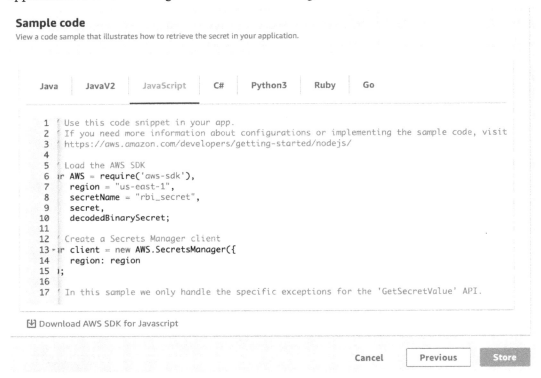

Figure 7.8 – Code snippets for retrieving our secret

Though not necessarily an IAM service, AWS Secrets Manager helps us manage and rotate our app credentials more securely.

Next, we will look at the services that are available for logging and auditing within AWS.

Logging and auditing

Unlike the identity services that merited their own chapters, or even the services we looked at earlier within this chapter, **AWS CloudTrail** and **Amazon CloudWatch** may not seem worth much of a mention. However, logging and auditing are essential components of **non-repudiation**. Non-repudiation is when we have assurances that something, such as an action, signature, or event, cannot be denied by a person. IAM ties the event, action, account, and others to the individual, and auditing and logging help prove that the event occurred.

We will quickly look at the two services AWS provides for audit and logging. The first is AWS CloudTrail, which captures the events that occur within an AWS account. The second is Amazon CloudWatch, which is a monitoring and logging service that can be used with AWS services and resources.

AWS CloudTrail

AWS CloudTrail captures events that occur within an AWS account to help us address compliance, governance, and operational and risk auditing. When an IAM object, such as a user, role, or a service, takes an action within the AWS account, the action is logged in detail. We can see the latest actions that have been taken by the `redbeardidentity` user account, including changing its password and starting and stopping an EC2 instance:

Event name	Event time	User name	Event source	Resource type
ChangePassword	March 09, 2021, 09:01:00 (UT...	redbeardidentity	iam.amazonaws.com	-
ConsoleLogin	March 09, 2021, 09:00:35 (UT...	redbeardidentity	signin.amazonaws.com	-
ConsoleLogin	March 04, 2021, 10:50:20 (UT...	redbeardidentity	signin.amazonaws.com	-
StopInstances	February 24, 2021, 11:24:48 (...	redbeardidentity	ec2.amazonaws.com	AWS::EC2::Instance
SendSSHPublicKey	February 24, 2021, 11:10:54 (...	redbeardidentity	ec2-instance-connect.amazonaws.com	AWS::EC2::Instance
AuthorizeSecurityGroupIngress	February 24, 2021, 11:09:49 (...	redbeardidentity	ec2.amazonaws.com	AWS::EC2::SecurityGroup
AuthorizeSecurityGroupIngress	February 24, 2021, 11:09:48 (...	redbeardidentity	ec2.amazonaws.com	AWS::EC2::SecurityGroup
RevokeSecurityGroupIngress	February 24, 2021, 11:09:48 (...	redbeardidentity	ec2.amazonaws.com	AWS::EC2::SecurityGroup
SharedSnapshotVolumeCreated	February 24, 2021, 11:05:35 (...	-	ec2.amazonaws.com	-
RunInstances	February 24, 2021, 11:05:31 (...	redbeardidentity	ec2.amazonaws.com	AWS::EC2::VPC, AWS::EC2::A...
AuthorizeSecurityGroupIngress	February 24, 2021, 11:05:29 (...	redbeardidentity	ec2.amazonaws.com	AWS::EC2::SecurityGroup
CreateSecurityGroup	February 24, 2021, 11:05:29 (...	redbeardidentity	ec2.amazonaws.com	AWS::EC2::VPC, AWS::EC2::S...
CreateKeyPair	February 24, 2021, 10:59:31 (...	redbeardidentity	ec2.amazonaws.com	AWS::EC2::KeyPair
ConsoleLogin	February 24, 2021, 10:57:50 (...	redbeardidentity	signin.amazonaws.com	-

Figure 7.9 – User events captured across services in AWS CloudTrail

It could be tempting to think of it as AWS' native **security information and event management (SIEM)** software, it does not provide advanced threat analytics or baselining. However, AWS CloudTrail events can be fed into other AWS services such as **Amazon GuardDuty**, which performs such analytic functions. Alternatively, AWS CloudTrail can be configured to output event data into an organization's existing SIEM. AWS CloudTrail's main value stems from maintaining a detailed record of events for security and operational auditing, which can then be used for compliance and operational troubleshooting.

AWS CloudTrail logs come in JSON format. The log messages provide information about a specific event, the time, the initiator, the source, and the resource that was manipulated. This information is useful for compliance purposes, as well as getting additional operational information in case something is not working within the environment. Here is an example of the JSON log output from when `redbeardidentity` signed into the Management Console:

```
{
    "eventVersion": "1.08",
    "userIdentity": {
        "type": "IAMUser",
        "principalId": "AIDAWSFPVONALTHHLBKLK",
        "arn": "arn:aws:iam::451339973440:user/
redbeardidentity",
        "accountId": "451339973440",
        "userName": "redbeardidentity"
    },
    "eventTime": "2021-03-09T14:00:35Z",
    "eventSource": "signin.amazonaws.com",
    "eventName": "ConsoleLogin",
    "awsRegion": "us-east-1",
    "sourceIPAddress": "108.4.89.93",
    "userAgent": "Mozilla/5.0 (Macintosh; Intel Mac OS
X 10_15_6) AppleWebKit/605.1.15 (KHTML, like Gecko)
Version/14.0.3 Safari/605.1.15",
    "requestParameters": null,
    "responseElements": {
        "ConsoleLogin": "Success"
    },
    "additionalEventData": {
```

```
        "LoginTo": "https://console.aws.amazon.com/console/
home?state=hashArgs%23&isauthcode=true",
        "MobileVersion": "No",
        "MFAUsed": "Yes"
    },
    "eventID": "1849ad94-a60d-47c1-bc3c-0e1ffe3fe9fd",
    "readOnly": false,
    "eventType": "AwsConsoleSignIn",
    "managementEvent": true,
    "eventCategory": "Management",
    "recipientAccountId": "451339973440"
}
```

By default, AWS CloudTrail maintains event logs for 90 days, and those logs can be searched for from the Management Console. However, the typical AWS CloudTrail usage pattern involves creating something called a **trail**. Trails allow event logging to be exported into an Amazon S3 bucket for more durable storage and searching. Additionally, trails enable integration into other AWS services, such as Amazon CloudWatch, where monitoring, alarming, and notifications can be configured based on certain event types. Trails can be configured for a single region, multiple regions, or for every AWS account under an AWS Organization.

The events that AWS CloudTrail captures come in three different varieties. The first set of events are known as **management events**. These are the events that manipulate the services and objects of AWS, such as what we can see in the preceding screenshot. Another way to think of management events are events that occur on AWS' control plane through either the Management Console, CLI, or other interfaces such as APIs or SDKs.

The second are **data events**. These are events that operate at the data plane, such as when we're manipulating objects within an S3 bucket or calling a Lambda function. This event type acts similarly to application-level logging in terms of its verbosity and specificity as to what was manipulated, what action was taken, and who or what performed the action. As such, it is not natively enabled when creating a trail.

The last event type is **Insights events**. Insights are the bare minimum in terms of the threshold analytics capabilities provided through AWS CloudTrail. With Insights enabled, AWS CloudTrail will flag Insights by placing them into a unique folder within the default Management Console interface, or preface the events with a unique prefix when exporting to a bucket as part of a trail to highlight the historical aberration of the events. Insights are not enabled by default.

AWS CloudTrail provides an essential service, but its capabilities, such as anything other than its event logger, only shine through when it is combined with other services, such as Amazon CloudWatch. Next, we will look at Amazon CloudWatch, which is, among other features, AWS' logging service.

Amazon CloudWatch

We may be underselling Amazon CloudWatch by referring to it simply as "AWS' logging service." Whereas logging is one of its most visible and obvious use cases, it also performs various flavors of health and capacity monitoring and alerting, operational dashboarding, and anomaly detection. As availability is one of the three pillars of information security, this service can help with many of the operational challenges that come with running infrastructure in the cloud. However, for our purposes, we will stick to a very high-level overview of some of its capabilities.

Amazon CloudWatch offers real-time monitoring for AWS resources and applications by aggregating service metrics. An example of this is the CPU utilization in an Amazon EC2 instance or Amazon **Elastic Container Service (ECS)** cluster. By aggregating those metrics over time, Amazon CloudWatch can provide dashboarding to show historical trends, insights on how to improve the monitored service, and alerting when the metric either exceeds or dips below a specific threshold. We can then build additional actions based on the triggers and events provided by Amazon CloudWatch, such as scaling out an Amazon ECS cluster once CPU utilization crosses a high utilization threshold and stays there for a specific period.

Metrics can come from other places outside of AWS services. As we mentioned earlier, Amazon CloudWatch is the logging service for AWS services, but it can also be used for apps that have been deployed within AWS itself. We saw an example of the logging service in action when we imported our Amazon Cognito users in *Chapter 5, Introducing Amazon Cognito*:

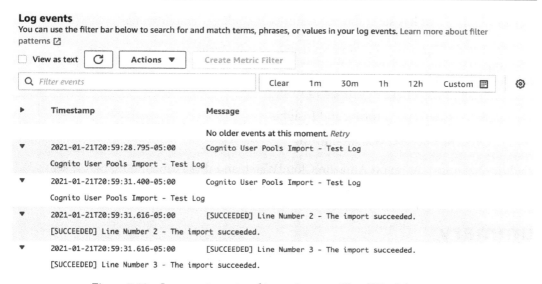

Figure 7.10 – Log events captured in an Amazon CloudWatch log group

With application logs going to Amazon CloudWatch, we can filter on certain strings to trigger events, such as sending an alert when a specific error message is found.

Amazon CloudWatch integrates with the Amazon Simple Notification Service to provide alerting based on events or alarms. Alarms can be triggered based on single actions, such as a server failing a health check, or based on a complex set of conditions known as a composite alarm. A composite alarm reduces the number of alerts by only alerting once a certain threshold of the monitored alarms trigger the composite alarm condition:

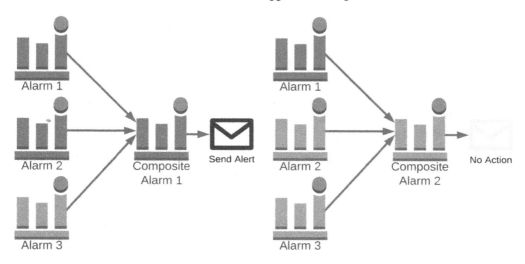

Figure 7.11 – Sample composite alarms

Amazon CloudWatch has several more features, including the following:

- Synthetic monitoring that mimics user access patterns using code snippets called canaries
- Container-specific monitoring and Insights
- Lambda-specific monitoring and Insights
- ServiceLens microservice monitoring

For additional information on Amazon CloudWatch and these capabilities, please see the *Further reading* section at the end of this chapter.

Summary

Now that you have finished this chapter, you should be familiar with some of the additional identity and identity-adjacent capabilities you can use to solve identity challenges on AWS. AWS Directory Service supports Active Directory workloads on AWS and extends an organization's AD footprint into AWS. AWS Secrets Manager allows programmatic secret storage and rotation, while AWS Key Management Service allows you to manage cryptographic keys that are used for encryption. Finally, AWS CloudTrail acts as the audit log for all actions taken on AWS services, while Amazon CloudWatch acts as a logging and resource monitoring service.

This concludes this section of this book, where we looked at specific AWS services. The next section will see us pivot toward practically applying these services to solve an enterprise-grade identity use case. In the next chapter, we will plan what we intend to accomplish with our practical implementation by using enterprise-grade tools and design patterns.

Questions

1. What version of AWS Managed Microsoft AD, either Standard or Enterprise, would be best suited for an organization consisting of 2,000 employees that requires support for full AD workloads?

2. An organization consisting of 2,000 employees only needs LDAP services for their applications in AWS. Which AWS Directory Service would be the most appropriate?

3. What is a customer master key?

4. AWS Secrets Manager can automatically rotate credentials for which three services, without requiring a custom Lambda function?

5. What is non-repudiation?

6. True/False: Amazon CloudWatch only provides logging aggregation.

Further reading

- AWS Key Management Service Developer Guide: `https://docs.aws.amazon.com/kms/latest/developerguide/overview.html`

- AWS Secrets Manager User Manual: `https://docs.aws.amazon.com/secretsmanager/latest/userguide/intro.html`

- AWS CloudTrail User Manual: `https://docs.aws.amazon.com/awscloudtrail/latest/userguide/cloudtrail-user-guide.html`

- Amazon CloudWatch User Manual: `https://docs.aws.amazon.com/AmazonCloudWatch/latest/monitoring/WhatIsCloudWatch.html`

Section 2: Implementing IAM on AWS for Administrative Use Cases

This section will help you conceptualize how to use the previously introduced AWS IAM services to solve common enterprise identity use cases, and will provide references and guidance on their implementation. Each chapter will iterate upon the capabilities of a hypothetical deployment, so you will be able to easily relate these concepts to your own environments, and see how they expand upon one another and interrelate.

This part of the book comprises the following chapters:

- *Chapter 8, An Ounce of Prevention – Planning Your Administrative Model*
- *Chapter 9, Bringing Your Admins into the AWS Administrative Backplane*
- *Chapter 10, Administrative Single Sign-On to the AWS Backplane*

8
An Ounce of Prevention – Planning Your Administrative Model

In a fast-paced enterprise setting, many practitioners find themselves building piecemeal solutions to complex business challenges in reaction to the changing demands and urgent deadlines imposed upon them by the business. As both **Identity and Access Management (IAM)** and the cloud are high-value, business-enabling technologies, it can be challenging to take the time before implementation to contemplate what a sustainable implementation pattern looks like for the business. Although engaging in this planning exercise may frustrate some stakeholders, organizations that fail to plan out their cloud administrative models often find themselves limited by their own short-term technical solutions. Solving use cases such as these should be a holistic, multi-step design and analysis process.

In this chapter, we will evaluate how we would like to apply an administrative model to the Redbeard Identity organization. We will do this by evaluating both business requirements and technical factors. By taking the time to consider our cloud administrative model upfront prior to implementation, our solution will be less likely to require rework as our business needs evolve. In other words, an ounce of prevention is worth a pound of cure.

By the end of this chapter, you will have done the following:

- Evaluated an organization's current-state IAM capabilities to inform the administrative model

- Evaluated the business structure, user attribute schema, and business requirements to inform the administrative model

- Determined the optimal **AWS Organizations** and account structure based upon the business requirements

Technical requirements

To get the most out of this chapter, you will need an AWS account.

Evaluating the organization's current IAM capabilities

Our objective over the next few chapters is to look for ways to link an organization's existing identity management infrastructure and the organization itself to AWS. More specifically, we want every administrator to have access to the backplane of the AWS account or accounts where appropriate, and for these existing user identities to become available to applications hosted on AWS. This means we will need to connect an existing org's IAM infrastructure to AWS and apply the appropriate provisioning, governance, and authorization models to ensure that appropriate access is granted. As we just completed a review of the AWS identity services, next we must look at our organization's IAM capabilities.

First, we must take an inventory of the current identity management landscape, capabilities, and maturity for the organization as that will help inform our administrative model. In order to make these examples comparable to scenarios found in real enterprise environments, we have designed an organization and configured enterprise-grade IAM capabilities to pair with our AWS environments. Let's start by taking a look at our organization's enterprise architecture and capabilities.

For the use cases that we will be exploring through the remaining chapters of this book, we will use a fictional company that we've predictably named *Redbeard Identity*. This organization has its own account management, identity provider, directory service, and strong authentication capabilities.

The Redbeard Identity organization wants to continue to use its own identity infrastructure as much as possible as part of its AWS deployment. However, it is fine with individual AWS accounts being bound by authorization rules managed within the AWS platform. Since we need to both address administrative authentication into AWS backplanes and provide our standard user identity information to apps deployed within AWS while still referring to the existing Redbeard Identity IDP and directory as the authoritative source of identity information, we recommend designing an AWS Organization to manage the accounts.

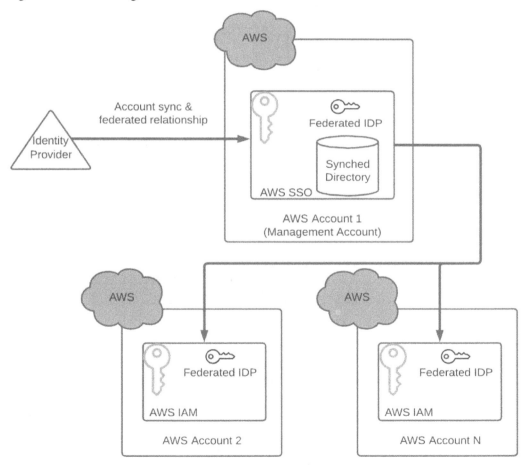

Figure 8.1 – AWS Organization and an external IDP

This will allow AWS SSO to manage administrative authentication into each account's backplane using the existing IDP for authentication, and also load the accounts into the local AWS SSO user store.

Next, we'll take a look at the business units and user accounts for guidance on how to design the AWS Organization.

Evaluating the business structure and account schema

As part of this exercise and others in the following chapters, we will be creating several user accounts inside a directory that we can use with various AWS services. Let's take a look at a sample account and its available attributes:

Attribute	Value
Login	redbeardidentity+ceo@gmail.com
firstName	Redbeard
lastName	Identity
title	CEO
displayName	Redbeard Identity
userType	Staff
employeeNumber	S17001
costCenter	10001
organization	Business Operations
division	Global Operations
department	Office of the CEO
managerId	redbeardidentity+ceo@gmail.com
Manager	Redbeard Identity

Table 8.1 – One user record from the organization

We've built this organization to include several users with diverse attribute values in order to set up several example scenarios that we are likely to see in an enterprise use case. To help us stay focused on how we are solving identity challenges on AWS rather than concerning ourselves with needing to remember names, each employee record is named for their job title. Furthermore, each account will track to a specific AWS use case or environment. We will dive into the details of the AWS organizational design in the next section. For now, let's look at all the accounts inside of our directory and a few key attributes that will help us determine how we can build a durable administrative model:

displayName	title	costCenter	organization	department	userType
Redbeard Identity	CEO	10001	Business operations	Office of the CEO	Staff
Sales Prod	Sales Associate	20001	Sales	Large Accounts	Staff
Sales Dev	Sales Engineer	20002	Sales	Large Accounts	Staff
Iam Prod	IAM Engineer	30001	Information Technology	Identity Operations	Staff
Iam Dev	IAM Developer	30002	Information Technology	Identity Development	Staff
Network Prod	Network Operations	30011	Information Technology	Network Operations	Staff
Network Dev	Network Developer	30012	Information Technology	Network Development	Staff
Cloud Prod	Cloud Engineer	30013	Information Technology	Cloud Operations	Staff
Cloud Dev	Cloud Developer	30014	Information Technology	Cloud Development	Staff
Admin Assistant	Administrative Assistant	10011	Business Operations	Business Administration	Staff
Summer Intern	Intern	30002	Information Technology	Identity Development	Contractor

Table 8.2 – The employees we will use with our examples

> **Tip**
> The CSV we used to populate our directory with user records is available here: `https://github.com/jonlehtinen/ImplementingAWSIdentity/blob/main/RedbeardIdentity_csv_template.csv`.

By looking at the example accounts and the full record from *Tables 8.1* and *8.2*, we can see a few patterns emerging that can help us plan our AWS integration. First, there appears to be a distinction between development and operational roles across both the *Sales* and *Information Technology* orgs. Next, the `costCenter` attribute also makes a distinction between those who are titled *Engineer* and those who are titled *Developer*, in addition to those two roles reporting to distinct departments under the broader *Information Technology* organization. This suggests a need for distinct budget reconciliation across those broader business functions, which may mean those two departments may require separate AWS accounts.

Though we can get valuable insights by looking at the existing IAM capabilities and account schema, at the end of the day we are implementing an AWS identity administration strategy in order to fulfill a business need. As such, any implementation includes partnering with the business stakeholders to understand the business goals and objectives of the program and folding those requirements into the technical capabilities we have discovered. For our purposes, the business objectives for Redbeard Identity are to maintain separate accounts for each department, maintain separation of duties between the development functions and the operations functions, and ensure that former employees cannot access any AWS resources once they have been terminated.

With this information in hand, let's begin drafting our AWS Organizations structure.

Designing the AWS organizational structure

Now that we have ascertained our organization's IAM capabilities, its business requirements for AWS integration, and the account schema, we can begin to lay the groundwork for how we will manage our organization's AWS accounts. While small organizations may be able to address their cloud workloads within a single account, enterprise-grade organizations often need to have additional regulatory and compliance requirements that demand additional segmentation between business units, job functions, and workloads. A well-planned multi-account structure will provide these benefits without increasing the administrative overhead.

Mapping business functions to OUs

We will do this through an AWS organization, OUs, and organizational SCPs. Before we begin the work of configuring all these things in the **Management Console**, it will be helpful to first come up with and document our plan for the organizational hierarchy. First is our management organization. This organization will be used only for managing the organization, OUs, and accounts, and providing a single place for consolidated billing. This account will be at the root of our organization's hierarchy:

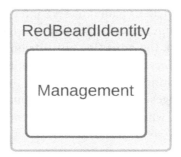

Figure 8.2 – The organization with management account

Next, we will create the OUs. It may be tempting to build an organizational structure that mimics that of the business, but that could introduce unnecessary administrative complexity when applying authorization policy. Consider the diversity of roles, business functions, and separation of duty concerns that are found in a standard business unit. Now imagine writing a corresponding authorization policy where those diverse requirements are scattered up and down the OU hierarchy.

Instead, it is best to build our OUs based on common business functions or controls that apply to a wide swath of business functions. An example of this is centralized services that service the entire business, such as IT, Security, or Audit. Let's create an OU for one such shared service that we saw when we looked at the accounts in directory, the **Information Technology** organization:

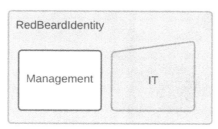

Figure 8.3 – Adding the IT OU

We could begin creating accounts for each of the IT services within the IT OU if our organization has no policy distinctions between their non-production and production workloads. Let's follow separation of duties guidelines and configure nested Production and Non-Production OUs beneath our IT OU. By doing this, all of our IT policies can be applied and managed through an organizational SCP applied exclusively on the IT OU. Further policy distinctions between the production and non-production accounts within the IT OU are then applied and managed separately upon the production and non-production OU respectively:

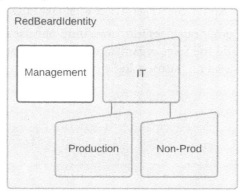

Figure 8.4 – Nested production and non-production OUs

From there, we can create individual accounts for each of the departments within IT. Each department's account will be independent of the other services, but still governed under the organizational SCP that constrains its environmental OU and the broader IT OU:

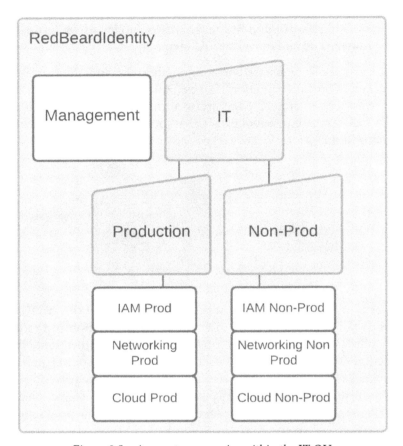

Figure 8.5 – Accounts per service within the IT OU

We repeat this process with other business functions until our use case is satisfied. For simplicity's sake, we'll only add the Sales organization for the purpose of demonstrating how we can restrict access based on business logic:

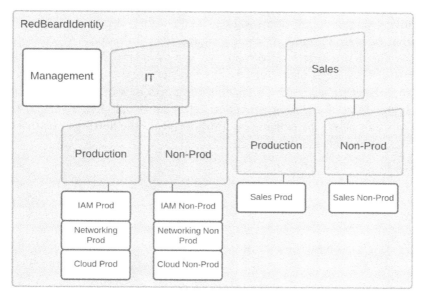

Figure 8.6 – Additional OUs and their accounts

Besides OUs that align to business functions, we can also consider other AWS account use cases where a distinct policy would be required. For example, consider the life cycle of the AWS accounts themselves. A sandbox OU could allow anyone to safely learn and experiment under their own accounts. Similarly, when an account is ready to be retired, an OU that applies an organizational SCP that disables all functionality could be used as a pre-deletion buffer to ensure resources remain available, but access is removed:

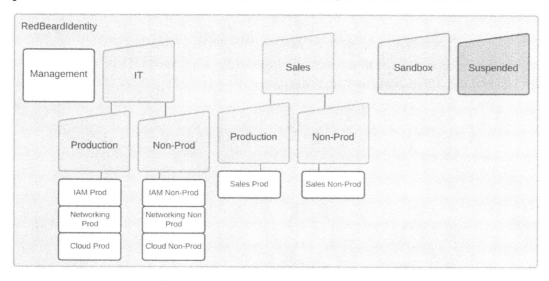

Figure 8.7 – Our organization's OU hierarchy

Our comparatively simple organization will be sufficient for our purposes. Now that we understand what we want to build, we can configure this within our AWS console. As we had already created an organization from our work in *Chapter 6, Introduction to AWS Organizations and AWS Single Sign-On*, we can simply create the additional OUs under that existing root. For detailed instructions on setting up a new organization, please see that chapter. If you do not have an organization defined or would like to start with a fresh organization, you can delete the existing organization from the **Settings** menu within the AWS Organizations service and begin fresh:

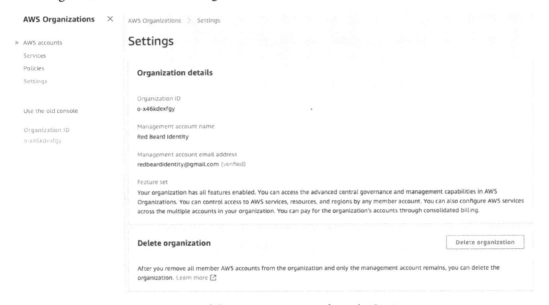

Figure 8.8 – We can delete our organization from the Settings menu

Once we are satisfied with the organization, we can add our top-level OUs by selecting the Root OU and selecting **Create new** from the menu:

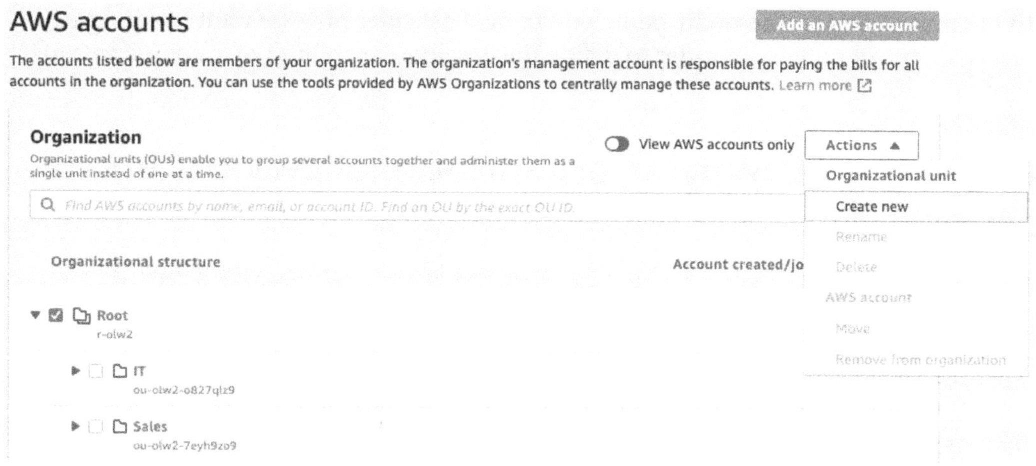

AWS accounts

Add an AWS account

The accounts listed below are members of your organization. The organization's management account is responsible for paying the bills for all accounts in the organization. You can use the tools provided by AWS Organizations to centrally manage these accounts. Learn more

Organization

View AWS accounts only | Actions ▲

Organizational units (OUs) enable you to group several accounts together and administer them as a single unit instead of one at a time.

Organizational unit

Q *Find AWS accounts by name, email, or account ID. Find an OU by the exact OU ID.*

Create new

Rename

Organizational structure

Account created/jo | Delete

AWS account

▼ ☑ 🗁 Root

Move

r-olw2

Remove from organization

▶ ☐ 🗁 IT

ou-olw2-o827qlz9

▶ ☐ 🗁 Sales

ou-olw2-7eyh9zo9

Figure 8.9 – Adding OUs to the Root OU

We repeat until our organization is as we wish:

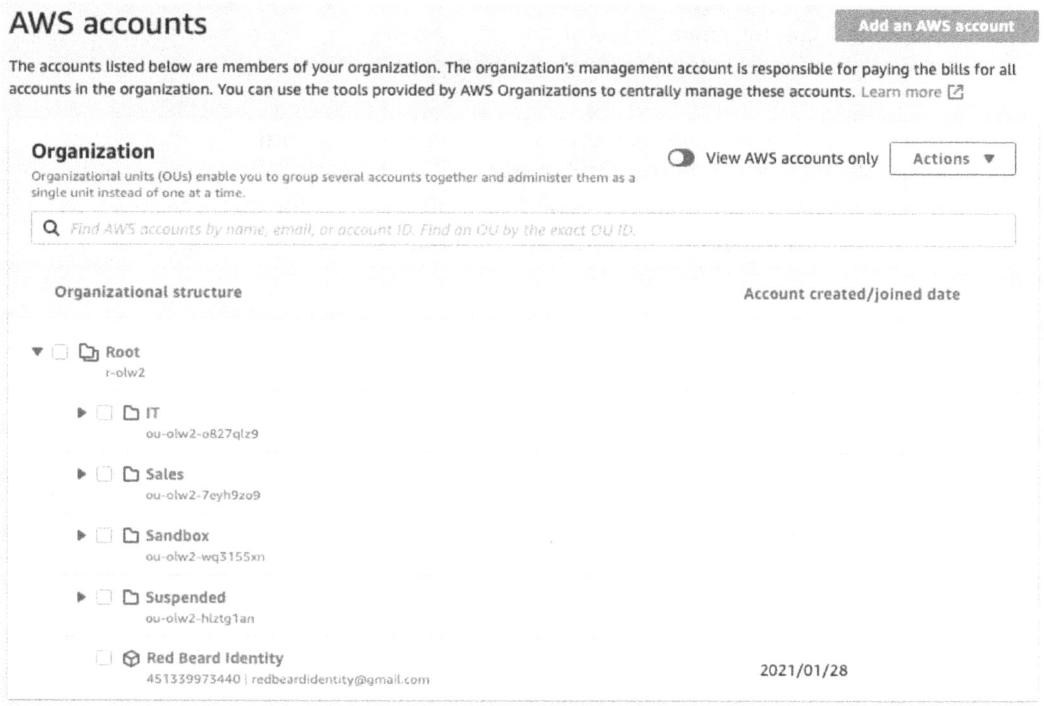

AWS accounts

Add an AWS account

The accounts listed below are members of your organization. The organization's management account is responsible for paying the bills for all accounts in the organization. You can use the tools provided by AWS Organizations to centrally manage these accounts. Learn more

Organization

View AWS accounts only | Actions ▼

Organizational units (OUs) enable you to group several accounts together and administer them as a single unit instead of one at a time.

Q *Find AWS accounts by name, email, or account ID. Find an OU by the exact OU ID.*

Organizational structure | Account created/joined date

▼ ☐ 🗁 Root

r-olw2

▶ ☐ 🗁 IT

ou-olw2-o827qlz9

▶ ☐ 🗁 Sales

ou-olw2-7eyh9zo9

▶ ☐ 🗁 Sandbox

ou-olw2-wq3155xn

▶ ☐ 🗁 Suspended

ou-olw2-hlztg1an

☐ 🔊 Red Beard Identity

451339973440 | redbeardidentity@gmail.com | 2021/01/28

Figure 8.10 – The completed organization hierarchy

With the organizational hierarchy built, we can now consider how we want to configure the service control policies in order to define the baseline capabilities of accounts within each of these OUs.

Designing and applying organizational service control policies

When we went through the exercise of designing our OUs, we classified each one by similarity of business functions. This was because common business functions will typically need to be governed by similar policies. By writing a single policy that we can apply at the OU, we simplify our policy administration for all the accounts within our organization. Now we'll write and apply those policies that will apply at the OU level for each of our organization's OUs.

> TIP
>
> **Service Control Policies (SCPs)** will vary wildly from organization to organization based on factors such as the local regulatory environment, the organization's cost tolerance, technical maturity, and so on. These are meant to be examples and not a comprehensive list of best-practice SCPs.

We will write each of these policies using the **Create new service control policy** wizard found under the **Policies** menu in the AWS Organizations service and save them with a descriptive name. We will then have each of these policies available for us to attach to any given OU by selecting that OU from the AWS Organizations service and opening the **Policy** tab.

First, we'll create a policy that will functionally disable the root account user for each account. For similar reasons as to why it is recommended to never use the root account outside of creating an IAM user account, we do this because the root account user is outside of the governance model for AWS Organizations. The root user account represents a risk for account takeover and even detachment from the organization. A service control policy blocking the root user from performing any functions within its own account greatly diminishes the blast radius in the event it is compromised.

We'll name this policy `Deny_RootUser_Actions`. From the **Create new service control policy** wizard we can either edit the existing statement or clear it out and drop in our own JSON:

```
{
    "Version": "2012-10-17",
    "Statement": {
```

```
    "Sid": "DenyRootUserActions",
    "Effect": "Deny",
    "Action": "*",
    "Resource": "*",
    "Condition": {
      "StringLike": { "aws:PrincipalArn": "arn:aws:iam::*:root"
}
    }
  }
}
```

Note the `Condition` statement that looks for the root account in the principal ARN. That condition is what prevents this SCP from disabling the entire account. We click the **Create policy** button and the wizard validates that our policy is valid, and assuming it is, our new service control policy is available to attach to our OUs:

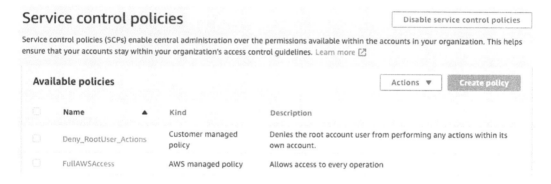

Figure 8.11 – The new SCP is now available for use

While we are addressing the simple security policies, let's also create the policy that we will use to disable all access and activity on the accounts that get placed into our *Suspended* OU. The objective of this policy is to ensure that all activity, keys, and processes are stopped within that account, but that the account and its contents remain available to the business in the event that the account needs to be reinstated. This will be a very simple policy:

```
{
  "Version": "2012-10-17",
  "Statement": [
    {
      "Sid": "DisableAccount",
      "Effect": "Deny",
```

```
        "Action": "*",
        "Resource": "*"
    }
   ]
}
```

The next policy we'll make will make sure that an account can't simply bypass whatever policies we apply to them by leaving the organization. This policy is the first one we have made where a specific service and action are referenced:

```
{
  "Version": "2012-10-17",
  "Statement":
    {
      "Sid": "DontForgetYoureHereForever",
      "Effect": "Deny",
      "Action": "organizations:LeaveOrganization",
      "Resource": "*"
    }
}
```

We now have four SCPs defined and available for use within our organization's management account. These SCPs include the three we just created, and the default **FullAWSAccess** that is created by default when we created the organization. As we have already seen how to attach policies using the Management Console and CLI in *Chapter 6, Introduction to AWS Organizations and AWS Single Sign-On*, we won't repeat that process for each OU and policy.

Let's map out which policies we would like to apply to which OU:

	Root	IT	Sales	Sandbox	Suspended
FullAWSAccess	X	X (inherited from Root)	X (inherited from Root)	X (inherited from Root)	X (inherited from Root)
Deny_RootUser_Actions		X	X	X	X
Full_Disablement					X
No_Leaving_Organization		X	X	X	X

Table 8.3 – OU SCPs and where they apply

Note that we don't have a choice but to allow the FullAWSAccess policy as it applies to the Root OU, and every OU and account will inherit that policy by virtue of being beneath it in the organizational hierarchy.

Now we apply the policies to our OUs using either the administrative console or CLI. Once this is done, we can validate that they are attached by looking at each OU within the AWS Organizations service, or by using the following command in CLI:

```
aws organizations list-policies-for-target --filter SERVICE_
CONTROL_POLICY --target-id <organization target id value>
```

Now that we have our organization configured, we are ready to configure AWS SSO to connect to our IDP so we can begin solving for our specific administrator and standard user authentication and authorization use cases. We will build that connection and onboard our administrative accounts in the next chapter.

Summary

In this chapter, we took a few moments to plan out our administrative model for an enterprise AWS deployment. We did this so we could accommodate the business requirements with a full understanding of the organization's IAM maturity and current-state capabilities, which will help us design administrative patterns that will be supportive in the long term. Once we had thought through the use cases, requirements, and capabilities, we designed and applied some high-level OU SCPs that will govern all the AWS accounts that we will be administrating moving forward.

We will build upon this foundational work in the next two chapters. First, in the next chapter, we will connect Redbeard Identity's external IDP and provision our administrative users into AWS SSO. Then, in *Chapter 11*, *Bringing Your Users into AWS*, we will address authentication and authorization use cases for those users into the AWS backplane.

Questions

1. Which of these did we not consider when designing our administrative model?

    a. The number of employees in the organization

    b. Current-state IAM capability

    c. Business objectives

    d. Business function alignment

2. Why did our *Suspended* OU have the `FullAWSAccess` SCP when we did not attach it to that OU?

Further reading

*   AWS best practices for organizations: `https://aws.amazon.com/organizations/getting-started/best-practices/`

*   Best practices for OUs with organizations: `https://aws.amazon.com/blogs/mt/best-practices-for-organizational-units-with-aws-organizations/?org_product_gs_bp_OUBlog`

9
Bringing Your Admins into the AWS Administrative Backplane

In the previous chapter, we took the time to put some thought into how we would like to administrate our AWS environments. Now that we know how our AWS accounts will be structured, and the criteria for separating each environment, we can begin bridging our example organization's IAM infrastructure to AWS.

The next two chapters will focus on implementing administrative access to the AWS backplane. First, we will address getting our organization's identity information into AWS SSO. This will entail connecting our AWS SSO service to our organization's identity provider. With our organization's IDP set as the external identity provider, we will then look at a couple of strategies for account management within AWS SSO when using an external IDP. The first will be manual **account linking**. The second will be automated provisioning and life cycle management through the **System for Cross-domain Identity Management protocol**, or **SCIM**. Finally, we will define the conditions for provisioning administrative users into the backplanes for their appropriate AWS accounts using groups.

In this chapter, we will be covering the following topics:

- Defining our organization's identity source
- Provisioning administrative accounts in AWS – account linking
- Provisioning administrative accounts in AWS – SCIM provisioning

Technical requirements

To get the most out of this chapter, you will need the following:

- An AWS account
- An SAML2 and SCIM compliant identity provider, such as Okta Identity Cloud, PingOne, or Azure Active Directory
- A populated user directory to act as the user store for that identity provider

Defining our organization's identity source

In the previous chapter, we were introduced to the Redbeard Identity organization. Based upon that organization's business requirements, organizational structure, current identity capabilities, and account attribute schema, we designed an administrative model for our AWS organization:

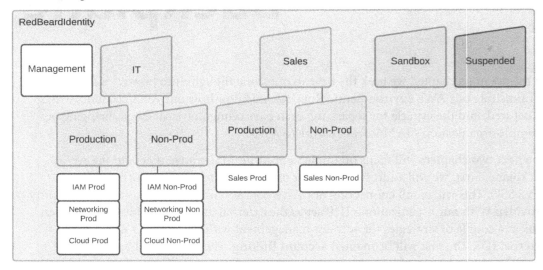

Figure 9.1 – The Redbeard Identity AWS organization admin model

Now that we know how we want AWS accounts to be created and managed, it is time to connect the existing identity provider to the AWS SSO service so that we can bring our administrators into their respective AWS administrative backplanes. Though we won't dive into the details of administrative user authentication and authorization until *Chapter 10, Administrative Single Sign-On to the AWS Backplane,* connecting an external identity provider to the AWS SSO service is a prerequisite for either a just-in-time provisioning and account linking model or for using SCIM.

> **Tip**
>
> The examples used in this book will be shown using the **Okta Identity Cloud**. Whereas the configuration within AWS will be the same regardless the IDP technology used within an organization, be advised that the requirements, steps, and options available for user account administration from within the IDP may be different, depending on your IDP platform.

The last time we worked within AWS SSO, we configured AWS SSO itself to be its own identity source. AWS SSO became our own SAML2 IDP, and we could provision the accounts that we created and managed within its own directory. This way, AWS SSO could act as the IDP for managed AWS accounts within our AWS organization. It could also be the IDP for our SaaS apps and any other SAML2 compliant **service providers (SPs)** where we wanted to use those identities:

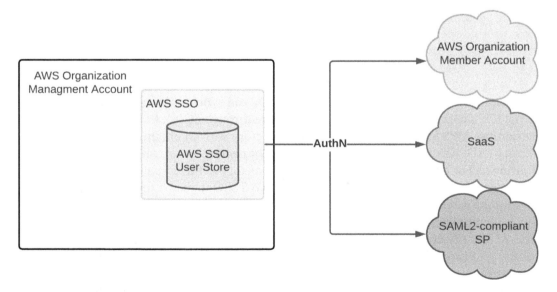

Figure 9.2 – AWS SSO as an IDP and user store with users created and managed within AWS SSO

Our use case is different this time. While AWS SSO remains the IDP and user store for administrative accounts for member accounts within our AWS organization, AWS SSO is no longer the authoritative source for that identity information – at least as far as our example Redbeard Identity company is concerned. AWS SSO will issue authentication tokens to each connected AWS account, but only upon receiving a successful authentication token from an external IDP:

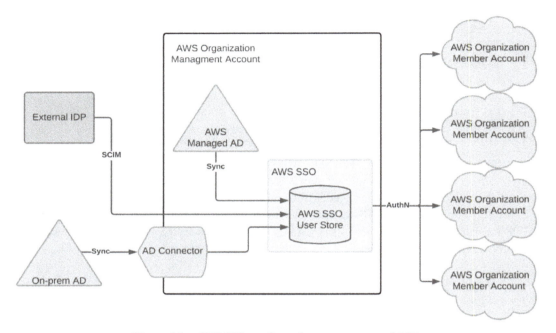

Figure 9.3 – AWS SSO configured to use an external IDP

The preceding diagram shows how the external IDP can synchronize accounts from the external IDP's user store into AWS SSO's user store through SCIM. In addition to SCIM provisioning, AWS SSO can populate its local directory with an organization's Active Directory (AD) accounts through either an **AD Connector**, or through an **AWS Managed AD** instance. For organizations that use Active Directory as the user store for their identity provider, both these methods will allow AWS SSO to match the authenticated user from the external IDP's SAML2 assertion to its corresponding record in AWS SSO. It is not a requirement that accounts be directly synchronized from the IDP's user store; Redbeard Identity does not use Active Directory, but it could still match accounts that are manually provisioned with matching values from its IDP's directory. That said, doing so would introduce significant administrative overhead. We will investigate the advantages of using a directory synchronization strategy such as SCIM or an AD Connector over manual provisioning as we move through this chapter.

Connecting our IDP to AWS SSO

Let's connect our IDP to AWS SSO:

1. From the AWS Management Console, we go to the AWS SSO service and select **Settings**.

2. Assuming our configuration remains the same from the activities we did in *Chapter 6, Introduction to AWS Organizations and AWS Single Sign-On*, we may already have values defined for our **Identity source** similar to what can be seen in the following screenshot:

AWS SSO > Settings

Settings

ARN ❶ arn:aws:sso:::instance/ssoins-7223ec67c031315d 🗐

Identity source

Your identity source is where you administer your users and groups, and where AWS SSO authenticates your users. You can choose between AWS SSO, SAML 2.0-compatible identity provider (IdP), or Active Directory (AD). Learn more

Identity source	AWS SSO | Change
Authentication	AWS SSO
Provisioning ❶	AWS SSO
Identity store ID ❶	d-9067650dfa 🗐
Attributes for access control ❶	Enabled | View details

Figure 9.4 – Identity source settings in AWS SSO

3. The preexisting configuration will not impact anything we need to do; however, if we want to start fresh, we can click the **Delete AWS SSO configuration** button at the bottom of the page to clear out all the settings. Regardless, once the settings have been cleared, we will return to this page and click **Change** next to **Identity source**.

4. This takes us to a wizard where we can define where our identities are sourced. As we intend to use an external IDP for our Redbeard Identity example, we will select the **radial** button next to **External Identity Provider**.

5. This populates some URLs required to configure the SAML2 relationship with the external IDP. It also provides a link to download the AWS SSO SAML metadata:

Change identity source

①————②

Choose identity source   Review

Choose where your identities are sourced

Your identity source is the place where you administer and authenticate identities. You use AWS SSO to manage permissions for identities from your identity source to access AWS accounts, roles, and applications. Learn more

AWS SSO
You will administer all users, groups, credentials, and multi-factor authentication assignments in AWS SSO. Users sign in through the AWS SSO user portal.

Active Directory
You will administer all users, groups, and credentials in AWS Managed Microsoft AD, or you can connect AWS SSO to your existing Active Directory using AWS Managed Microsoft AD or AD Connector. Users sign in through the AWS user portal.

● **External identity provider**
You will administer all users, groups, credentials, and multi-factor authentication in an external identity provider (IdP). Users sign in through your IdP sign-in page to access the AWS SSO user portal, assigned accounts, roles, and applications.

Configure external identity provider

AWS SSO works as a SAML 2.0 compliant service provider to your external identity provider (IdP). To configure your IdP as your AWS SSO identity source, you must establish a SAML trust relationship by exchanging meta data between your IdP and AWS SSO. While AWS SSO will use your IdP to authenticate users, the users must first be provisioned into AWS SSO before you can assign permissions to AWS accounts and resources. You can either provision users manually from the Users page, or by using the automatic provisioning option in the Settings page after you complete this wizard. Learn more

Service provider metadata

Your identity provider (IdP) requires the following AWS SSO certificate and metadata details to trust AWS SSO as a service provider. You may copy and paste, or type this information into your IdP's service provider configuration interface, or you may download the AWS SSO metadata file and upload it into your IdP.

AWS SSO SAML metadata	Download metadata file
AWS SSO Sign-in URL	https://redbeardidentity.awsapps.com/start
AWS SSO ACS URL	https://us-east-1.signin.aws.amazon.com/platform/saml/acs
AWS SSO Issuer URL	https://us-east-1.signin.aws.amazon.com/platform/saml/d-0

Hide Individual metadata values

Figure 9.5 – Configuration values for AWS SSO to become a service provider in our IDP

6. From our IDP, we will build the connection to our AWS Single-Sign On service by populating that connection information. We also need to define our **Application username format**. AWS SSO uses **Email** as its unique identifier, so we must set that as the unique identifier to be sent by the IDP. This process will look different depending on the IDP we use, but the basic details will be the same across each one:

Advanced Sign-on Settings

These fields may be required for a AWS Single Sign-on proprietary sign-on option or general setting.

AWS SSO ACS URL

    https://us-east-1.signin.aws.amazon.com/platform/saml

    Enter your AWS SSO ACS URL. Refer to the Setup
    Instructions above to obtain this value.

AWS SSO issuer URL

    https://us-east-1.signin.aws.amazon.com/platform/saml

    Enter your AWS SSO issuer URL. Refer to the Setup
    Instructions above to obtain this value.

Credentials Details

Application username format

    Email   ⌄

Update application username on

    Create and update   ⌄

Password reveal

    ☐ Allow users to securely see their password
       (Recommended)

    ┌───┐
    │ ❶   Password reveal is disabled, since this app is │
    │   using SAML with no password.   │
    └───┘

                                              Save

Figure 9.6 – The IDP side configuration for AWS SSO

7. Once we save the connection information inside our IDP, we can download our
    IDP's metadata file. This metadata file provides the IDP's endpoints, username
    identifier format information, and most importantly, the public signing key that
    AWS SSO can use to validate that an incoming assertion is genuinely from the IDP.
    Once we have the IDP's metadata, we can go back to AWS SSO and upload it in the
    Change identity source wizard. Alternatively, we could also populate the values
    directly in AWS SSO if our IDP did not support metadata:

Identity provider metadata

──

AWS requires specific metadata provided by your identity provider (IdP) to establish trust. You may copy and paste from your IdP, type the metadata in
manually, or upload a metadata exchange file that you download from your IdP.

        IdP SAML metadata*   [RBIIDPmetadata.xml]   [Browse...]

                    If you don't have a metadata file, you can manually type your metadata values

Figure 9.7 – Uploading the external IDP metadata into AWS SSO

8. Next, we are prompted to confirm that we want to change our identity source for AWS SSO. This page details several warnings about the additional responsibilities we will take on by doing so. Type ACCEPT and move on:

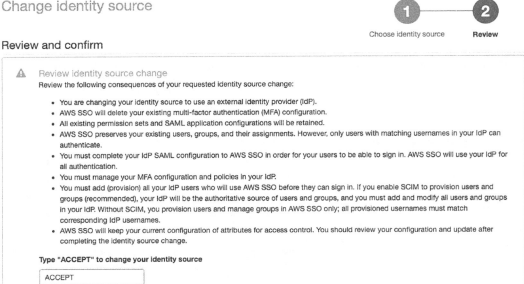

Figure 9.8 – Confirming the change of our identity source

9. AWS SSO processes the information and confirms the change:

Complete

We have successfully configured your AWS SSO

Return to settings

✅ Creating external identity provider configuration

✅ Enabling external identity provider

Figure 9.9 – New identity source configured

10. With the IDP configured, our AWS organization will now use our external IDP for user authentication. However, we must still manage the accounts that will be used with the AWS accounts:

AWS SSO > Settings

Settings

ARN ℹ arn:aws:sso:::instance/ssoins-7223ec67c031315d 🗐

Identity source

Your identity source is where you administer your users and groups, and where AWS SSO authenticates your users. You can choose between AWS SSO, SAML 2.0-compatible identity provider (IdP), or Active Directory (AD). Learn more

Identity source	External Identity Provider | Change
Authentication	SAML 2.0 | View details
Provisioning ℹ	Manual | Enable automatic provisioning
Identity store ID ℹ	d-9067650dfa 🗐
Attributes for access control ℹ	Enabled | View details

Figure 9.10 – Our provisioning option is currently set to manual

Now that we have an external IDP configured, we need to populate AWS SSO's user store with our administrative accounts. First, we will manually provision some accounts to demonstrate account linking functionality.

Provisioning administrative accounts in AWS – account linking

Account linking is when a service provider correlates a locally managed account with the subject of an external IDP's federated token. The local account may get created in a just-in-time fashion from the information contained within the IDP's authentication token, or the account may have been created earlier and was correlated by matching on a unique identifier, such as an email address. Arguably, when both AWS SSO and the IDP use Active Directory as their account stores, but the IDP itself does not manage the accounts, this is also an example of account linking. Though all the data ultimately stems from the same Active Directory instance, there is no explicit link between the account, as presented by the IDP, and the account stored within the AWS user store.

However, our example company is not using Active Directory. As such, we need to manually create some matching user records inside AWS SSO for our administrators. The AWS CLI does not have a function for bulk importing accounts, so administrating accounts in this fashion can be tedious. Let's create an account for our `Iam Dev` user:

1. From the **AWS SSO** service, under the **Users** section, we start by hitting the **Add user** button. This takes us to a wizard where we can enter various attribute values for our user:

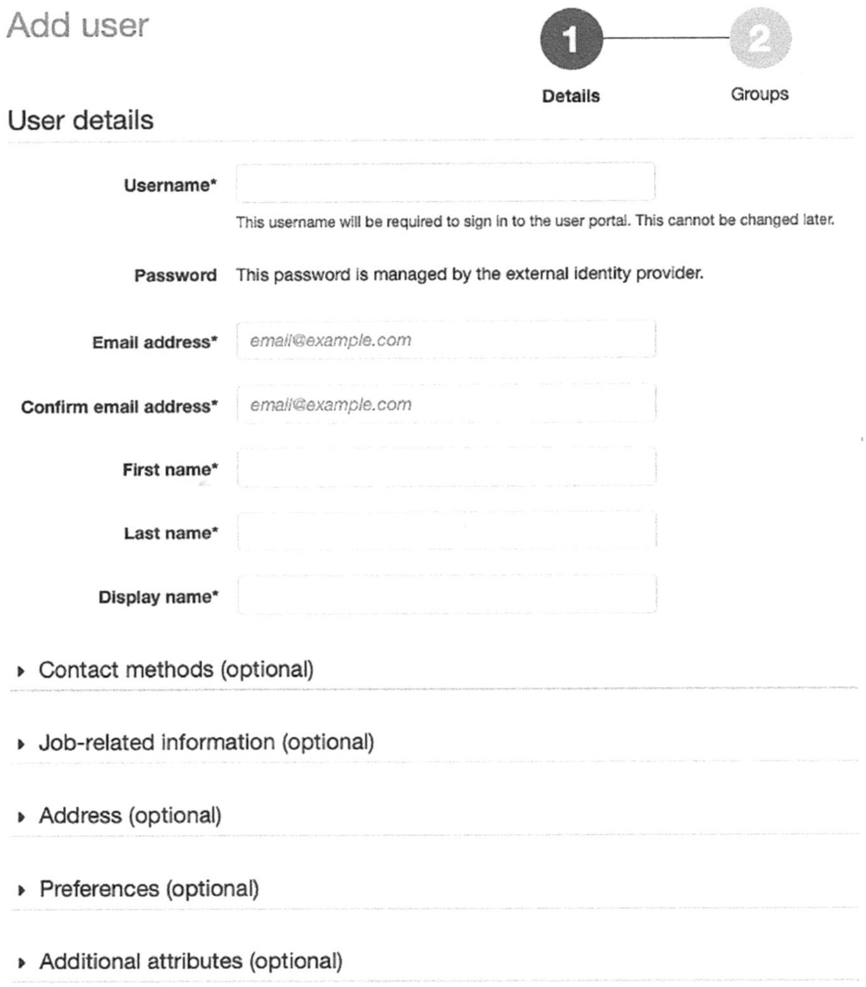

Add user

	1	2
	Details	Groups

User details

Username*	
	This username will be required to sign in to the user portal. This cannot be changed later.
Password	This password is managed by the external identity provider.
Email address*	email@example.com
Confirm email address*	email@example.com
First name*	
Last name*	
Display name*	

▸ Contact methods (optional)

▸ Job-related information (optional)

▸ Address (optional)

▸ Preferences (optional)

▸ Additional attributes (optional)

Figure 9.11 – The AWS SSO Add user wizard

2. There are several sections available for us to populate information. However, only the first section is required to make a fully functional AWS SSO user record. We can reference the values inside our external IDP's directory for `Iam Dev` and populate the required attributes:

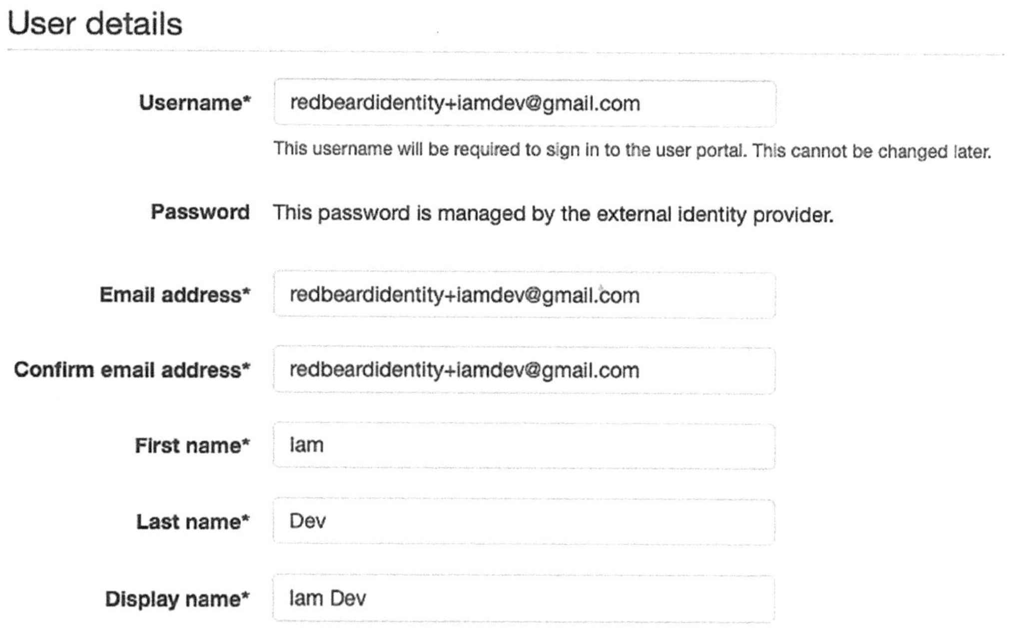

User details

Username*	redbeardidentity+iamdev@gmail.com
	This username will be required to sign in to the user portal. This cannot be changed later.
Password	This password is managed by the external identity provider.
Email address*	redbeardidentity+iamdev@gmail.com
Confirm email address*	redbeardidentity+iamdev@gmail.com
First name*	Iam
Last name*	Dev
Display name*	Iam Dev

Figure 9.12 – Populating essential attributes for our user record

3. While it is not required that we fully populate the record with the remaining attributes from our external IDP's user store, we will go ahead and do it in the interest of keeping our data consistent across data stores. Strictly speaking, we could populate these attributes with distinct values compared to those found inside our external IDP; so long as the username value and any other attributes that are sent from the external IDP are the same, the remaining attributes do not need to match. That said, we do not have any need to change our attribute values, so we will keep them consistent and fill in the rest of the record. Once this is completed, we will be ready to proceed.

4. Next, we are asked to add the user to some groups. Our AWS SSO has not been configured with any groups, but we are using groups to manage access to AWS SSO within our external IDP. We will continue that pattern within AWS SSO, so we will click **Create group**:

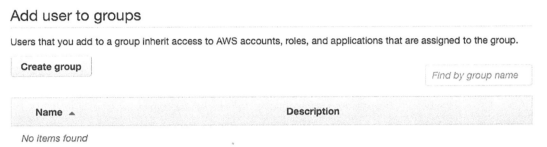

Figure 9.13 – The Create group option in AWS SSO

5. The Iam Dev user is a member of two groups inside our external IDP. The first is AWS_IT_IAM_Dev, while the second is AWS_Sandbox. We must enter the group's name and a description to create the first group and then repeat this for the second:

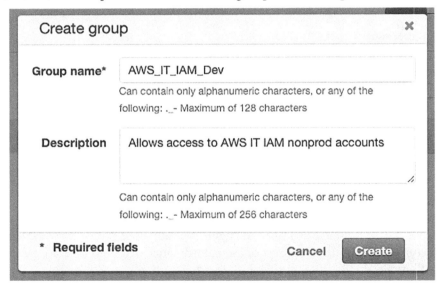

Figure 9.14 – Creating groups in AWS SSO

6. Now that we have two groups available to select, and they are the two groups we want Iam Dev to be a member of, we can tick the boxes shown in the following screenshot and click **Create user**:

Add user to groups

Users that you add to a group inherit access to AWS accounts, roles, and applications that are assigned to the group.

Name ▲	Description
☑ AWS_IT_IAM_Dev	Allows access to AWS IT IAM nonprod accounts
☑ AWS_Sandbox	Allows access to AWS Sandbox

Figure 9.15 – Adding the groups to Iam Dev

7. Once processing is complete, we can return to our **Users** section in AWS SSO. There, we will find our **Iam Dev** user record, identical to how it exists in our external IDP. We can also verify that it is a member of the two groups that we created:

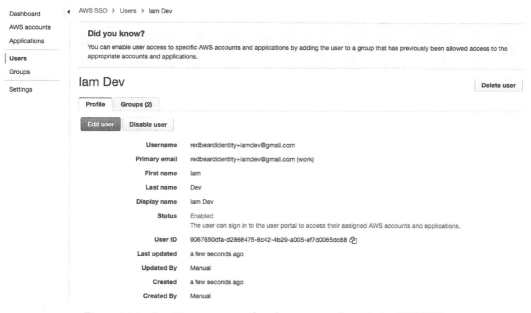

Figure 9.16 – Iam Dev user record and group membership in AWS SSO

With `Iam Dev` complete, the next step is to repeat this process for every AWS admin that we want to bring over from the Redbeard Identity IDP. This would be a tedious undertaking, though, since we only have a handful of accounts that exist within that user store. Rather, we'll focus on the better administrative method for the Redbeard Identity use case, before we begin using these accounts and groups for authentication and authorization in the next chapter.

Limitations of manual provisioning and account linking

Setting aside the tedium of manually creating those accounts, manual provisioning has other significant drawbacks. We are focusing on getting our administrators into the administrative backplane, but auditors are much more interested in evidence showing that terminated administrators have been removed from critical systems within a reasonable timeframe. Whereas an organization could build a manual process around joiner, mover, and leaver flows that ensure all compliance requirements are met, it would be costly to operate and prone to human error.

Fortunately, most enterprise deployments will not use manual provisioning. Active Directory connectors to on-premises deployments ensure that terminated accounts are also disabled in the AWS SSO user store. More modern organizations can also leverage the IDP as its provisioner through SCIM, which will ensure that any status change on a user record or group will be immediately updated to AWS SSO. Let's look at a SCIM implementation for Redbeard Identity's use case.

Provisioning administrative accounts in AWS – SCIM provisioning

System for Cross-domain Identity (SCIM) provisioning is a standards-based RESTful account provisioning service that sends account information in a standardized JSON format. When we enable automatic provisioning with SCIM, the directory objects that we specify for our IDP to synchronize in the user store for our AWS SSO service will automatically be created, updated, and deleted, in tandem with their counterparts inside the user store of our external IDP.

How SCIM works

Before we enable SCIM for our example use case, let's take a quick look at how SCIM operates:

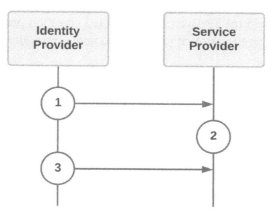

Figure 9.17 – SCIM create and update flows

The SCIM provisioning flows for creating and updating accounts are rather straightforward:

1. The IDP that acts as the authoritative source for provisioning in the service provider's user store pushes the accounts and attributes based on that service provider's predefined account schema. This is where things such as **attribute transformation** can occur to manipulate the data between the IDP and SP to accommodate the SP's specific attribute requirements.

2. With the accounts and groups now provisioned in the SP's user store, the service provider will be able to link the subject of the authentication token it receives from the IDP with up-to-date directory information from that same IDP's authoritative user store.

3. As attributes are updated or group memberships change, the IDP intermittently pushes the changes to the affected objects to maintain synchronization. This includes **account deprovisioning**.

SCIM is mostly a push-based process from the IDP to the service provider. That said, there are flows where the trigger for synchronizing a directory object is predicated on a request from the service provider. This trigger could be something such as an SP-initiated authentication request to the IDP for an account that hasn't been found yet within the SP's user store. This is a common practice, where the aim is to minimize the proliferation of identity data across providers:

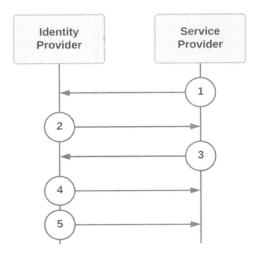

Figure 9.18 – SSO-triggered SCIM provisioning event

4. The service provider reaches out to the IDP to authenticate a user.

5. The IDP sends the authentication token to the service provider.

6. Upon not finding a corresponding account within its local user store, the SP makes a request to the IDP to provision that account in its user store.

7. The IDP responds by provisioning the account in the SP's user store.

8. Now that the account has been added to the SP's user store, the IDP will push any changes to that account into the SP's user store.

Compared to the administrative overhead of manual account administration, SCIM is a good way to ensure our existing organization's joiner, mover, and leaver flows are quickly reflected in downstream third-party applications.

Enabling automatic provisioning in AWS SSO

Let's connect the Redbeard Identity organization's external IDP to our AWS SSO service's user store using SCIM:

1. First, from the AWS Administrative Control Panel, we must go to our **AWS SSO** service and select the **Settings** menu.

2. Under **Identity source**, click the link that says **Enable automatic provisioning**. This opens the **Inbound automatic provisioning** wizard.

3. We will be presented two values: our **SCIM endpoint** and **Access token**. We will use these to configure the SCIM connection inside our IDP. Similar to other secrets issued inside AWS, this will be the one opportunity we will have to capture the access token value, so we must make sure to document it someplace safe:

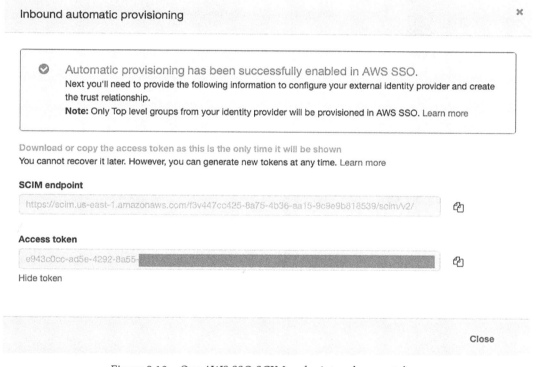

Figure 9.19 – Our AWS SSO SCIM endpoint and access token

4. Now, we will go to the AWS SSO app configuration inside our IDP. For the Redbeard Identity organization's IDP, this can be found under the **Provisioning** tab. There, we must click the button labeled **Configure API Integration**:

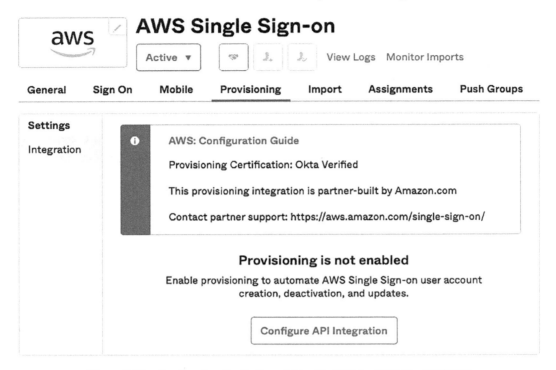

Figure 9.20 – Configuring the Redbeard Identity IDP for SCIM to AWS SSO

5. From there, we must tick the box to enable API integration, which will present us with the fields we need to populate with our SCIM endpoint and API token values. We will drop the values we got from AWS SSO into the field and test the connection. If all is well, we will see a confirmation message stating that the connection was verified successfully:

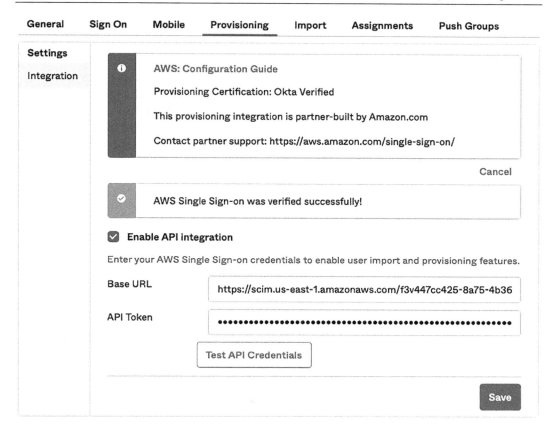

Figure 9.21 – The Redbeard Identity IDP is now connected to the AWS SSO user store via SCIM

6. Now that our external IDP and AWS SSO's user store are connected, we have some additional options inside our IDP for what operations we wish to allow, and how we want to map our IDP's attributes to the user store schema of AWS SSO. We'll click **Edit** and make sure that we are configured to create our users, update user attributes, and deactivate users inside AWS SSO when there is a change inside our IDP. Fortunately, our IDP is natively supported by AWS SSO, so a preconfigured mapping of attributes has already been configured for us. If we want to customize our IDP's user schema, or want to adjust mappings, this is where we could make those changes. Once satisfied with the settings, we can click on **Save** and move on:

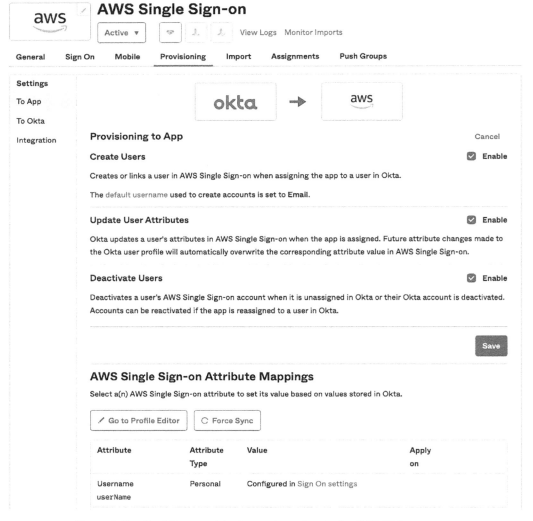

Figure 9.22 – Enabling provisioning and syncing in the AWS SSO user store

7. Now that our users have been assigned to groups inside our IDP's directory, we will assign access to AWS SSO through membership to those groups from our IDP. Whereas only a handful of users are found in each one of these groups based upon their job function, every individual user who is a member of one of those groups is now in scope for provisioning in AWS SSO's user store:

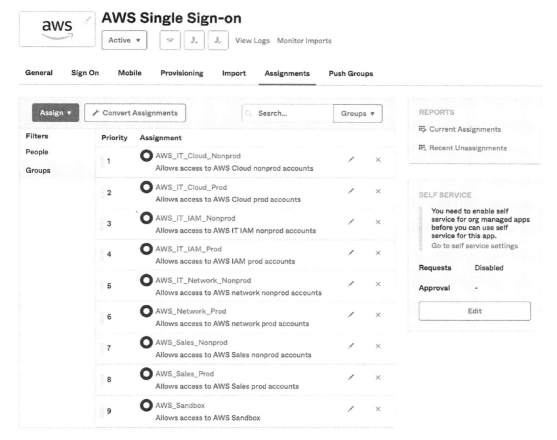

Figure 9.23 – Group-based access control used to control provisioning in AWS SSO from the IDP

8. We can confirm that this logic is working by refreshing the user list inside AWS SSO. Within seconds, our users will be provisioned:

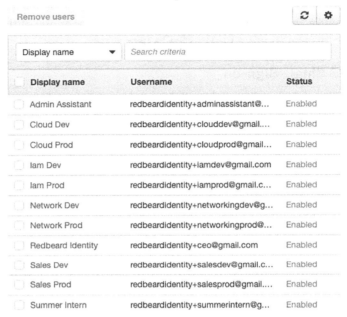

Figure 9.24 – SCIM-provisioned users appearing within the AWS SSO user store

9. We may optionally choose to also push specific groups from our IDP into AWS SSO. This will provide us with additional options for authorization, so we will do so. Similarly, those groups and their memberships will be populated within the AWS SSO directory moments later:

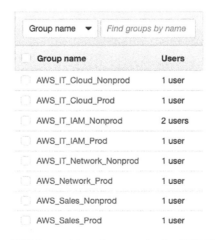

Figure 9.25 – SCIM-provisioned groups in the AWS SSO directory

With that, we can now manage AWS SSO users and groups centrally using our organization's authoritative identity system.

> **TIP**
>
> The specifics of the IDP configuration will vary, basedpon the IDP used. The Redbeard Identity IDP is built upon the Okta platform, which is available for free for developer use cases. Please review the *Further reading* section for references on using other major identity platforms.

SCIM in action

Let's observe how SCIM works with AWS SSO by creating a new user within the Redbeard Identity IDP and adding it to the **AWS_Sandbox** group. We will call this new user New User, and it will be a contractor reporting to **Sales Dev** within the Sales organization. We will bootstrap this user by updating the CSV file we used in *Chapter 8, An Ounce of Prevention – Planning Your Administrative Model*. This will quickly create a richly featured account while leaving existing ones untouched (unless we want them to be updated):

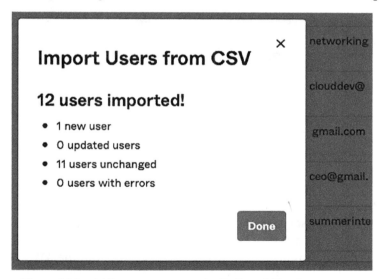

Figure 9.26 – Adding a new user to the Redbeard Identity user store

Once **New User** has activated their account, they become live in our IDP's directory, but they won't be found inside AWS SSO's user store:

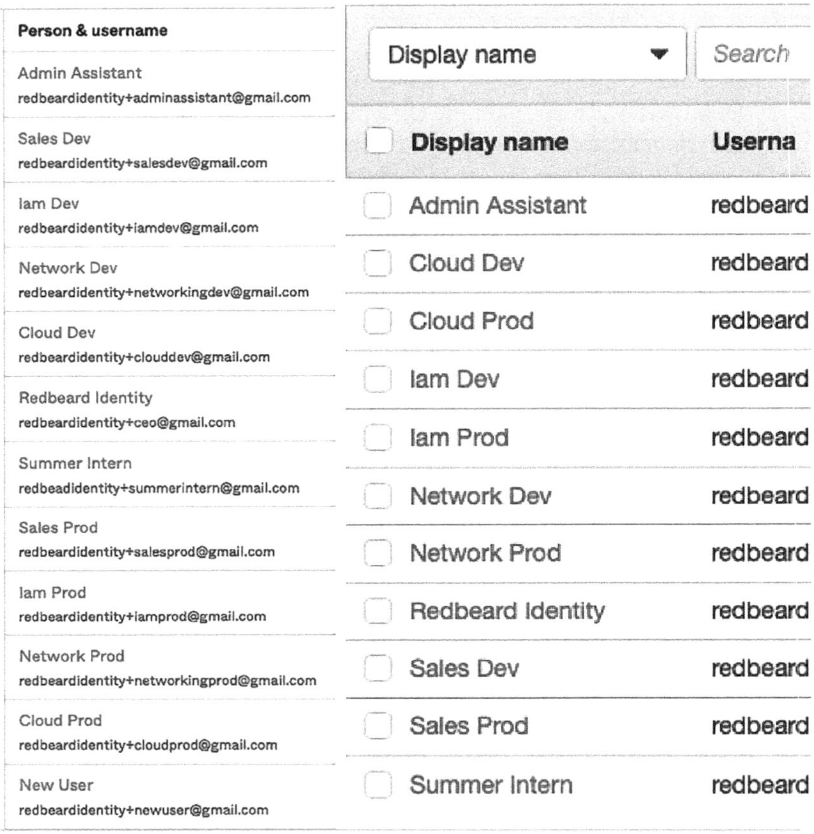

Figure 9.27 – RBI's user store on the left; AWS SSO's user store on the right

This is because the Redbeard Identity organization assigns AWS SSO access to group membership. To get **New User** into AWS SSO, we will need to assign it to a group that governs access in AWS SSO on the IDP side. As everyone within Redbeard Identity is allowed to access the sandbox accounts, we will place **New User** in the **AWS_Sandbox** group, within the IDP, and refresh our AWS SSO user store. If we select that group to view its membership, we will see **New User**:

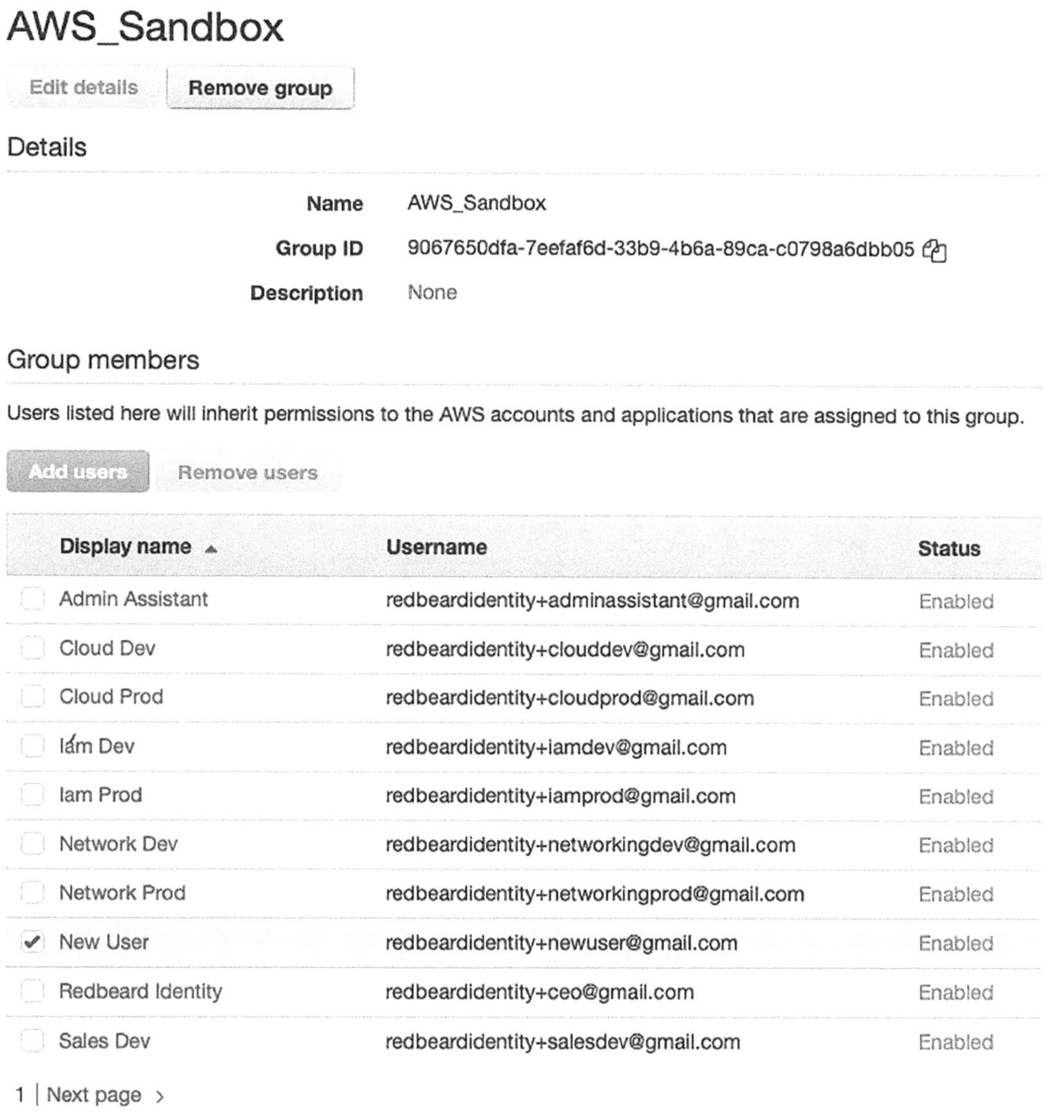

AWS_Sandbox

Edit details | **Remove group**

Details

Name	AWS_Sandbox
Group ID	9067650dfa-7eefaf6d-33b9-4b6a-89ca-c0798a6dbb05
Description	None

Group members

Users listed here will inherit permissions to the AWS accounts and applications that are assigned to this group.

Add users | Remove users

Display name ▲	Username	Status
Admin Assistant	redbeardidentity+adminassistant@gmail.com	Enabled
Cloud Dev	redbeardidentity+clouddev@gmail.com	Enabled
Cloud Prod	redbeardidentity+cloudprod@gmail.com	Enabled
Iam Dev	redbeardidentity+iamdev@gmail.com	Enabled
Iam Prod	redbeardidentity+iamprod@gmail.com	Enabled
Network Dev	redbeardidentity+networkingdev@gmail.com	Enabled
Network Prod	redbeardidentity+networkingprod@gmail.com	Enabled
✔ New User	redbeardidentity+newuser@gmail.com	Enabled
Redbeard Identity	redbeardidentity+ceo@gmail.com	Enabled
Sales Dev	redbeardidentity+salesdev@gmail.com	Enabled

1 | Next page >

Figure 9.28 – New User inside the AWS_Sandbox group

Whereas automatic provisioning is convenient, the bulk of the security and compliance value that SCIM offers comes from automatic deprovisioning. Let's deactivate **New User** from our IDP and watch what happens.

First, he is deactivated within our IDP's user store and automatically loses access to the applications he is assigned from his group memberships. This means he loses access to AWS SSO at the IDP:

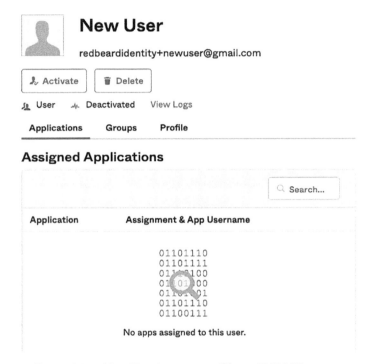

Figure 9.29 – New User is now cut off from AWS SSO access

Checking the AWS SSO user store, we can now see that the corresponding account for **New User** has been disabled there as well. It has also been cut off from the AWS SSO side:

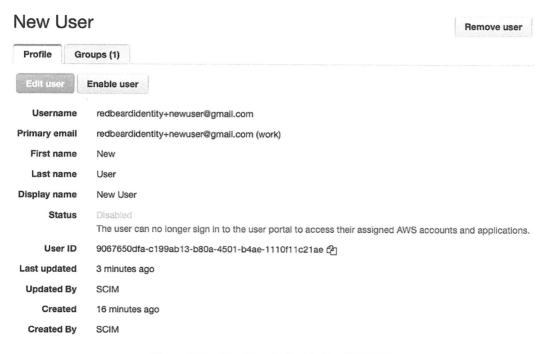

Figure 9.30 – New User is disabled in AWS SSO

Depending on our organization's account management and compliance policies, we may optionally delete the record entirely from AWS SSO at this point.

SCIM provides powerful and automated account management capabilities when leveraging an external IDP. For organizations that either do not use Active Directory as their user store or would prefer not to extend their AD footprint into the cloud, SCIM provides modern, API-driven identity life cycle management capabilities to ensure that only the right accounts and groups are provisioned in AWS for administrative access.

Summary

In this chapter, we looked at how we can bring our administrative accounts into the AWS administrative backplane. First, we connected our external identity provider to our AWS SSO service. Then, we looked at two different methods to manage administrative accounts. The first was manual account linking, where an administrator must provision, deprovision, and monitor account and group membership for changes inside the external IDP's user store, to then mimic those changes inside AWS SSO's own user store. The second was SCIM, a RESTful, API-based identity provisioning protocol that automatically synchronizes accounts, attributes, and groups between the external IDP and AWS SSO.

Now that we have our user stores synchronized using SCIM, we are positioned to leverage those accounts and groups, along with their attributes, to address administrative authentication and authorization to AWS resources. We will explore that topic in detail in the following chapter.

Questions

1. What are manual provisioning and account linking?
2. What is SCIM?
3. What are some of the advantages of SCIM over manual provisioning?

Further reading

Take a look at the following link to find out more about the supported identity providers and configuration instructions for AWS SSO SCIM: `https://docs.aws.amazon.com/singlesignon/latest/userguide/supported-idps.html`.

Code samples

The following is the updated Redbeard Identity CSV file for this chapter: `https://github.com/jonlehtinen/ImplementingAWSIdentity/blob/main/RedbeardIdentity_csv_template_new_scim_user.csv`.

10
Administrative Single Sign-On to the AWS Backplane

In the previous chapter, we built out the provisioning and account synchronization processes between our **Amazon Web Services** (**AWS**) environment and the Redbeard Identity organization's existing **identity provider** (**IDP**). Our administrative users are now synchronized to the AWS **single sign-on** (**SSO**) user directory from our external IDP using the **System for Cross-domain Identity Management** (**SCIM**). Of course, populating the AWS SSO user store is only half of the administrative access equation. Next, we will address administrative user authentication and authorization to ensure that each administrator can only access the environment that is appropriate for them.

The following topics will be covered in this chapter:

- Why use federation for AWS administrators?—Learn why identity federation is a good pattern for managing administrator access into the AWS control plane

- Assigning access to AWS accounts—Assign accounts and authenticate administrative users into AWS accounts through AWS SSO using an external IDP

- Implementing fine-grained access management for administrators—Set the limits of administrator access within an account through **permission sets**

- Administrative SSO using the AWS **command-line interface (CLI)**—Use SSO with an external IDP to obtain temporary credentials to use within the AWS CLI, and issue commands with those temporary credentials

Technical requirements

To get the most out of this chapter, you will need the following:

- An AWS account

- A **Security Assertion Markup Language 2 (SAML2)** and an SCIM-compliant IDP such as Okta Identity Cloud, PingOne, or **Azure Active Directory (Azure AD)**

- A populated user directory to act as the user store for that IDP

- A workstation running the AWS CLI

- A text editor or **integrated development environment (IDE)** to edit **JavaScript Object Notation (JSON)/YAML Ain't Markup Language (YAML)** files, such as Microsoft **Visual Studio Code (VS Code)**

The code samples used in the chapter can be found at the following links:

- Updated Redbeard Identity **comma-separated values (CSV)** file: `https://github.com/jonlehtinen/ImplementingAWSIdentity/blob/main/RedbeardIdentity_csv_template_new_scim_user.csv`

- `ITS_ec2_policy.json` document: `https://github.com/jonlehtinen/ImplementingAWSIdentity/blob/main/ec2_policy.json`

Why use federation for AWS administrators?

Before we dive into the mechanics of connecting our AWS environment with our external IDP, let's take a moment to revisit our assumptions around why we would choose to use an external IDP for AWS access in the first place. As we have seen throughout this book, AWS has multiple services capable of addressing user authentication and authorization. It could be argued that given the AWS **Identity and Access Management (IAM)** service itself already evaluates every transaction and has the capability to handle user management, authentication, and authorization, daisy-chaining additional components to that service unnecessarily complicates matters.

The argument for using identity federation with administrative accounts echoes the same arguments for identity federation with most other third-party applications. Identity federation, especially automated provisioning and deprovisioning, helps control the proliferation of user and company data on third-party systems. Additionally, federation also simplifies the sign-on experience to the third-party applications by allowing those applications to trust users authenticated by an external IDP. This improves the **user experience** (**UX**) and reduces the surface area of attack for credential theft. Users also don't have to manage an additional set of credentials.

Finally, consider how other abstraction layers simplify powerful, yet complicated toolsets. We wouldn't expect software to be written in machine code. Instead, software frameworks and languages expand the accessibility and functionality of the powerful base language that executes inside every **central processing unit** (**CPU**). Whereas abstraction layers necessarily involve additional systems, services, and processing, that additional tooling complexity is designed to make the underlying service simpler to understand and consume. The leap in complexity for setting up the same level of administrative oversight using AWS IAM compared to accomplishing the same task using AWS SSO is nowhere near as severe as writing any program using machine code instead of JavaScript, but it is an example of how some things are simplified through additional abstraction and complementary services.

Now that we have had a refresher on why we are implementing things the way we are, let's begin testing administrative authentication between AWS and our external IDP.

Federated sign-in using an external IDP

We set up the connection between our IDP and AWS SSO as a prerequisite to synchronizing our administrative accounts from our IDP's user store into AWS SSO's user store in *Chapter 9, Bringing Your Admins into the AWS Administrative Backplane*. Now, let's verify if our users can sign in.

Depending upon the IDP platform used, there may be a preliminary step required before any users will be able to sign in to AWS using the external IDP. **Identity as a Service (IDaaS)** platforms such as Okta and Azure AD require applications to be assigned to users within the platform before the IDP will issue an authentication token for that user.

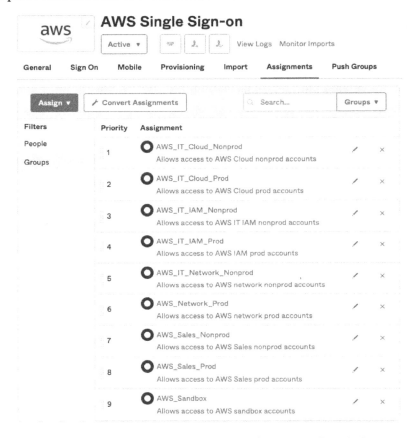

Figure 10.1 – Groups assigned access to AWS SSO from the Redbeard Identity IDP

The Redbeard Identity IDP allows direct assigning of applications to users. However, that isn't the best administrative model. Instead, we assigned access to AWS SSO from the IDP to various directory groups found within our IDP's directory. This means that any users found within those groups will be authorized to access the AWS SSO instance we configured inside our IDP. This is an example of **coarse-grained authorization**. Coarse-grained authorization is a binary yes/no application access decision that is managed by either the IDP (through a mechanism such as group membership or application assignment) or the application itself (through a comparable mechanism). This makes the Redbeard Identity IDP the first enforcement point for administrative access into our AWS environment.

The groups shown in *Figure 10.1* cover everyone within the IDP's directory. Whereas most of the groups only have one or two members, the AWS_Sandbox group covers everyone not previously covered by another group membership. Allowing the full set of users we created for the Redbeard Identity use case to access the Sandbox is fine for our purposes, but it would be prudent not to make such access a birthright entitlement if we had more than a dozen records in our directory.

Now that we have addressed the prerequisites for coarse-grained access to AWS through our IDP, we can attempt to access AWS in two different ways. The first is through our IDaaS platform's user application portal, and the second is through the SSO link provided by AWS SSO.

Let's start by walking through the IDaaS method, as follows:

1. First, we sign in to our IDP's portal as a Redbeard Identity user, as illustrated in the following screenshot:

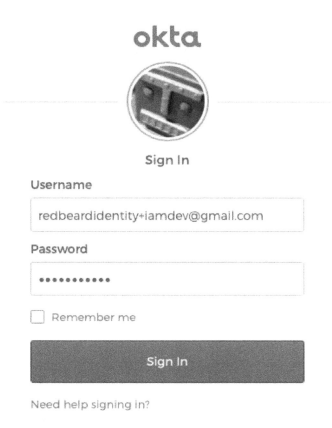

Figure 10.2 – Signing in as Iam Dev

2. Once signed in, we see the AWS SSO application among our approved applications, as illustrated in the following screenshot. This is because `Iam Dev` is a member of two qualifying groups that grant access at the IDP level, `AWS_IT_IAM_Nonprod` and `AWS_Sandbox`:

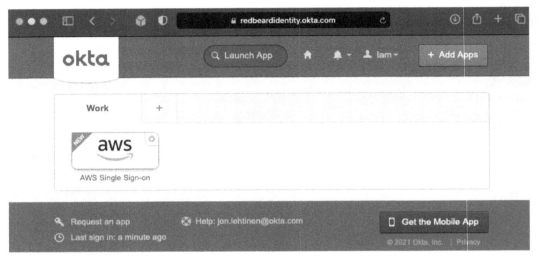

Figure 10.3 – AWS SSO within the IDaaS app launcher

3. We can click the AWS icon and start the SSO transaction to our AWS environment.

   The alternate path for authentication is through the AWS-provided sign-on link. This link is found in the **Settings** section of our AWS SSO service, under **User portal**, as illustrated in the following screenshot:

User portal

The user portal is a central place where your users can see and access their assigned AWS accounts, roles, and applications. Share this URL with your users to get them started with AWS SSO.

User portal URL   https://redbeardidentity.awsapps.com/start

Figure 10.4 – User portal Uniform Resource Locator (URL) in the AWS SSO settings

4. If we go to that address, AWS SSO redirects us to our IDP for authentication, as illustrated in the following screenshot:

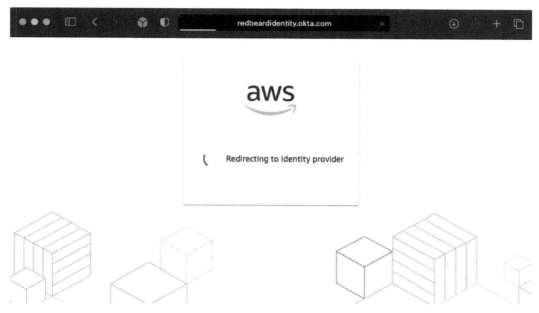

Figure 10.5 – Redirecting to the external IDP for user authentication

5. We arrive back at our IDP, this time with an AWS banner. We will sign in using the
    `Iam Dev` account once more.

Unfortunately, at this point, it doesn't matter which of the two paths we choose as we
still need to configure our authorization policies within AWS itself. The IDP may control
coarse-grained access to AWS SSO, but AWS will manage authorization to accounts and
resources within AWS once the federated user makes it into AWS.

Next, we'll take a look at our options for assigning access to our federated users inside
AWS.

Assigning access to AWS accounts

Now that we can sign in to AWS SSO with our external IDP, we need to assign accounts
to users within AWS SSO in order to close the loop between the authorization controlled
by the IDP and the authorization controlled by AWS. If we considered the IDP's
authorization control coarse-grained, AWS SSO provides options for fine-grained control
through a variety of mechanisms. Let's start with some basic authorization controls and
refine the permissions further as we go.

We can see all of our AWS accounts listed in the **AWS accounts** menu inside AWS SSO, as illustrated in the following screenshot:

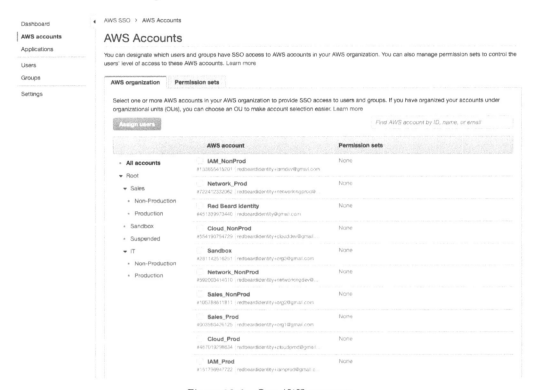

Figure 10.6 – Our AWS accounts

Presently, we have no users assigned to any of them. We also do not have any permission sets assigned to any of the accounts. A **permission set** defines what an AWS user can do within an AWS account when signing in through AWS SSO. A permission set is stored as an AWS IAM role that is assumed by the federated user within the member AWS account. Users can be assigned multiple permission sets, but they may only assume one of those permission sets at a time when signing in to a member AWS account. We will explore permission sets in greater detail in the following section.

Let's start by assigning users to the IAM_NonProd account. We do this by selecting that account and clicking the **Assign users** button, as illustrated in the following screenshot:

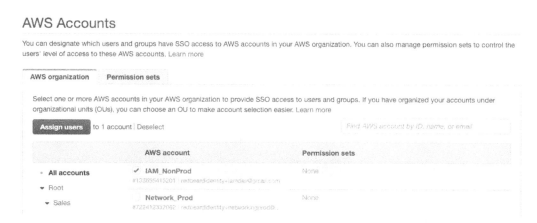

Figure 10.7 – Selecting an account to assign users to it

This takes us to a wizard with which we can manually assign either individual users or groups to that account. Whereas we only have a handful of users in our example, best practice recommends against managing access on a per-account basis. We took the time to build our IDP-side app provisioning logic to be group-based because it simplifies **user access management (UAM)**. We will assign each account to its corresponding group's name, starting with this one. Let's tick AWS_IT_IAM_Nonprod, as illustrated in the following screenshot, and move on:

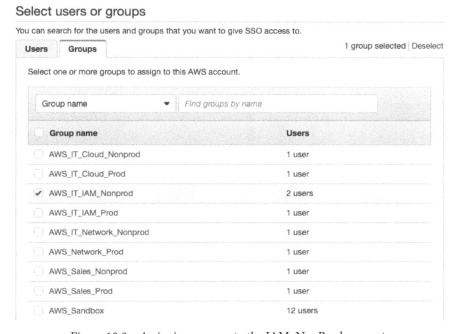

Figure 10.8 – Assigning a group to the IAM_NonProd account

Next, we assign permission sets. As AWS SSO configures itself as the IDP within each managed account's AWS IAM instance, we need to define a role or roles that will be available for our users to assume upon successfully federating into that account's AWS IAM service. We still have the two policies we created in *Chapter 6, Introduction to AWS Organizations and AWS Single Sign-On*, so we will enable both of those permission sets to start. We can now finish the wizard. The process is illustrated in the following screenshot:

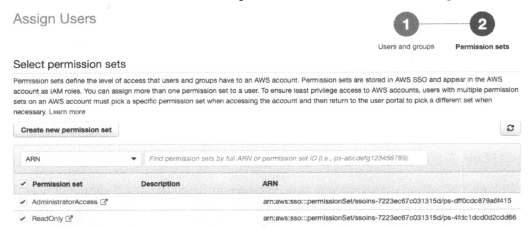

Figure 10.9 – Assigning permission sets

The wizard processes the request, and if all goes well, we get a readout of everything that was created inside that account, as illustrated in the following screenshot:

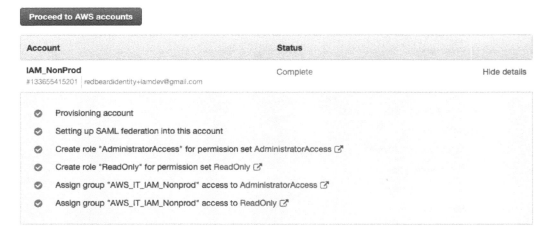

Figure 10.10 – Provisioning accounts, SAML federation, and assumable roles

We can now go back to our AWS accounts, where we see the permission sets defined next to the IAM_NonProd account, as illustrated in the following screenshot:

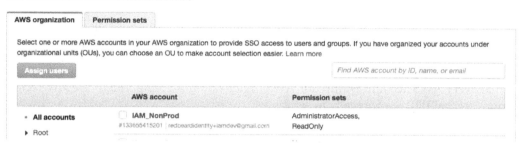

AWS Accounts

You can designate which users and groups have SSO access to AWS accounts in your AWS organization. You can also manage permission sets to control the users' level of access to these AWS accounts. Learn more

| AWS organization | Permission sets |

Select one or more AWS accounts in your AWS organization to provide SSO access to users and groups. If you have organized your accounts under organizational units (OUs), you can choose an OU to make account selection easier. Learn more

Assign users   Find AWS account by ID, name, or email

	AWS account	Permission sets
• **All accounts**	☐ **IAM_NonProd**	AdministratorAccess,
▶ Root	#133655415201 | redbeardidentity+iamdev@gmail.com	ReadOnly

Figure 10.11 – The account is now configured with permission sets

We'll repeat that exercise for each account. Once we are done, each account will be accessible by every administrator account assigned to that account's corresponding access control group back at our external IDP, as illustrated in the following screenshot:

| AWS organization | Permission sets |

Select one or more AWS accounts in your AWS organization to provide SSO access to users and groups. If you have organized your accounts under organizational units (OUs), you can choose an OU to make account selection easier. Learn more

Assign users   Find AWS account by ID, name, or email

	AWS account	Permission sets
• **All accounts**	☐ **IAM_NonProd**	AdministratorAccess,
▶ Root	#133655415201 | redbeardidentity+iamdev@gmail.com	ReadOnly
	☐ **Network_Prod**	AdministratorAccess,
	#722412332062 | redbeardidentity+networkingprod@...	ReadOnly
	☐ **Red Beard Identity**	None
	#451339973440 | redbeardidentity@gmail.com	
	☐ **Cloud_NonProd**	AdministratorAccess,
	#554190754729 | redbeardidentity+clouddev@gmail....	ReadOnly
	☐ **Sandbox**	AdministratorAccess,
	#281142516251 | redbeardidentity+org3@gmail.com	ReadOnly
	☐ **Network_NonProd**	AdministratorAccess,
	#592003414010 | redbeardidentity+networkingdev@...	ReadOnly
	☐ **Sales_Nonprod**	AdministratorAccess,
	#103788611811 | redbeardidentity+org2@gmail.com	ReadOnly
	☐ **Sales_Prod**	AdministratorAccess,
	#003980426125 | redbeardidentity+org1@gmail.com	ReadOnly
	☐ **Cloud_Prod**	AdministratorAccess,
	#467019298634 | redbeardidentity+cloudprod@gmail...	ReadOnly
	☐ **IAM_Prod**	AdministratorAccess,
	#151796947722 | redbeardidentity+iamprod@gmail.c...	ReadOnly

Figure 10.12 – Each account configured with assigned group and permission sets

The only aberration is the `Red Beard Identity` AWS account, which in this scenario is acting as the management account. So far, we have been signing in using a local AWS IAM account to perform all of the administrative functions. However, in the spirit of truly delegating authentication and authorization to an externalized IDP, we should assign access to our management AWS account to an account controlled by the external IDP as well. Since we have used an AWS IAM account named `redbeardidentity` for most of what we have done so far, we'll create a corresponding account in our IDP in the same name and set it as an assigned user for our management account with full administrator permissions.

Now that every AWS account within our AWS organization is configured to use AWS SSO with our external IDP, we are ready to sign in.

Signing in to the administrative console

As it stands, we have 10 different accounts that we could sign in to using our external IDP. Let's proceed as follows:

1. Let's start by signing into our external IDP's user portal as the `Iam Dev` user, as illustrated in the following screenshot:

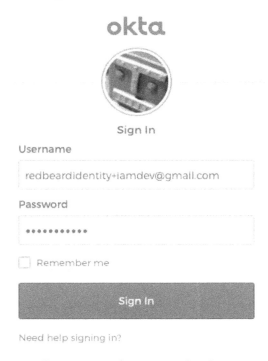

Figure 10.13 – Signing in as Iam Dev

2. Once signed in, we again see the AWS SSO application among our approved applications, as we did last time. This is illustrated in the following screenshot:

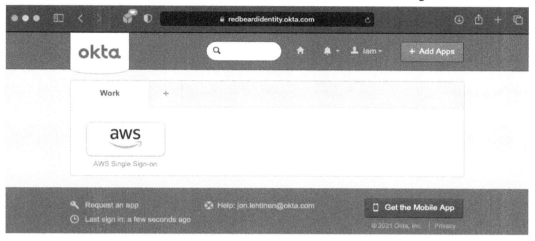

Figure 10.14 – AWS SSO within the IDaaS app launcher

3. We can click the AWS icon and start the SSO transaction to our AWS environment. After some redirects, we are back at the AWS SSO application launcher, except this time we actually have AWS accounts available for us to use, as we can see here:

Figure 10.15 – IAM_NonProd and Sandbox accounts available to the Iam Dev user

4. We will select the `IAM_NonProd` environment. When we click it, we see a pair of options appear beneath the account name, as illustrated in the following screenshot:

Figure 10.16 – Two links for IAM_NonProd

5. The names for each of these options correspond to the permission sets that we attached to the account inside AWS SSO. Each of these links is for assuming a specific role within the managed AWS account that corresponds to the permission set we assigned to that account. Let's open the management console using the `ReadOnly` role, as illustrated in the following screenshot:

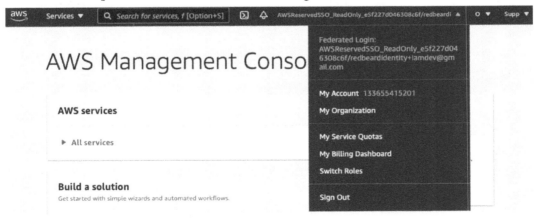

Figure 10.17 – Signed in to IAM_NonProd under the ReadOnly role

We can see the name of our federated login's assumed role once we are in the environment. Attempting to do anything—such as creating a **Simple Storage Service (S3)** bucket—throws an `Access Denied` error, which confirms the policy is operating as expected.

> **Tip**
> Always validate the logic of a permission set to verify what it does or does not do. These environments and resources were set up and used exclusively for this book, so I took a shortcut by reusing the permission set that we created in *Chapter 6, Introduction to AWS Organizations and AWS Single Sign-On*. You should never make the same assumption in a live environment.

6. Launching the `AdministratorAccess` role shows a different assumed role upon landing in the console, as we can see here:

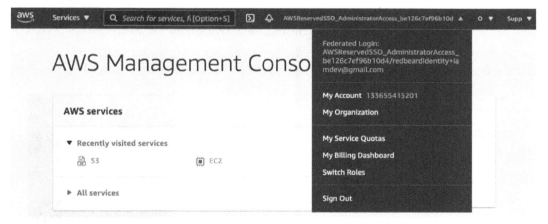

Figure 10.18 – The AdministratorAccess assumed role in the IAM_NonProd account

Unlike the previous role, this one is unrestricted. However, unrestricted roles are unusual in real deployments. We will need to revisit our permission sets to fine-tune the access each administrator gets within each account.

7. But before we do that, let's first sign in as the Redbeard Identity user through our external IDP to get back into the administrative console for our management AWS account. You can see this user account in the following screenshot:

Figure 10.19 – The Sandbox and management AWS account are available

As with most users, this one has access to the Sandbox. But more importantly and uniquely, this account can access the management AWS account, as we can see here:

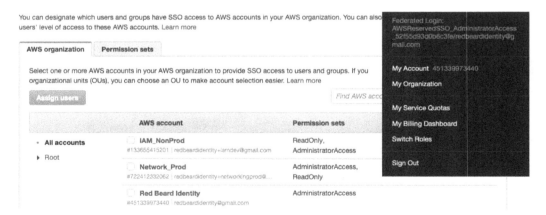

Figure 10.20 – Signing in as an admin on the management account through federation

We now have options to address some additional security best practices, such as removing the `redbeardidentity` AWS IAM user account that we created and have used for all administrative tasks up to this point. This will also remove the long-lived programmatic credentials we have assigned to that user, which represent a potential security risk to our environment. Now that we have connected everything through an external IDP, we are free to further harden our AWS environments.

> **Tip**
>
> In this chapter, we will use the *Redbeard Identity* user account as the example account that matches the organization, division, and department attributes of another user. The CSV file referenced in this chapter (and in *Chapter 8, An Ounce of Prevention – Planning Your Administrative Model,* and *Chapter 9, Bringing Your Admins into the AWS Administrative Backplane*) includes an account that has all the same attributes as the *Redbeard Identity* user, but it has been renamed as *Super User* for ease of identification and disambiguation from the *Redbeard Identity, CEO* user account.

Next, we'll look into refining the user permissions into each environment beyond the basic read-only and administrator access that we used to test out our configuration.

Implementing fine-grained access management for administrators

So far, we only have two levels of access for our administrators inside our AWS accounts once those administrators are placed inside a group that allows them to sign in to AWS SSO: AdministratorAccess and ReadOnly. If we defined group-based access that determines if a user is permitted to even access AWS SSO as coarse-grained access management, then the access granted by these two permission sets represents a very rudimentary example of **role-based access control** (**RBAC**). By layering on additional concepts, we can further refine our authorization model into something that is only allowed access to specific resources based upon the assumed role and the user's attributes, to achieve fine-grained access management through **attribute-based access control** (**ABAC**).

Permission sets and managed authorization policies

To achieve fine-grained access management through ABAC, we will need to marry an improved set of permission sets with a customized authorization policy. Let's start our journey there by taking a look at the first two permission sets we created and used with AWS SSO, AdministratorAccess and ReadOnly.

We can check the defined permission sets from the AWS SSO service's console under **AWS accounts**. We can see all of the permission sets that we have defined, regardless of whether they are in use or not, by clicking on the **Permission sets** tab, as illustrated in the following screenshot. We can click on a permission set to see its properties:

AWS Accounts

You can designate which users and groups have SSO access to AWS accounts in your AWS organization. You can also manage permission sets to control the users' level of access to these AWS accounts. Learn more

AWS organization	**Permission sets**

Permission sets define the level of access that users have to their assigned AWS accounts. Choose from the following predefined permission sets or create custom ones. Users that have been assigned multiple sets can choose a permission set at the time they sign in to the user portal. Learn more

Create permission set   Delete

ARN	▼	Find permission sets by full ARN or permission set ID (i.e., ps-abcdefg123456789).

Permission set	Description	ARN
● AdministratorAccess		arn:aws:sso:::permissionSet/ssoins-7223ec67c031315d/ps-dff0cdc879a6f
○ ReadOnly		arn:aws:sso:::permissionSet/ssoins-7223ec67c031315d/ps-4fdc1dcd0d2c

Figure 10.21 – Permission sets available for use with AWS SSO

First, we have `AdministratorAccess`. We'll click on this and see how it is configured. You can see the configuration in the following screenshot:

AdministratorAccess

Permissions	AWS accounts

General

Name	AdministratorAccess
Description	*Not provided*
ARN	arn:aws:sso:::permissionSet/ssoins-7223ec67c031315d/ps-dff0cdc879a6f415
Session duration	1 hour
Relay state	*Not provided*

Edit

AWS managed policies

AWS managed policies are policies in IAM that you can attach to this permission set when you need to grant predefined permissions that are job related (such as AdministratorAccess) or service specific (such as AmazonCloudDirectoryFullAccess). You can attach up to 10 AWS managed policies to this permission set. Learn more

Attach managed policies

IAM policy
AdministratorAccess ☐   Detach

Figure 10.22 – The AdministratorAccess permission set's configuration

Here, we can adjust things such as the duration of the federated session. This is useful for allowing different periods of session validity based upon the sensitivity of the permission set. A superuser may require a shorter session before reauthentication to the external IDP, whereas a read-only permission set could allow a much longer session duration.

More germane to the topic of authorization, however, is that this is where we can attach managed policies to our permission set. In the case of `AdministratorAccess`, we see that it has the `AdministratorAccess` managed policy attached. If we click on that managed policy, we are taken to the AWS IAM service where we can see the policy document itself, as illustrated in the following screenshot:

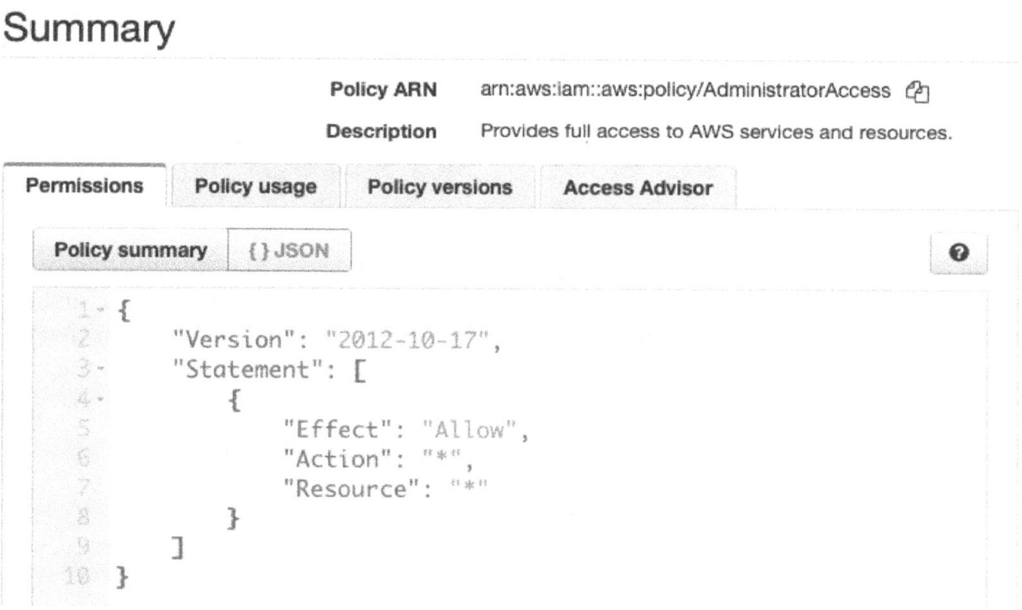

Summary

Policy ARN	arn:aws:iam::aws:policy/AdministratorAccess
Description	Provides full access to AWS services and resources.

Permissions	**Policy usage**	**Policy versions**	**Access Advisor**

Policy summary	{ } JSON		❓

```json
1 {
2     "Version": "2012-10-17",
3     "Statement": [
4         {
5             "Effect": "Allow",
6             "Action": "*",
7             "Resource": "*"
8         }
9     ]
10 }
```

Figure 10.23 – The AdministratorAccess managed policy that makes up the AdministratorAccess permission set

If we repeat that exercise with the `ReadOnly` permission set, we see that it has the `ViewOnlyAccess` managed policy attached. Clicking on that managed policy also shows us the policy document. As it lists every AWS service and every command that lists, describes, or gets values from those services, it is quite long. Here is a snippet from the document:

```json
{
    "Version": "2012-10-17",
    "Statement": [
        {
```

```
    "Action": [
        "acm:ListCertificates",
        "athena:List*",
        "aws-marketplace:ViewSubscriptions",
        "autoscaling:Describe*",
        "batch:ListJobs",
        -- (truncated for space) --
        "workmail:Describe*",
        "workspaces:Describe*"
    ],
    "Effect": "Allow",
    "Resource": "*"
}
]
}
```

We have validated the policies behind our permission sets and seen that we can attach managed policy objects to our permission sets, but none of that is going to help us further refine access from either the full `AdministratorAccess` or `ReadOnly` options that we currently have available. However, if we look further down on the properties of the permission set, we have the option to edit permissions on these policies, as we can see here:

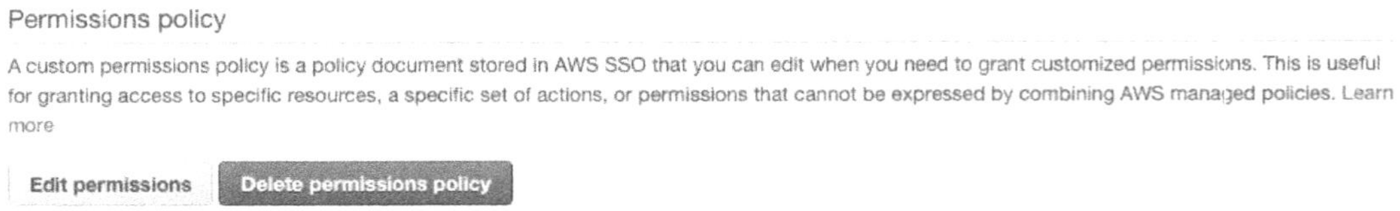

Permissions policy

A custom permissions policy is a policy document stored in AWS SSO that you can edit when you need to grant customized permissions. This is useful for granting access to specific resources, a specific set of actions, or permissions that cannot be expressed by combining AWS managed policies. Learn more

Edit permissions   Delete permissions policy

Figure 10.24 – Option to customize permission set policy on an existing permission set

We can refine these existing permission sets to be more restrictive. Alternatively, we can create custom policies on new permission sets. The end result will be the same—a custom authorization policy that we can use to apply fine-grained access control to the resources within our AWS accounts.

Permission sets and custom authorization policies for fine-grained access control

Before we set about building our ABAC policies, let's look again at our AWS organization and the administrative user attributes in the directory, and list out the use cases we need to address to determine an optimal way to solve the business problem in front of us.

First, here is a map of our AWS organization:

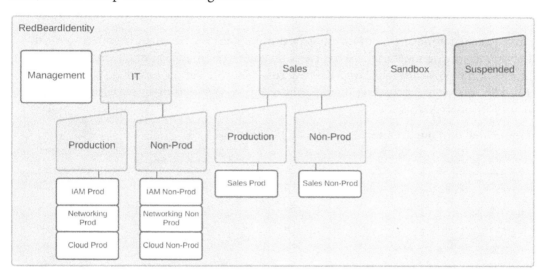

Figure 10.25 – The Redbeard Identity AWS organization map

Whereas a real organization would have stakeholders informing the requirements, for this exercise we are going to make some inferences based on the organizational layout that will provide our access control requirements. Based on the layout, we can assume the following:

- Only the superuser (redbeardidentity@gmail.com in this case) may have access to the **Management** account.

- Only members of the **IT** organization should have access to **IT** organization AWS accounts. The same applies to the **Sales** organization.

- Only workers in operational departments should have read/write access to **Production** AWS accounts. They may optionally have read-only access to **Non-Prod** AWS accounts.

- Only workers in development departments should have read/write access to **Non-Prod** AWS accounts. They may optionally have read-only access to **Production** AWS accounts.

- Each division's operational and development departments may only access their own division's AWS accounts.

- IT Support staff require access to stop and start **Elastic Compute Cloud** (**EC2**) instances inside AWS accounts within the IT **organizational unit** (**OU**) and the Sandbox.

- Anyone may access the **Sandbox** AWS account.

- Nobody may access accounts in the **Suspended** OU.

This may seem complicated, but fortunately, our preplanning has already gotten us most of the way to solving most of these use cases through a combination of the coarse-grained authorization policy through group-based access controlled at the external IDP. We now just need to marry that coarse-grained policy with a corresponding fine-grained policy managed through AWS SSO.

Mapping it all out, our architecture will look like this:

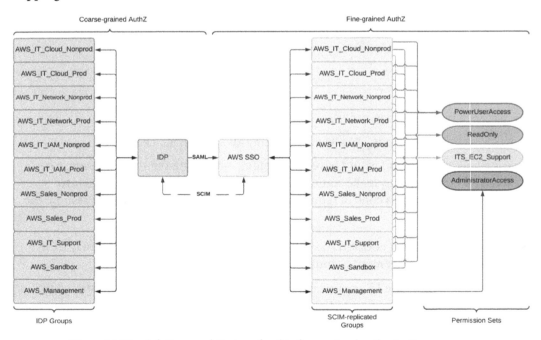

Figure 10.26 – Solutions architecture for this fine-grained authorization use case

Since we can use the group membership to fulfill all of our organizational requirements, we just need to identify an attribute that will allow us to further refine the level of access that will work for limiting the IT Support department to only affecting EC2 instances owned by the **IT** organization. The `costCenter` will be a good candidate for this for when we start writing our permission sets.

Writing our fine-grained authorization policies

In order to enable our fine-grained authorization use cases, we will need to make a few adjustments to our configuration within AWS SSO. First, we need to customize the authorization policy documents that we will use within each of the four permission sets that we will be creating. Let's return to our AWS SSO service within the management AWS account and create a new permission set by hitting the **Create permission set** button under the **Permission sets** tab on the **AWS accounts** menu. We will start by writing one of our new permission sets for the administrators within each of the AWS accounts. The process is illustrated in the following screenshot:

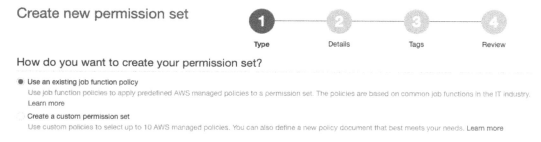

Figure 10.27 – Selecting a canned role or custom permission set

The first option we have is to either select an existing job function policy or create a custom permission set. AWS provides several predefined policies based upon popular job descriptions, as illustrated in the following screenshot. We should take a look at these offerings before deciding to write our policy from scratch:

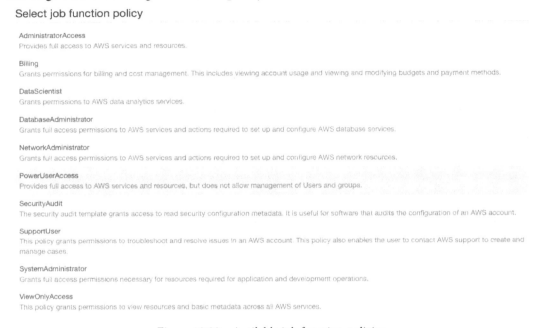

Figure 10.28 – Available job function policies

The next step lists the various job function policies, along with a brief description of what they can do. Since we don't want our administrative users to be able to remove our IDP configuration from each AWS account's AWS IAM instance, nor remove the account from the AWS organization entirely, we should look for something down-scoped from the full `AdministratorAccess` policy. Fortunately, there is a `PowerUserAccess` policy. We can click it to see its **JavaScript Object Notation (JSON)** policy in AWS IAM, as reproduced here:

```
{
    "Version": "2012-10-17",
    "Statement": [
        {
            "Effect": "Allow",
            "NotAction": [
                "iam:*",
                "organizations:*",
```

```
            "account:*"
        ],
        "Resource": "*"
    },
    {
        "Effect": "Allow",
        "Action": [
            "iam:CreateServiceLinkedRole",
            "iam:DeleteServiceLinkedRole",
            "iam:ListRoles",
            "organizations:DescribeOrganization",
            "account:ListRegions"
        ],
        "Resource": "*"
    }
  ]
}
```

This policy will bar administrative users from all AWS IAM functions except creating or deleting service-linked roles and listing roles, generally speaking. It also keeps them out of AWS Organizations and Account services. This seems like a winner, so we can proceed.

Once we have clicked through the rest of the wizard and have our new `PowerUserAccess` permission set listed among the other two that were previously there, we can click it and open its properties. Let's adjust the session duration to 9 hours to accommodate a full working day. We have to do this by entering a custom duration in seconds, as 9 hours is not an option in the drop-down menu. The following screenshot illustrates the process:

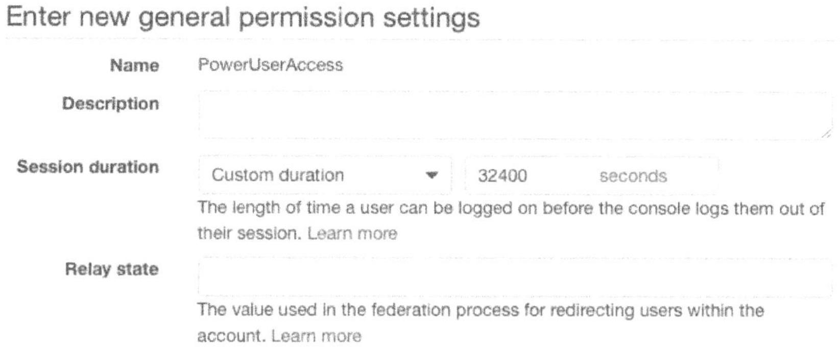

Enter new general permission settings

Name	PowerUserAccess
Description	
Session duration	Custom duration ▼ 32400 seconds
	The length of time a user can be logged on before the console logs them out of their session. Learn more
Relay state	
	The value used in the federation process for redirecting users within the account. Learn more

Figure 10.29 – Setting a custom session duration

With that done, we may proceed. The next screen gives us an option to update the role the permission set creates in the AWS accounts it is attached to. When modifying a permission set that is attached to an AWS account, we must update that role for the changes we make to be reflected in the assumed role in the member account's AWS IAM service. You can see an example of this in the following screenshot. As we haven't assigned this permission set to any of our accounts, we can move on:

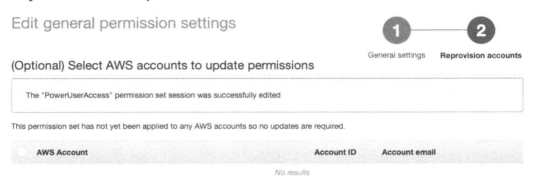

Figure 10.30 – Option to reprovision accounts

For ReadOnly, we can follow the same process we used for PowerUserAccess, as there is a default ViewOnlyAccess job function policy available for us to select. We won't repeat the steps here. This means we now have the AdministratorAccess, PowerUserAccess, and ReadOnly permission sets defined. Now, we need to write a custom policy for our IT Support administrators to ensure they can only modify EC2 instances tagged with a costCenter that belongs to the **IT** organization across AWS accounts within the **IT** OU of our AWS organization.

To do this, we once more create a new permission set. However, this time, we select the **Create a custom permission set** option, which will take us to the following screen:

Figure 10.31 – Creating a custom permission set

Next, we set our new policy's name, description, and session duration. We will keep it at the same 9 hours as the rest of our policies since our risk tolerance already accepts 9-hour sessions for policies with much greater levels of access. Beneath those settings, we will tick the option to create a custom permissions policy. This will open up a policy editor for us to use, as illustrated in the following screenshot:

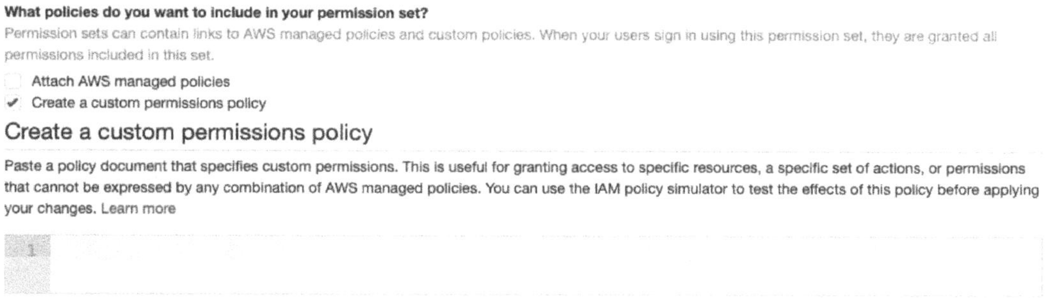

What policies do you want to include in your permission set?

Permission sets can contain links to AWS managed policies and custom policies. When your users sign in using this permission set, they are granted all permissions included in this set.

- Attach AWS managed policies
- ✔ Create a custom permissions policy

Create a custom permissions policy

Paste a policy document that specifies custom permissions. This is useful for granting access to specific resources, a specific set of actions, or permissions that cannot be expressed by any combination of AWS managed policies. You can use the IAM policy simulator to test the effects of this policy before applying your changes. Learn more

Figure 10.32 – The policy editor for writing our custom policy

Let's start by adding the essential components of our policy document—the version and the statement, as follows:

```
{
    "Version": "2012-10-17",
    "Statement":
```

Our IT Support admins will need to view EC2 instances, so that will be our first statement. As this is a multi-effect policy, we will need to mind our JSON syntax. You can view the code we're using in the following snippet:

```
{
    "Version": "2012-10-17",
    "Statement": [
        {
            "Effect": "Allow",
            "Action": [
                "ec2:DescribeInstances"
            ],
            "Resource": "*"
        },
```

Next, we want them to be able to start and stop instances. So, we write our next effect, as follows:

```
{
    "Version": "2012-10-17",
    "Statement": [
        {
            "Effect": "Allow",
            "Action": [
                "ec2:DescribeInstances"
            ],
            "Resource": "*"
        },
        {
            "Effect": "Allow",
            "Action": [
                "ec2:StartInstances",
                "ec2:StopInstances"
            ],
            "Resource": "*"
```

As written (and setting aside the malformed JSON), this policy is too permissive. We need to apply a `condition` statement that limits the specific EC2 instances they can manipulate within shared environments, such as the Sandbox. We can use resource tags and the `ec2:ResourceTag` variable to enforce this. We can look at our IDP's user store for a list of cost centers to use within our `condition` statement. The code is illustrated in the following snippet:

```
{
    "Version": "2012-10-17",
    "Statement": [
        {
            "Effect": "Allow",
            "Action": [
                "ec2:DescribeInstances"
            ],
            "Resource": "*"
        },
```

```
        {
            "Effect": "Allow",
            "Action": [
                "ec2:StartInstances",
                "ec2:StopInstances"
            ],
            "Resource": "*",
            "Condition": {
                "StringEquals": {
                    "ec2:ResourceTag/CostCenter": ["3001",
        "30002", "30011", "30012", "30013", "30014", "31013", "31000"]
                }
            }
        }
    ]
}
```

We drop that into the editor for validation, and we can move on. We are given a chance to review, and then our final permission set is created.

> **Tip**
>
> Whereas it is alluring to look to `ResourceTag` and `RequestorTag` variables to apply fine-grained authorization in sweeping policy statements, the reality is that there is tremendous variation between the actions that support `ResourceTag/RequestorTag` variables. In fact, there are even several different types of `ResourceTags` variables depending upon the service, including `s3:ResourceTags`, `ec2:ResourceTags`, and the general `aws:ResourceTags` variable. Some services and actions do not support `ResourceTags` variables at all. This is partially why we leaned more heavily on using group membership, permission sets, and differing AWS accounts for accomplishing the bulk of our authorization policy objectives.

Now that we have all of our policies aligned to our use cases, we are now ready to put them to use.

Putting it all together for administrative authorization

We have completed building all the pieces that will enable our administrative authentication and authorization model in our AWS accounts, but there isn't anything intrinsic to any of the managed policies we attached to the permission sets that will limit access to specific AWS accounts. That authorization will come from assigning specific groups to the account, and then assigning which permission sets (the assumed role within that account) each of those groups will have access to. Let's begin doling out our access to fulfill our eight business objectives, as follows:

1. Let's start with the `IAM_NonProd` AWS account. We tick the box next to that account and hit the **Assign users** button, as illustrated in the following screenshot:

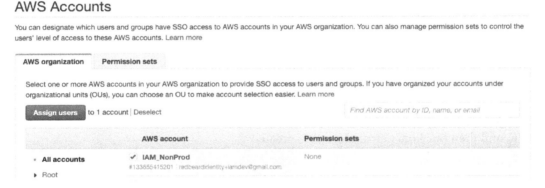

Figure 10.33 – Assigning users to IAM_NonProd

2. We open the **Groups** tab and select all of the groups that should have access to this account based upon our eight authorization use cases. Based upon those use cases, the `AWS_IT_IAM_Nonprod` group gets access. The `AWS_IT_IAM_Prod` and `AWS_IT_Support` groups will also be in scope for access, but in order to limit access of the `SuperUserAccess` permission set to just the `AWS_IT_IAM_Nonprod` group, we will only select that for now. We tick just that group and move on to assign permission sets to the groups.

3. We match the permission set to the group. This is how we constrain access within the environment. After we have selected the `PowerUserAccess` and `ReadOnly` permission sets, we can click **Finish**.

4. It will then provision the role into the account. We then repeat the process with the other groups that need access.

5. Let's do the AWS_IT_IAM_Prod group next. That group will only be allowed to have the ReadOnly permission set. We complete the process and start the cycle once more for the AWS_IT_Support group.

6. The AWS_IT_Support group gets the ITS_EC2_Support permission set. We complete that process, and we are done setting up access for the IAM_NonProd AWS account.

7. We can take a look at the groups and their permission sets to validate everything looks OK by clicking on the IAM_NonProd account, as illustrated in the following screenshot:

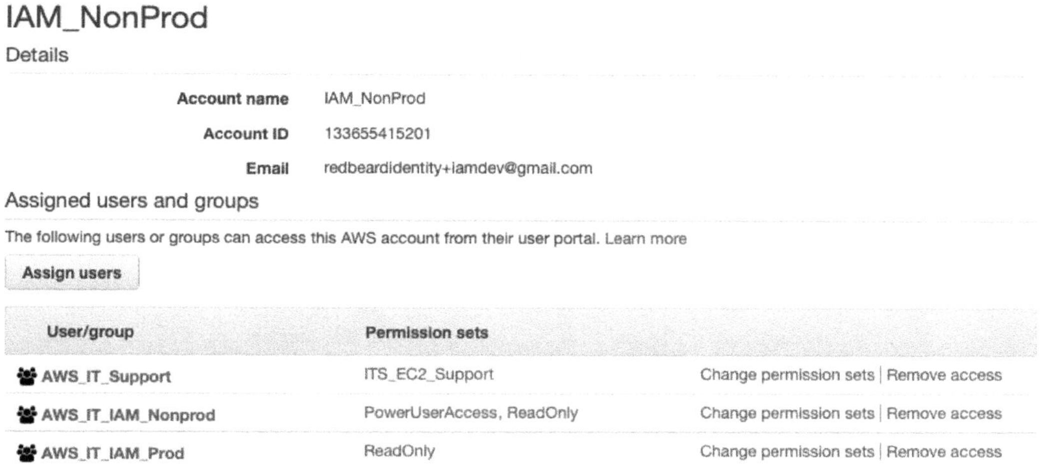

IAM_NonProd

Details

Account name	IAM_NonProd
Account ID	133655415201
Email	redbeardidentity+iamdev@gmail.com

Assigned users and groups

The following users or groups can access this AWS account from their user portal. Learn more

Assign users

User/group	Permission sets	
👥 AWS_IT_Support	ITS_EC2_Support	Change permission sets | Remove access
👥 AWS_IT_IAM_Nonprod	PowerUserAccess, ReadOnly	Change permission sets | Remove access
👥 AWS_IT_IAM_Prod	ReadOnly	Change permission sets | Remove access

Figure 10.34 – The groups and permission sets assigned to IAM_NonProd

We then iterate through all the remaining accounts, repeating this process until each account has the appropriate groups and permission sets assigned to fulfill our authorization use cases. Once we are done, our AWS accounts will look like this:

	AWS account	Permission sets	
• All accounts	IAM_NonProd #133655415201 · redbeardidentity+iamdev@gmail.com	PowerUserAccess, ReadOnly	and 1 more
▶ Root	Network_Prod #722412332062 · redbeardidentity+networkingprod@...	PowerUserAccess, ReadOnly	and 1 more
	Red Beard Identity #451339973440 · redbeardidentity@gmail.com	AdministratorAccess	
	Cloud_NonProd #554190754729 · redbeardidentity+clouddev@gmail....	ITS_EC2_Support, PowerUserAccess	and 1 more
	Sandbox #281142516251 · redbeardidentity+org3@gmail.com	ITS_EC2_Support, PowerUserAccess	
	Network_NonProd #592003414010 · redbeardidentity+networkingdev@...	ITS_EC2_Support, PowerUserAccess	and 1 more
	Sales_NonProd #105788611811 · redbeardidentity+org2@gmail.com	PowerUserAccess, ReadOnly	
	Sales_Prod #003980426125 · redbeardidentity+org1@gmail.com	PowerUserAccess, ReadOnly	
	Cloud_Prod #467019298634 · redbeardidentity+cloudprod@gmail...	ITS_EC2_Support, PowerUserAccess	and 1 more
	IAM_Prod #151796947722 · redbeardidentity+iamprod@gmail.c...	ITS_EC2_Support, PowerUserAccess	and 1 more

Figure 10.35 – Groups and permission sets assigned to AWS accounts

We can now see how this looks through a user's eyes to validate that our policies work as expected. Signing in as IT Support, we see our eight accounts available to us, as illustrated in the following screenshot:

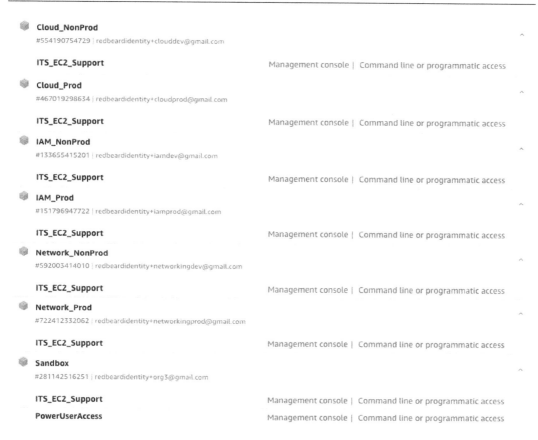

Figure 10.36 – IT Support's available accounts and roles

As our most granular policy was the `ITS_EC2_Support` policy, let's validate that this is working as intended. The `Iam Dev` user created a pair of EC2 instances inside the `IAM_NonProd` account. Let's sign in to that account as the IT Support user to see what happens when we try to stop them. You can see the outcome here:

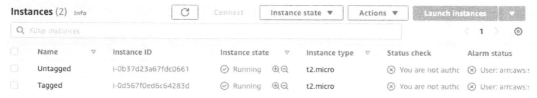

Figure 10.37 – IT Support can see the instances

The IT Support user can see the two instances. We will start by first trying to stop the untagged instance. Fortunately, we receive an error, indicating our policy appears to be working so far. Note in the following screenshot that it is enciphered; our user does not have the appropriate permissions to read the error message:

Figure 10.38 – Failure to stop the untagged EC2 instance

Next, we will try the tagged instance, as follows:

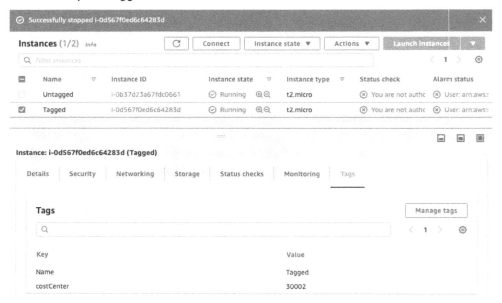

Figure 10.39 – Successfully stopping the tagged EC2 instance

The It Support user was able to stop this instance. We can see the tagging that made this possible in the preceding screenshot.

Now that we have validated our fine-grained authorization policies are working the way we want for our administrative use case, let's take a look at our administrative SSO and authorization configuration for AWS CLI access.

Administrative SSO using the AWS CLI

One of the primary benefits of using AWS SSO for administrative access is the issuance of temporary credentials. Whereas we have used durable programmatic credentials for AWS CLI access in the past, we can now use a browser for SSO and instantiate a temporary session without needing to issue or store those credentials on our workstation. We do this by selecting the command-line or programmatic access link after signing in to AWS SSO from our external IDP, as illustrated in the following screenshot:

Get credentials for PowerUserAccess   ✕

AWS account 133655415201 (IAM_NonProd)

Use any of the following options to access AWS resources programmatically or from the AWS CLI. You can retrieve new credentials as often as needed. Learn more

macOS and Linux | Windows | PowerShell

Option 1: Set AWS environment variables
Option 1: Set AWS environment variables Learn more

```
export AWS_ACCESS_KEY_ID="ASIAR6HT3ZGQSR2EEQ74"
export AWS_SECRET_ACCESS_KEY="OvpVYIZP5bt/yb7FI+X13+M7coCd9YxxRkvRxWa+"
export AWS_SESSION_TOKEN="IQoJb3JpZ2luX2VjEJv//////////wEaCXVzLWVhc3QtMSJHMEUCICFTGRZ4AobBpzvJEs
```

Option 2: Add a profile to your AWS credentials file
Paste the following text in your AWS credentials file (typically found at ~/.aws/credentials). Learn more

```
[133655415201_PowerUserAccess]
aws_access_key_id = ASIAR6HT3ZGQSR2EEQ74
aws_secret_access_key = OvpVYIZP5bt/yb7FI+X13+M7coCd9YxxRkvRxWa+
aws_session_token = IQoJb3JpZ2luX2VjEJv//////////wEaCXVzLWVhc3QtMSJHMEUCICFTGRZ4AobBpzvJEsRFhb7DKN
```

Option 3: Use individual values in your AWS service client (Learn more)

AWS Access Key Id	ASIAR6HT3ZGQSR2EEQ74	Copy
AWS Secret access key	OvpVYIZP5bt/yb7FI+X13+M7coCd9YxxRkvRxWa+	Copy
AWS session token	IQoJb3JpZ2luX2VjEJv//////////wEaCXVzLWVhc3QtMSJHMEUCICFTGRZ4/	Copy

Figure 10.40 – Our temporary AWS CLI credentials through AWS SSO

We will sign in as the `Iam Dev` user once again and copy the commands to export the variables we need to use the AWS CLI with our temporary credentials. These credentials are valid for the duration of the session we defined within the permission set for this assumed role. For this particular role, these credentials are good for 9 hours. Once we enter the values, we can validate that they work by entering a basic command. The process is illustrated in the following screenshot:

```
jonlehtinen@ ~ % export AWS_ACCESS_KEY_ID="ASIAR6HT3ZGQSR2EEQ74"
export AWS_SECRET_ACCESS_KEY="OvpVYIZP5bt/yb7FI+X13+M7coCd9YxxRkvRxWa+"
export AWS_SESSION_TOKEN="IQoJb3JpZ2luX2VjEJv//////////wEaCXVzLWVhc3QtMSJHMEUCICFTGRZ4AobBpzvJEsRFhb7DKMy32sW9j4QMK/OHJXUeAiEA9XThJDWGS
usT43d0Gbxz/jpokfjJNwW7fhJOEizqZxcq5AMINBAAGgwxMzM2NTU0MTUyMDEiDGBag29zS2wBWCMFFyrBAyorkKdJc9UN5sYaYyygu/EJ0b6PKxztf/XLNPSOr+c8vU8LiV92
J/1ydvNL7hbPU6ICJ5u4J5qhTwJW64ED6Z8WvnpaJXW5qv8VTvScfAZi+U1kxfFFvDYTwQ9+raw5yez8Seqp3mTWMw6MS8kyhxO20EIQa3ZMTXcmMs1dfULD2Olbwpy64ApLjNI
JQ8sTSXKjwJJD49q4QR29BFXxLsEaBUt1MtkrD3i8BFd30B08z99rUdUgQMqbL2T4dAwrekZDClLccgN7krsqd401hZvCSx7TRGTz5nh2W+k5QzZtFGuK1+k/fi92hvCtRasWsx+
dhvwfxOIET6hRuVC85pzZ7Rgq5jq0CV0Na/XbpMWj0l4v5hPbdJKccWPUP43Q/15pN9veVXaG3x92TFbxQwNWAeB9PgozBeCd7+mWgKeO5LNrjvqcp++2yHhKbjL6Q3HQOVr5yQ
cFjQvYqtHyR7At5z1f9YiUByLWY3EwQCp4y97i+qsN9wyx9y1JCdHPXTkuDjLqdId2bVNEzWAwIy4M8hwiSDoMeYmXtkky3xyUA7vAK+4zSLdcV81KbdCmkHsfDfJ2CXT7jnNMv
WRXWJErMMK/ThYUGOqYB/ZqXSBPjBuIxhhjxrJp+flhg4rmUnaqSTiiFsaF8masST1T+Vx8zNa5f+io5gK7joiHx/6RjwmP0KPqHdxhlfNvnXiOm9qPm4mQRnXT4EOG6yevXjYC
+QFyrS6g7nhdttuBEbdZ5pW+hJhgs2SBvcg1P5B8+31dHwUa+TaLz3R4MjpB03d2iv6Z+ZPCD/Kwj+X1Yn80B646BFUfKwg9mNn5EjMe1FA=="
jonlehtinen@ ~ % aws ec2
> aws ec2 describe-instances
Reservations:
- Groups: []
  Instances:
  - AmiLaunchIndex: 0
    Architecture: x86_64
    BlockDeviceMappings: []
    CapacityReservationSpecification:
      CapacityReservationPreference: open
    ClientToken: ''
    CpuOptions:
      CoreCount: 1
      ThreadsPerCore: 1
    EbsOptimized: false
    EnaSupport: true
    EnclaveOptions:
      Enabled: false
    HibernationOptions:
      Configured: false
```

Figure 10.41 – Setting temporary credentials through AWS SSO and validating they work

After creating a tagged and untagged EC2 instance, let's sign in to the AWS CLI as the IT Support user and attempt to stop both instances once again. We can get the instance **identifiers (IDs)** using the `describe-instances` command once again. `i-0b3670c6354bed31d` is untagged, and `i-01296365cf6d834e6` is tagged.

First, we will try the untagged instance, as follows:

```
jonlehtinen@ ~ % aws ec2 stop-instances --instance-id i-0b3670c6354bed31d

An error occurred (UnauthorizedOperation) when calling the StopInstances operation: You are not authorized to perform this operation. E
ncoded authorization failure message: WT9Gl1QS52jRRq2s5DfeTQF53s5o3qPXelqn3dHBJjKjO3_odokh1WJwM_kwb4PqnpNoiiLAMp4tm4a9ZvPFNgCKC1jyzV2eT
dX3jm9pra72CL7KZ1RR-r4adQdQeIBHu2ktV5h8pG9csaUrw4g-54G2oZGC7eBy_ooBket3I2JKZUlc3MVQ1fMPKiB4RgoLZZuobl Qil6SxsFKwmCHygdb78gXP4O0pEMKMMdwy
lvsgSDgT-0Jlvz DJ8o7Tyim3sfv5EKpd1lP0nvirg8FRt1Y_6vtMzoSb3mDlSdtnIZkGD4Xh-D4n1espdpVduE4qy-SSiOC9SPZa1pkJLs3Dfre8HrWIBcFdEBs0PH8K_UQkBQC
9NKlbcRqDZOSb0JQOS_qaRR7WvHdzPhTu25SdzNMm8lma2LZ2HNaqqe_p_Es86Mh5eH2hDhIsO5uaDKF7ds_vUp0We5pFRjwhyAimcmUGGJeQUqwA1wvIzkQV50X_5bViQc9bbZ
oAGwVgFy76zJdsWBZrrtVKAS5yubw8eZf8nIoxhVttoAF9xdCPiQu8UR57NJIZ-1g6-YKpc_5dXm2nIIKlsdRy-fxXKjUbUnMM1fUL z9GR_ig4XTypU_lhdlCj9ZGQwi0xIVHXS
orzEL3SEmTRDDN-5pVqX01ZY9t0UjL8xSEcX7J3n-WXM57jlRujPT1fdXy5Eiinw11zW_oH117DNgN7NaIDH4wL3SPAhJ7PXCI1KxB2mbkkNm3MeBajP6pu-z72s_Bj8QdAKZWp
3t0
jonlehtinen@ ~ %
```

Figure 10.42 – Untagged instance cannot be stopped by IT Support user from the AWS CLI

As predicted, we cannot stop it, given our entitlements. Next, we will try the tagged instance, as follows:

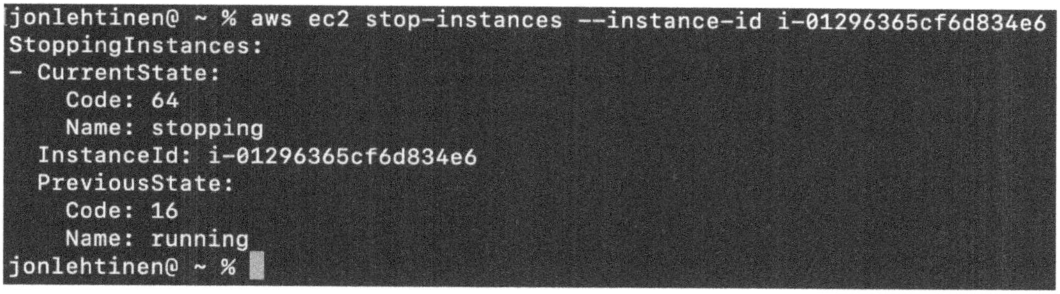

```
jonlehtinen@ ~ % aws ec2 stop-instances --instance-id i-01296365cf6d834e6
StoppingInstances:
- CurrentState:
    Code: 64
    Name: stopping
  InstanceId: i-01296365cf6d834e6
  PreviousState:
    Code: 16
    Name: running
jonlehtinen@ ~ %
```

Figure 10.43 – Tagged instance is stopped

This time, it works. Our administrative authentication and authorization model using an external IDP is working as expected for all our use cases.

With AWS SSO and permission sets, we are able to perform administrative functions from the CLI without creating or storing a long-lived programmatic credential and while using a federated identity.

Summary

In this chapter, we put into practice what we have learned across several AWS services to design and apply an administrative account authentication and authorization model. By using an external IDP, we were able to quickly deprovision access for administrators. Synchronizing our external IDP's users and groups into AWS SSO via SCIM laid the foundation for us to pair coarse-grained authorization control managed at the IDP with a fine-grained authorization policy controlled by AWS to fulfill our administrative authorization business objectives. We wrote a custom authorization policy for our permission sets using conditional operators. Finally, we saw how that model extends into the AWS CLI as well and improves security by eliminating long-lived programmatic credentials.

In the next chapter, we will switch our focus to application-centric identity using Amazon Cognito. We will address making our enterprise user accounts available for our applications hosted in AWS.

Questions

1. What is coarse-grained authorization?

2. What is fine-grained authorization?

3. What is a permission set?

4. How does a permission set grant access to the AWS accounts to which it is attached?

    a. It provisions users into that account.

    b. It automatically creates an assumable AWS IAM role for the federated user to assume in the managed AWS account.

    c. It doesn't.

5. AWS CLI access through AWS SSO removes the need to use long-lived programmatic credentials managed through AWS IAM.

    a. True

    b. False

Further reading

To learn more on the subject:

- *Manage SSO to your AWS accounts:* https://docs.aws.amazon.com/singlesignon/latest/userguide/manage-your-accounts.html

- *Attribute-based access control:* https://docs.aws.amazon.com/singlesignon/latest/userguide/abac.html

- *Actions, resources, and condition keys for AWS services:* https://docs.aws.amazon.com/service-authorization/latest/reference/reference_policies_actions-resources-contextkeys.html

Section 3: Implementing IAM on AWS for Application Use Cases

Modern organizations often have a hybrid cloud/data center strategy for their internal application portfolio. Whereas AWS-deployed applications could reach back into the on-premises data center to look up user attributes, replicating that information to the cloud using AWS Directory Services provides a better user experience and increased developer flexibility. Additionally, organizations can leverage Amazon Cognito to facilitate application identity use cases for AWS-hosted applications, all while retaining their existing identity provider as the authoritative source of user identity information.

This part of the book comprises the following chapters:

- *Chapter 11, Bringing Your Users into AWS*
- *Chapter 12, AWS-Hosted Application Single Sign-On Using an Existing Identity Provider*

11
Bringing Your Users into AWS

In the previous chapter, we implemented the authentication and authorization components of the administrative user model, which we initially conceptualized back in *Chapter 8, An Ounce of Prevention – Planning Your Administrative Model*. We accomplished our objectives through a combination of service control policies from AWS Organizations, AWS **Single Sign-On** (**SSO**) permission sets, and group-based access controlled by an external **identity provider** (**IDP**). Our requirements for administrative user access focused on gaining access to AWS accounts and the resources within those accounts. However, what are our options for providing user identity information to those applications that our organization intends to host on AWS?

In this chapter, we will review how administrative and non-administrative identity use cases differ, examine several possible solution architectures to solve this challenge (some using AWS services and some not), and then build a solution that will enable AWS-deployed applications to access user information through our organization's authoritative sources.

In this chapter, we will cover the following topics:

- Distinguishing administrative users from non-administrative users
- Solutions to non-administrative user use cases for apps on AWS
- Using Managed AD and trusts
- Creating a trust between AWS Managed AD and on-premises AD

Technical requirements

To get the most out of this chapter, you will need the following:

- An AWS account
- A SAML2 or SCIM-compliant IDP, such as Okta Identity Cloud, PingOne, or Azure Active Directory
- An Active Directory domain
- A remote desktop client to sign in to remote Windows servers, such as Microsoft Remote Desktop

Distinguishing administrative users from non-administrative users

We already have a connection to our AWS accounts via AWS SSO and our external IDP for the user accounts that are entitled to access our AWS environments. However, what can we do to ensure that all user accounts are available to applications in AWS, even if the users never need administrative access to an AWS account? To answer that question, first, we need to clarify how we define each of these types of accounts. In the broadest terms, administrative accounts are accounts that have enhanced permissions to modify system settings, create other accounts, and change the permissions for what other accounts can do. For our Redbeard Identity AWS use case, we classified administrative accounts as those accounts that had access to and could manipulate resources within an AWS account. We made distinctions as to where a given user had their administrative privileges via group memberships to specific accounts and permission sets that limited their abilities once they federated into their available AWS account.

So far, in the Redbeard Identity example, we have not created a distinction between administrative accounts and standard user accounts within the Redbeard Identity organization. The best practice is to separate highly privileged accounts from standard user accounts. Whereas some organizations might not consider the day-to-day work that occurs within an AWS account to be highly privileged, for the purposes of our Redbeard Identity example, we will say that access to the AWS administrative backplane is a privileged administrative function and that the accounts used when accessing AWS are distinct from common user accounts. However, even if we were to assume that Redbeard Identity had a single set of accounts for all of its use cases, the next question is whether that single connection from the external IDP to AWS SSO would be sufficient for all administrative and non-administrative use cases that the organization has for both AWS administration and AWS-hosted applications.

There are a few application-centric use cases that are not well-served by that single-connection model. For example, users within a group that grants access to an AWS account are currently available within the AWS SSO user directory. Applications hosted on AWS that need to be accessible to everyone within an organization require their IDP to have access to the full user population within its identity store, not just the subset of users who are members of an AWS administrative group.

AWS SSO might be a fully featured IDP capable of handling the authentication and authorization of applications that use it for federated identity, but organizations might not want to provide two distinct authoritative sources of identity within their applications. The Redbeard Identity organization already has a strategic IDP that it uses as its authoritative source for its workforce identity. Although AWS SSO was leveraged to facilitate account access and administrative authorization, the strategic external IDP remains the authoritative source of identity for AWS SSO.

Next, we will decide how to solve the challenge of exposing the full population of Redbeard Identity users to AWS-hosted applications.

Solutions to non-administrative user use cases for apps on AWS

Let's consider some of the solution architectures that are available to us when providing access to non-administrative user identity information to applications hosted within AWS. We will start with a baseline where we do not leverage any AWS services at all in order to access our user identities:

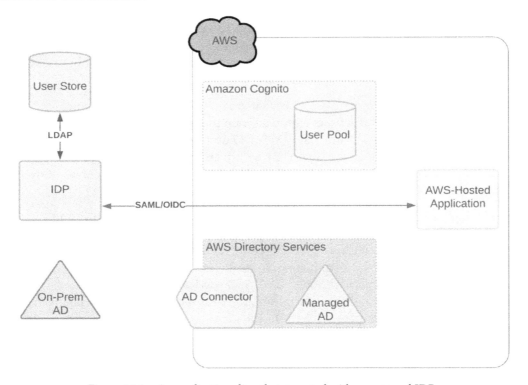

Figure 11.1 – An application directly integrated with an external IDP

In this configuration, the application, or its web server, is configured to operate as either a SAML service provider or an **OpenID Connect** (**OIDC**)-reliant party. Previously, we mentioned how services such as Amazon Cognito offer SDKs and code samples to facilitate application integration with those services. Standards bodies and open source communities offer similar plugins, SDKs, and web server modules that are designed to facilitate the adoption of standards-based identity protocols, such as SAML2 and OIDC. While this reduces the barrier for adoption, this model still requires the application team to maintain either the plugin or the web components that handle authentication and authorization. The web app either does not maintain a user store (that is, a single-page app) or provisions local accounts in a just-in-time fashion.

Tip

Organizations such as the OpenID Foundation maintain a library of certified (meaning the submitted has paid for independent compliance validation) and uncertified libraries and plugins to help app and web servers consume OIDC. The Shibboleth Consortium maintains Shibboleth SP, which is a popular open source SAML2 service provider plugin. Another popular open source SAML2 implementation is SimpleSAMLphp, which is maintained by Uninett. Links to all of these resources can be found in the *Further reading* section of this chapter.

A minor variation on this design includes an additional connection to the external IDP's **System for Cross-domain Identity Management (SCIM)** endpoint to address user provisioning and attribute synchronization into the application's user store:

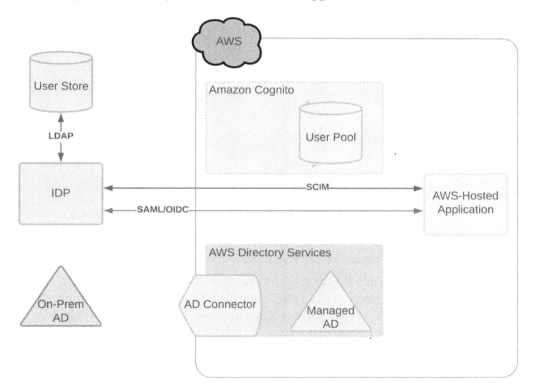

Figure 11.2 – An application with direct IDP authentication and SCIM provisioning

While this is better than the previous architecture, it still places significant administrative responsibility on the app team to maintain and harden the identity components of the application. However, this design does address provisioning, deprovisioning, and attribute synchronization.

Continuing with the architectures that do not use AWS components, the application could use the external IDP for user authentication and just-in-time provisioning, and then perform an additional attribute lookup through an LDAP lookup against either an external IDP's user store or a proxied view of that user store:

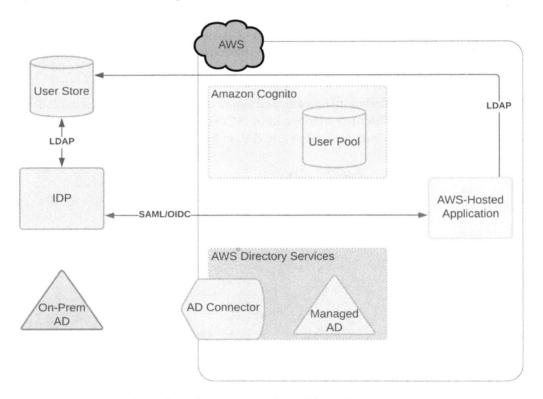

Figure 11.3 – An external IDP for authentication with an additional attribute lookup via an LDAP bind

We do not recommend this pattern for several reasons. First, securely executing an LDAP bind across not just security domains but also from AWS to an IDaaS platform or our own on-premises user store requires significant additional network and security architecture to execute. Secondly, an application that cannot get all the necessary user attributes from the OIDC ID token or SAML assertion at login time has authorization requirements that require bringing the application user store and authoritative user store closer in sync to avoid any data discrepancy issues that could lead to unauthorized access issues.

For applications where directory lookup is required for user authentication and authorization, theoretically, we could bind directly against either an on-premises AD or our external IDP's user store:

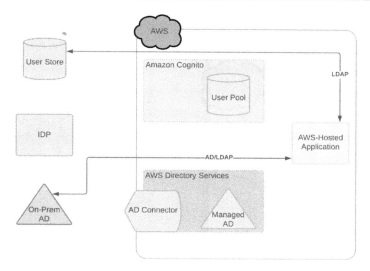

Figure 11.4 – An ill-advised directory-centric application architecture

However, the same caveats apply as before; this requires significant infrastructure to ensure secure communication between the external directories and the AWS-hosted application. For Active Directory workloads, this is particularly ill-advised as AD uses the full port range on the network for certain functions.

A far more secure solution architecture involves leveraging AWS Directory Services for applications that require Active Directory support. We can use the AD Connector to connect the application to the on-premises Active Directory domain controller:

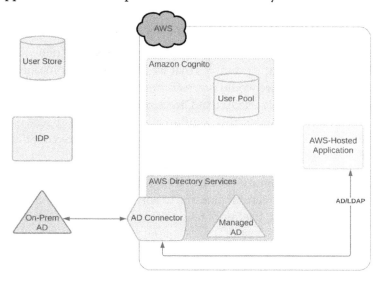

Figure 11.6 – An application using the AD Connector for AD workloads

Alternatively, we could extend the on-premises AD forest using a trust with a Managed AD instance with AWS Directory Services:

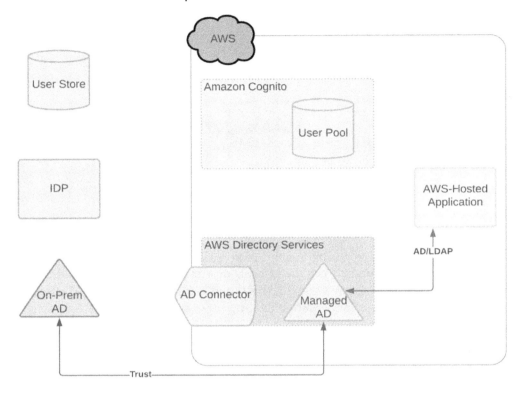

Figure 11.7 – Using Managed AD for AD workloads

While AWS Directory Services is fine for organizations that have AD dependencies, we have not needed AD with any of the examples that we have gone over in our Redbeard Identity organization. Although we would not recommend deploying AD infrastructure if an organization does not already use Active Directory, the pervasiveness of Active Directory within the enterprise makes this solution attractive. This is because it enables us to easily and securely expose user identity information to AWS-hosted applications.

The Redbeard Identity example focused on standards-based protocols and cloud-based implementations. If Redbeard Identity were a real organization, we would encourage it to continue following these architectural patterns to solve its use cases. As such, we could recommend that it uses a Cognito user pool to expose its user information to AWS:

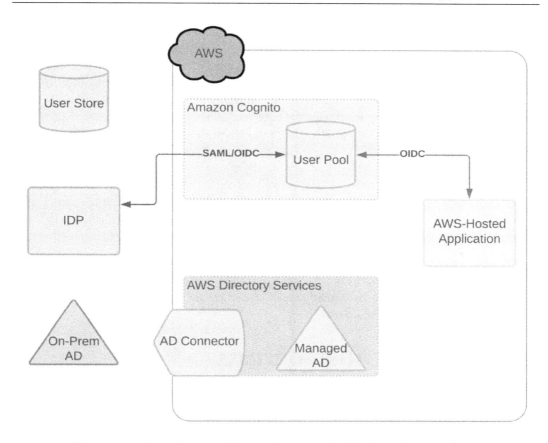

Figure 11.8 – An application references Cognito, which looks to the external IDP

In this design, the application will look to an Amazon Cognito user pool for its user information. The user pool will act as the application's user store, and detailed attributes will be provided at authentication time through the Cognito identity token. Since Amazon Cognito user pools provide a standards-compliant OIDC IDP, additional attributes can be accessed through the /userinfo endpoint, as needed, if the application is sufficiently entitled and scoped to have that access. To ensure that the Redbeard Identity organization is still able to control access to its applications despite the AWS-hosted application looking to Amazon Cognito for user authentication and attributes, we can configure the user pool to be populated through Redbeard Identity's strategic IDP. This way, we will be able to populate the pool with the full complement of employee identity records, whether they have access to an AWS account or not. Additionally, we will be able to ensure that access to AWS-hosted applications is revoked when the user is terminated inside the external IDP's user store.

So far, this solution appears to be the one that is most aligned with how the Redbeard Identity organization has operated based upon our previous implementation examples; however, we won't use it here. This design comes with a number of limitations that could make it less desirable compared to using something such as a Managed AD instance in a trust with an on-premises AD directory, chiefly Amazon Cognito user pools relying upon just-in-time provisioning when using an external IDP. We can ensure users will lose access to applications that look to the user pool when they are terminated at the external IDP, but this design won't disable the account object within the user pool. Additionally, applications that have directory requirements along with user authentication and authorization requirements would not be well served if this was the only way we exposed our user's identities to AWS. This is because they would only be able to query the directory for the subset of users that had already authenticated into the application via the user pool.

As such, we will introduce Active Directory to the Redbeard Identity organization. In the next section, we will create a Managed AD instance and build a trust with an existing AD forest. This will allow the AWS-hosted applications to have access to the on-premises user account information through that trust.

Using Managed AD and trusts

We will bring our non-administrative users into AWS using a Managed AD instance in AWS Directory Services. Strictly speaking, we don't even need to import our user's accounts into the Managed AD environment in order to accomplish our goal. We can arrange for the Managed AD instance to perform lookups and binds against our on-premises AD forest using a trust. A **trust** allows two or more AD domains to authenticate against resources that are available in the other:

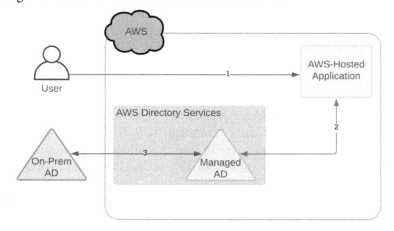

Figure 11.9 – A user signing in to an app through a domain trust

Consider the example in *Figure 11.9*. An AWS-hosted application that requires either AD or LDAP for user authentication or authorization is configured to look to an AWS Managed AD instance for user information. The Managed AD and the on-premises AD have a two-way trust:

1. The user signs in to the application.

2. The application looks to the Managed AD to verify the user's credentials. However, if the user's account is not found inside the Managed AD's domain, it will then look inside the on-premises AD for that account.

3. It finds the account there, validates the credentials, pulls out the appropriate user information, and the user has access to the application.

The name of the existing on-premises AD deployment that Redbeard Identity has is `example.local`. Note that it is important to keep the names straight as we build the trusts.

> **Tip**
>
> We will be addressing the critical steps that are needed to establish a trust between an on-premises AD forest and an AWS Managed AD forest. As organizations can have several different network architectures, we will not be diving deeply into the network connectivity considerations between our on-premises network and the AWS **Virtual Private Cloud** (**VPC**) where the AWS Managed AD instance will be deployed. More information on the required ports and VPC configurations can be found in the *Further reading* section at the end of this chapter.

Next, let's take a look at what it takes to create a Managed Microsoft AD instance using AWS Directory Service.

Creating a Managed Microsoft AD instance

The Redbeard Identity organization has an existing on-premises Active Directory that has a user account for every worker in the organization. Its **Identity Governance and Administration (IGA)** platform provides each new employee with an account inside that AD domain, and attributes are synchronized between that domain and the IDP's user store. Though individuals might have different accounts for SSO through the IDP and AD, they are all correlated as part of the same identity:

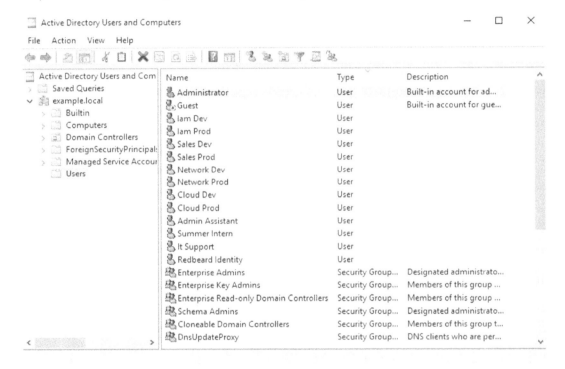

Figure 11.10 – The view of user accounts inside an on-premises AD

We need to build a matching AD instance using AWS Directory Service. To do this, perform the following steps:

1. Go to AWS Directory Service in the AWS Management Console. From the left-side menu, select **Directories**, which is beneath **Active Directory**.

2. Click on the button labeled **Set up directory**.

3. We will select the **AWS Managed Microsoft AD** option and then click on **Next**.

4. Here, we have options for sizing. As Redbeard Identity is a small organization, we can stick with the **Standard Edition** option:

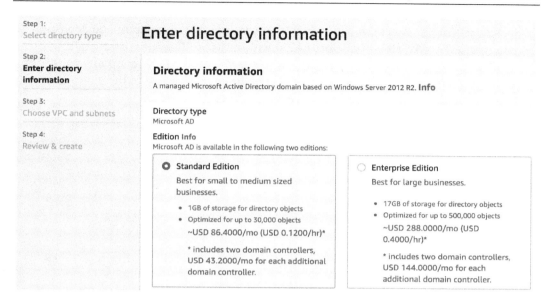

Figure 11.11 – The Standard Edition or Enterprise Edition sizing options

5. We also need to provide the domain name, NetBIOS name, description, and administrator password for the domain admin account in this section:

Directory DNS name
A fully qualified domain name. This name will resolve inside your VPC only. It does not need to be publicly resolvable.

corp.example.com

Directory NetBIOS name - *Optional*
A short identifier for your domain. If you do not specify a NetBIOS name, it will default to the first part of your Directory DNS name.

CORP

Maximum of 15 characters, can't contain the following characters: ` / : * ? " < > | ` . It must not start with ` . ` .

Directory description - *Optional*
Descriptive text that appears on the details page after the directory has been created.

RBI managed AD

Maximum of 128 characters, can only contain alphanumerics, and the following characters: ` _ @ # % * + = : ? . / ! - ` . It may not start with a special character.

Admin password
The password for the default administrative user named Admin.

•••••••••••

Passwords must be between 8 and 64 characters, not contain the word "admin", and include three of these four categories: lowercase, uppercase, numeric, and special characters.

Confirm Password

•••••••••••

This password must match the Admin password above.

Cancel   Previous   Next

Figure 11.12 – Additional configuration options

6. Next, we need to select the VPC where the managed AD instance will be deployed, along with two **Availability Zones (AZs)** within that VPC where the domain controllers will be deployed. The VPC, where a Managed AD is set up in a trust with an on-premises AD, will require some sort of direct connection to the enterprise network, either through a direct connection or VPN:

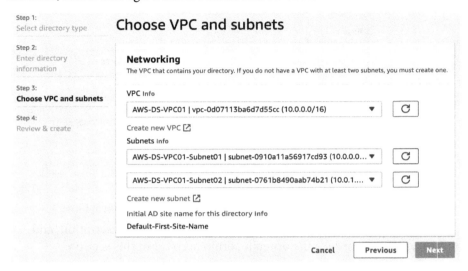

Figure 11.13 – The VPC and subnet settings

7. Finally, we verify that the configuration is what we want, and we create our Managed AD instance:

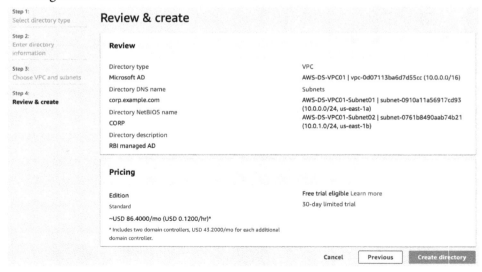

Figure 11.14 – Confirming the settings for our Managed AD instance

> **Tip**
>
> Most people do not have access to an on-premises AD environment for testing like this. Please refer to the *Further reading* section at the end of this chapter for tutorials on how to set up a lab in AWS EC2 to create a simulated on-premises AD environment that can be used to set up a trust with AWS Managed Microsoft AD.

The Managed AD instance will take some time to complete the setup. Once it has been completed, we can view the details of our new AD environment by clicking on the directory ID from the AWS Directory Service:

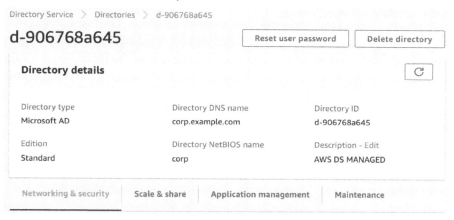

Figure 11.15 – Our Managed AD is complete

Now we have our AWS-hosted AD instance, called `corp.example.com`. Next, we need to make several configuration changes within our on-premises AD environment in preparation for establishing the trust.

Preparing the on-premises AD for a trust – conditional forwarders

Active Directory provides much more functionality beyond user and machine identity, such as DNS. As such, we will need to make some non-identity-related configuration adjustments to our AD environments before we can implement our trusts; specifically, we need to define **conditional forwarders**. Conditional forwarders are DNS servers that forward requests to specific DNS servers in a network based upon the DNS name that was queried. In AD trust topologies, we set up conditional forwarders so that requests to resolve DNS names belonging to a different forest within the trust can be fulfilled by the DNS servers managed by that domain. This configuration ensures the DNS servers that can correctly resolve a domain name to an IP address receive the DNS requests.

First, we need to create a conditional forwarder within `example.local` that will route all requests for resources on `corp.example.com` to the DNS servers hosted on `corp.example.com`. Please note that, for this example, we are working with Windows Server 2019:

1. Sign in as an admin to the `example.local` domain controller and open the **Server Management Console**.

2. From the top-level menu, select **Tools**, and then click on **DNS**. This will open the **DNS Manager** utility.

3. Expand the domain controller menu in the left-hand pane. This will reveal a folder called **Conditional Forwarders**.

4. Right-click on the **Conditional Forwarders** folder and select **New**.

5. We need the DNS domain values and the IP addresses of the DNS domain values for the AWS Managed AD instance. We can find those values by clicking on our Managed AD instance in AWS Directory Service. The DNS domain name is `corp.local.com`, and the DNS address values can be found underneath the **Networking & security** tab:

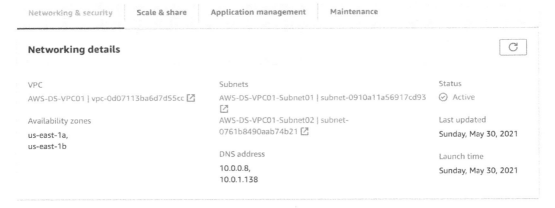

Figure 11.16 – The DNS addresses for the corp.example.com domain

6. Returning to the domain controller, we will enter the DNS domain value of `corp.local.com` into the **DNS Domain** field. Then, we will also enter each of the IP addresses into the IP addresses field. Finally, we will tick the **Store this conditional forwarder in Active Directory** box and select **All DNS servers in this forest** as the replication option:

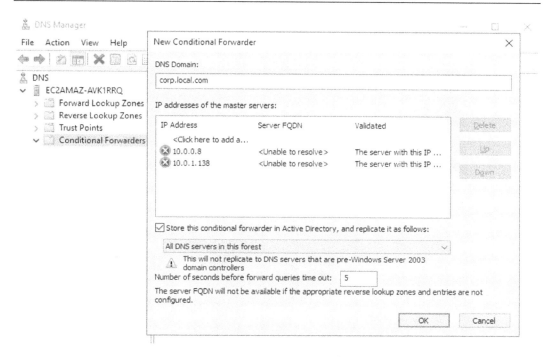

Figure 11.17 – Setting up the conditional forwarder on the on-premises domain

7. Click on **OK** and exit.

8. While we are here, we also want to capture the DNS server IPs used by `example.local` for when we need to build the conditional forwarder on `corp.example.com`. Right-click on the server object inside the **DNS Manager** window and then select **Properties**.

9. Click on the **Interfaces** tab to view the IP addresses of the DNS server. Make a note of the value for when we create the other conditional forwarder:

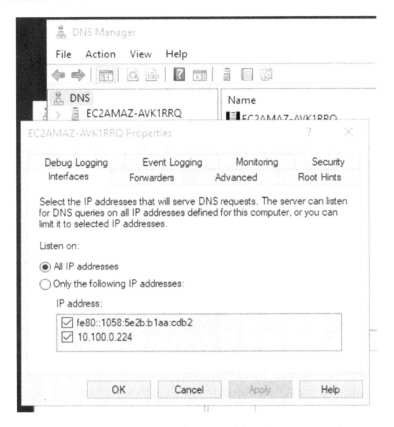

Figure 11.18 – Capturing the example.local DNS server IP

We are now ready to begin building the trusts.

Creating the trusts between on-premises and AWS Managed AD

In this section, we will continue working on the example.local domain controller. Open the **Server Manager** on the domain controller, and follow these steps:

1. From the top-level menu, select **Tools**.

2. Then, select **Active Directory Domains and Trust**.

3. In the **Active Directory Domains and Trusts** console, right-click on the example.local domain and select **Properties**. Then, navigate to the **Trusts** tab:

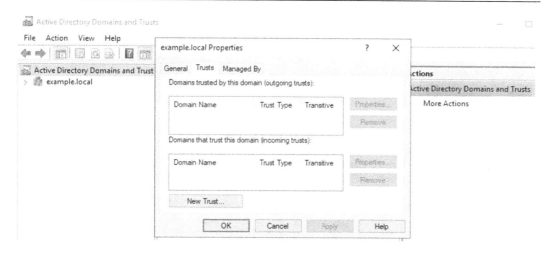

Figure 11.19 – The Trusts tab for the example.local domain

4. Click on the **New Trust** button to open the **New Trust Wizard**. Proceed past the first welcome screen and enter the name of the domain in which you want to establish the trust. Enter `corp.exaple.com` and then click on **Next**:

Trust Name
You can create a trust by using a NetBIOS or DNS name.

Type the name of the domain, forest, or realm for this trust. If you type the name of a forest, you must type a DNS name.

Example NetBIOS name: supplier01-int
Example DNS name: supplier01-internal.microsoft.com

Name:

corp.example.com

< Back   Next >   Cancel

Figure 11.20 – Naming the domain for the trust

5. We will select a forest trust for this use case as both domains are owned by a single organization and will be inhabited by the same population of users. Click on **Next**:

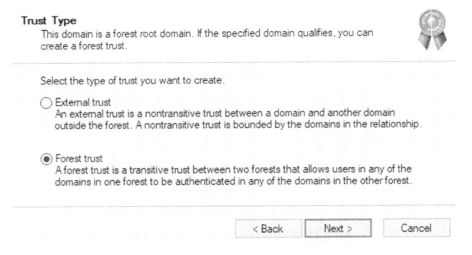

Figure 11.21 – Defining the trust type

6. Next, we will define the directionality of the trust relationship. We will stick with a two-way trust for this use case and then click on **Next**:

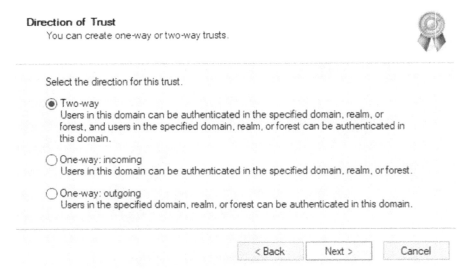

Figure 11.22 – Defining the trust directionality

7. For now, we will only create the trust on `example.local`. So, leave it set to **This domain only** and move on:

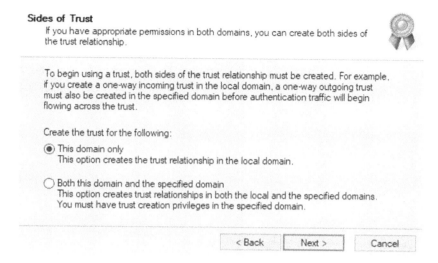

Figure 11.23 – Sides of Trust

8. Next, we will define a trust password that will be used to validate new trust relationships from the target domains. Enter and confirm the password and move on. We will use this password when we build the trust from the other domain inside this one.

9. Next, we get to confirm our settings. If we are satisfied, we can click on **Next**. We will see that our trust creation has been completed, at least on this side. Click on **Next** again to move on:

Figure 11.24 – Trust Creation Complete

10. The next two windows provide opportunities to confirm the incoming and outgoing trusts. We will skip this for now. That's because we have not built the other half of the trust on corp.example.com yet. Finally, we hit the end of the wizard; click on **Finish** to close the window. Now we can view the new trust in our **Trusts** tab:

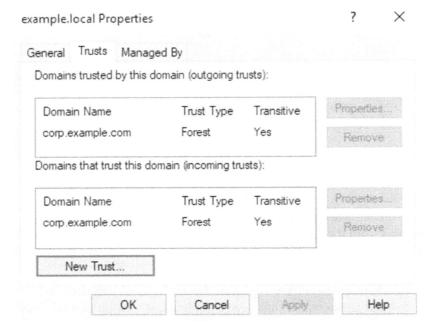

Figure 11.25 – The new trust with corp.example.com from example.local

With that out of the way, we are now ready to build the second half of the trust on the AWS Managed Microsoft AD instance.

Preparing the Managed AD for a trust – conditional forwarders

When using AWS Managed AD, we are not able to sign in to the domain controllers to make configuration changes to the forest. The AWS Directory Services control pane allows us to access many of the basic domain administrator tasks, and in theory, we shouldn't need to preconfigure anything in the Managed AD forest for the trust to be established. In practice, we can sidestep some trust verification issues by preconfiguring the conditional forwarders for the example.local domain on the corp.example.com domain controllers.

To do that, we will need to join an EC2 instance to the Managed AD domain, and sign in to it using the Managed AD's domain admin's credentials.

Joining an EC2 instance to the Managed AD domain

Before we create the EC2 instance, we will use it to manage the Managed AD domain controllers. First, we will create an AWS IAM role that will allow the EC2 service to automatically join EC2 instances to instances of AWS Managed AD. AWS offers managed policies for this function. We won't go through all of the steps required to launch the EC2 instance here, but we will highlight the essential steps to enable that EC2 instance to automatically join the domain.

We will need to reference this role when we create the EC2 instance, so let's create it:

1. From AWS IAM, navigate to the **Roles** menu and click on the **Create role** button.

2. As this role will be used by an AWS service, we will leave that option selected. We will select the **EC2** option underneath **Choose a use case**, as that is the service that will be using the role. We are ready to proceed:

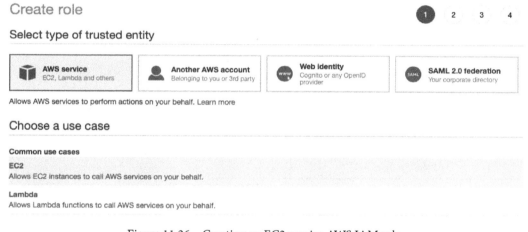

Figure 11.26 – Creating an EC2 service AWS IAM role

3. We will attach two managed policies to this role, that is, `AmazonSSMManagedInstanceCore` and `AmazonSSMDirectoryServiceAccess`. After we have found each one and selected it, we can proceed to the next step.

4. We could optionally tag the role if we want to. Once we have set the tags to our liking, we can proceed. Remember to take one last look at the role before creating it.

5. Finally, we can name our role and give it a description. We will call it `EC2DomainJoin`. Update the description and create the role.

6. The role is now available for us to assign to our EC2 instance:

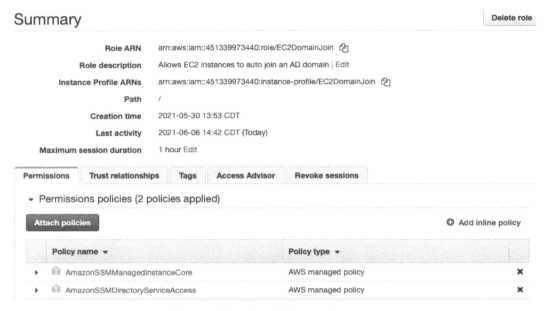

Figure 11.27 – Our EC2DomainJoin role

The other prerequisite for the EC2 instance is a security group that allows us to RDP into it and allows communication between itself and the AWS Managed AD domain controllers. Our Managed AD forest is deployed on its own VPC and two subnets with the following CIDR ranges:

Name	CIDR range
AWS-DS-VPC01	10.0.0.0/16
AWS-DS-VPC01-Subnet01	10.0.0.0/24
AWS-DS-VPC01-Subnet02	10.0.1.0/24

Table 11.1 – The Managed AD's VPC and subnet CIDR ranges

Our EC2 instance will need to be deployed inside that VPC as well; otherwise, modifications will need to be made to the VPC's route table to allow communication from an external VPC if we want to deploy it elsewhere. As the intricacies of VPC networking are beyond the scope of the topic at hand, we will simply deploy it to the same VPC as the Managed AD forest. Regardless, we will need to capture the VPC ID for future reference when creating the security group.

> **Tip**
>
> The VPC, subnet, routing table, and security group requirements for Managed AD will vary depending upon your organization's enterprise architecture. It is always a best practice to reduce the scope of a security group to the minimum access that it requires to operate, though additional consideration will be required should any future modifications be required.

We could safely operate by allowing connections from the subnet's IP ranges on the ports used by Active Directory, as we know this VPC will not see any further use. However, in the interest of operating securely, we have a comparatively simple option to constrain connectivity into the EC2 instance from just the Managed AD domain controllers. The AWS Directory Service automatically created a security group for its domain controllers when it deployed our Managed AD domain. We can build the security group for the EC2 instance to only allow connections from that security group ID on the ports used with Active Directory. The values we would need in this security group are as follows:

Protocol	Port range	Source
TCP	3389	My IP (RDP)
TCP	53	DC Security Group ID (DNS)
TCP	88	DC Security Group ID (Kerberos)
TCP	389	DC Security Group ID (LDAP)
TCP	464	DC Security Group ID (Kerberos password)
TCP	445	DC Security Group ID (SMB/CIFS)
TCP	135	DC Security Group ID (Replication)
TCP	636	DC Security Group ID (LDAPS)
TCP	49152-65535	DC Security Group ID (RPC)
TCP	3268-3269	DC Security Group ID (LDAP GC and LDAPS GC)
UDP	53	DC Security Group ID (DNS)
UDP	88	DC Security Group ID (Kerberos)
UDP	123	DC Security Group ID (Windows Time)
UDP	389	DC Security Group ID (LDAP)
UDP	464	DC Security Group ID (Kerberos password)

Table 11.2 – The security group settings for the EC2 instance

Let's create this security group so that we will have it available to attach to the EC2 instance when we create it. From the AWS Management Console, navigate to the EC2 service. From there, perform the following steps:

1. From the **Network & security** menu, select **Security Groups**.

2. First, we will search the list of existing security groups for the one that was created by AWS Directory Services for the Managed AD domain controllers. Once we have identified it, we will capture its security group ID:

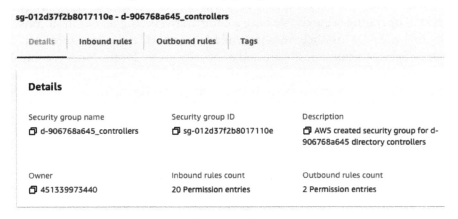

Figure 11.28 – The details of the security group

3. Now that we have the security group ID value, we can proceed by creating a new security group. After clicking on the **Create** button, give the security group a name, description, and select the VPC where you want it to be deployed. Give it a name and description that is memorable and descriptive, but most importantly, select the VPC ID where the Managed AD is deployed:

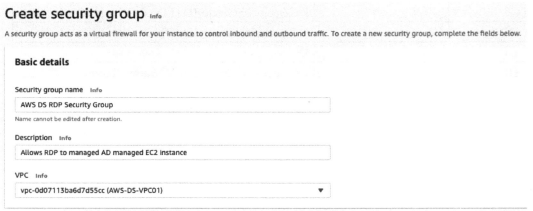

Figure 11.29 – The name, description, and VPC ID for the security group

4. Next, we will set up all of the inbound rules that we outlined earlier:

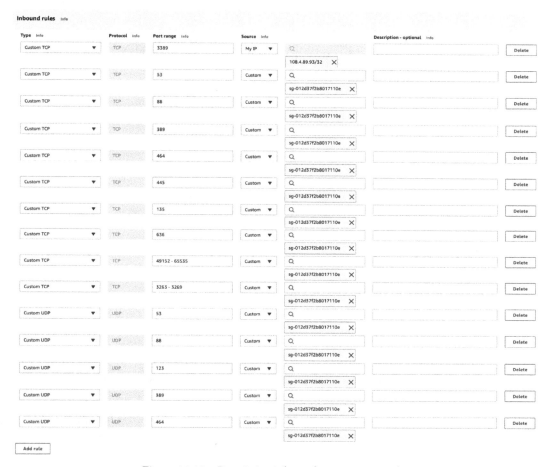

Figure 11.30 – Restrictive inbound connectivity rules

5. We will leave outbound rules in their default state, and we can add tags that we deem appropriate. Now, we can create the security group.

We now have the security group, IAM role, and VPC information. Next, we need to create the EC2 instance. The only additional requirements to create the EC2 instance are to select a Windows Server 2016 or later AMI, to select an instance type that has sufficient resources to run whichever version of Windows Server we choose, and to automatically assign the instance a public IP so that we can RDP into it. We will apply the majority of our selections during *Step 3* of the instance creation, as follows:

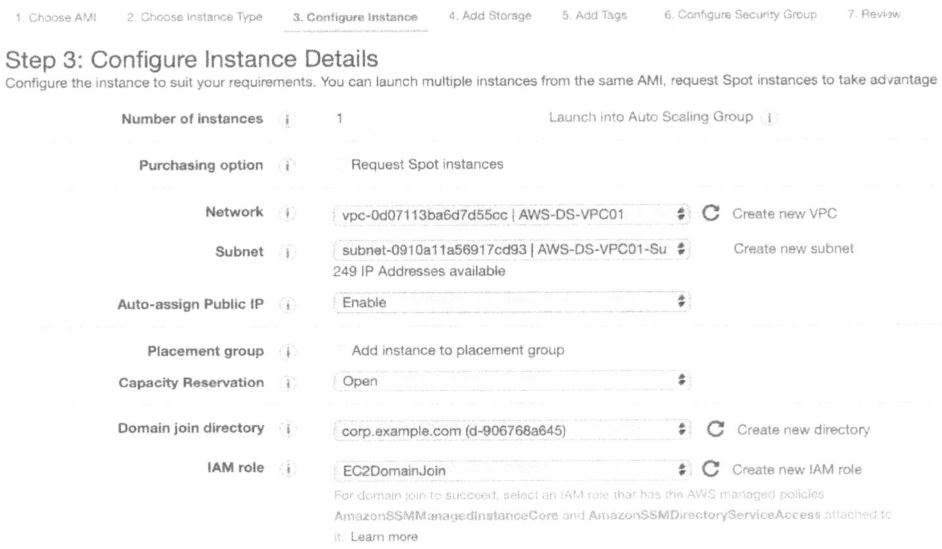

Figure 11.31 – The EC2 instance VPC, public IP, domain join, and IAM role selections

In the preceding screenshot, where we applied the IAM role we created, we also get to select the directory where the instance will be joined. Since this instance is meant to aid in the administration of the `corp.example.com` domain, we will select that one.

We can proceed through the wizard making whatever other customizations are required for our use case until *Step 6*, where we must select the security group that we created earlier:

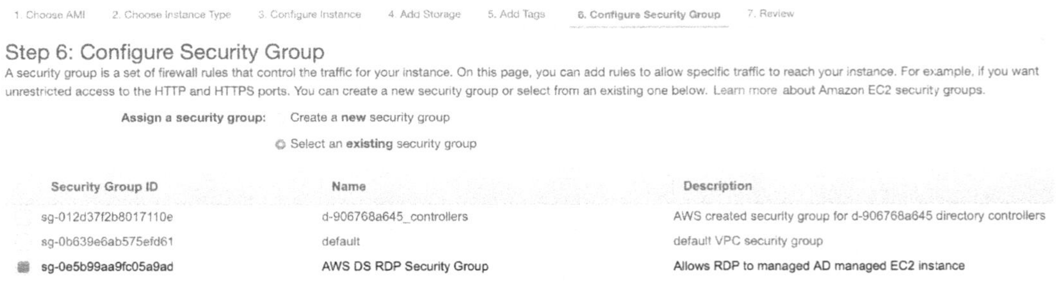

Figure 11.32 – Applying the security group to the EC2 instance

After reviewing all of our selections, we can create our instance. Once our instance is running and we have recovered the password of the administrator, we sign in to it using an RDP client. Although we have joined this instance to the Managed AD forest, we aren't done yet. There are still additional configuration steps that are needed within Windows Server itself to enable it to act as a remote domain controller management server.

Enabling AD management tools on the Managed AD management server

We have signed in to the EC2 instance as a local administrator. Though this server is joined to the domain, it is not configured to administrate any domain controllers within the domain where it is joined. We will install the AD management tools needed to enable us to administrate the domain controllers in `corp.example.com`, at which point we will sign in to this server using the domain admin account of the Managed AD domain.

From the Windows Server desktop, perform the following steps:

1. Open the **Start** menu, and select **Server Manager**.

2. From the dashboard, click on **Add roles and features**. This opens the **Add Roles and Features Wizard**.

3. We can click past the first screen and go straight to **Installation Type**. Make sure that **Role-based or feature-based installation** is selected, and then move on.

4. On the next screen, leave the default option of **Select a server from the server pool** selected, and make sure that your local server, identified by its internal IP address, is selected. Assuming that is in order, select **Next**:

Figure 11.33 – Selecting the local server for feature installation

5. We will skip the **Server Roles** by selecting next and moving straight to the **Features**. Here, tick the box next to **Group Policy Management**. Expand the drop-down menu next to **Remote Server Administration Tools** and then expand the drop-down menu next to **Role Administration Tools**. From there, tick the boxes next to **AD DS and AD LDS Tools** and **DNS Server Tools**. When those are all checked, click on **Next**:

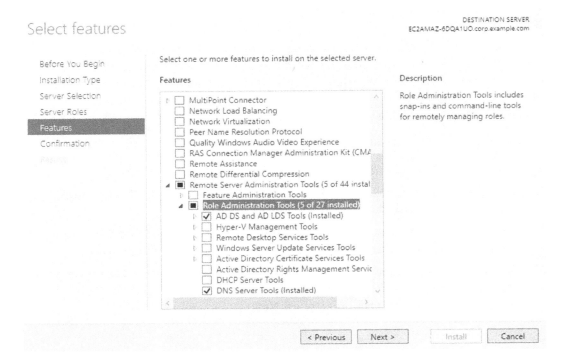

Figure 11.34 – Selecting the required AD management features

6. From the **Confirm Selections** screen, verify your selection and click on **Install**.

7. Once this has been completed, you will have the tools you need to administrate the Managed AD domain from the **Start** menu:

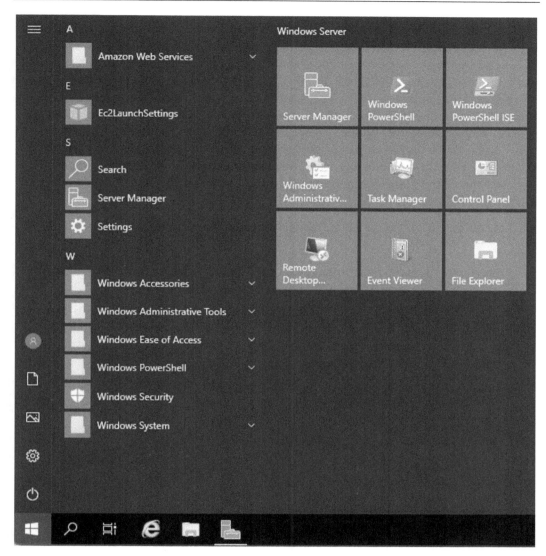

Figure 11.35 – The necessary tools have been installed

With that prework done, we can now sign out of the EC2 instance and sign back in using the username and password that we defined for the domain admin when we first set up our Managed AD instance.

Configuring the Managed AD conditional forwarder

Finally, we are ready to perform the task we mentioned that we were going to do several sections ago, that is, to configure the conditional forwarder for the on-premises AD domain in the Managed AD domain. Strictly speaking, this step, and all of the prework we performed to arrive at this step, is not considered necessary. However, it was the author's experience that the two domains could not verify their mutual trust until the conditional forwarder was preconfigured on the Managed AD domain. Furthermore, given that it is almost certainly an administrative necessity that we would build a management server for our Managed AD domain in any event, we recommend defining the conditional forwarders and all the prework required to get there, as ultimately, this will still benefit the reader despite not being *required*.

With that behind us, let's define a conditional forwarder so that we can finally build the trust:

1. After signing in to the EC2 instance using the domain administrator credentials for the Managed AD domain, open the **Server Manager** dashboard.

2. In the top-level menu, open the **Tools** menu, and select **DNS** from the drop-down menu:

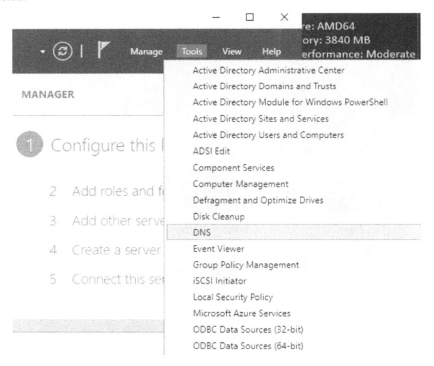

Figure 11.36 – The Managed AD Domain DNS configuration options

3. This opens up the **DNS Manager** tool. We will need to connect to one of the Managed AD domain's DNS servers. We can find those values in the **Networking Details** section of our Managed AD directory in AWS Directory Services. Once you have the IP address for one of them, enter that value and click on **OK**:

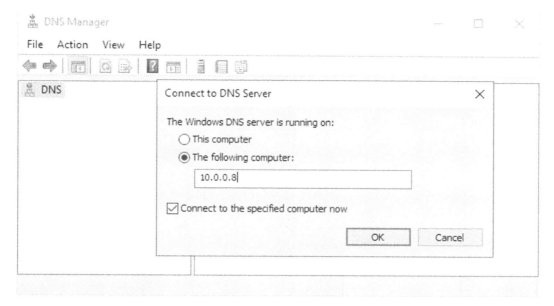

Figure 11.37 – Connecting to one of the Managed AD's DNS servers

4. Once connected, you can view the configuration details for the DNS server. We want to set up a conditional forwarder, so right-click on the **Conditional Forwarder** folder and select **New Conditional Forwarder** to launch the creation wizard.

5. We will enter the domain name and the IP address of the on-premises Active Directory forest's DNS server. We will also tick the box to store the conditional forwarder in the Managed AD and replicate it across DCs within that forest. Once the values have been entered, click on **OK**:

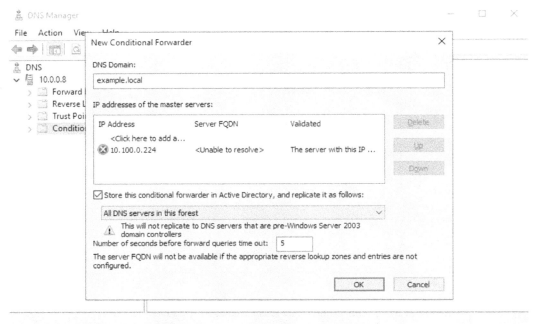

Figure 11.38 – Setting up the conditional forwarder

6. Now, the conditional forwarder for the on-premises domain has been set up in our Managed AD domain:

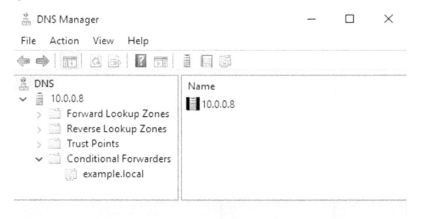

Figure 11.39 – The new conditional forwarder

With the conditional forwarder set up, we are now ready to return to the **AWS Directory Service** dashboard and finalize the trust between our two domains.

Creating the trust between AWS Managed AD and on-premises AD

As we have touched so many different AWS services and created so many resources throughout this chapter, we should take a moment to reflect upon why we went through all of this effort. The aim of this exercise was to provide a mechanism by which non-administrative user identity information could be made available to applications and resources hosted inside our AWS environment. We elected to make our on-premises Active Directory accounts available through AWS Managed AD care of a two-way trust. Once the trust has been established, the accounts in both domains will be able to access resources in each of the domains. Applications that use Active Directory for user authentication or attribute lookup will be able to look inside both domains for user information.

Now that we have done all of the necessary supporting work to get to this point, let's configure the forest trust between the AWS Managed AD and our on-premises AD:

1. From the **AWS Directory Service** dashboard, click on the directory to open the directory details screen.

2. We can view the existing trust relationships on this screen. Note that there shouldn't be any listed yet. Click on the **Add trust relationship** button to begin:

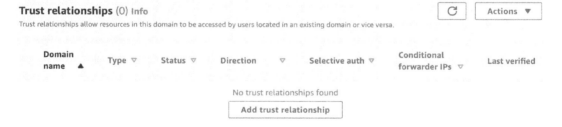

Figure 11.40 – The current trust relationships for corp.example.com

3. This opens a wizard where we can define the details for the new trust relationship. We will select a **Forest trust** and list the example.local domain as the target domain name:

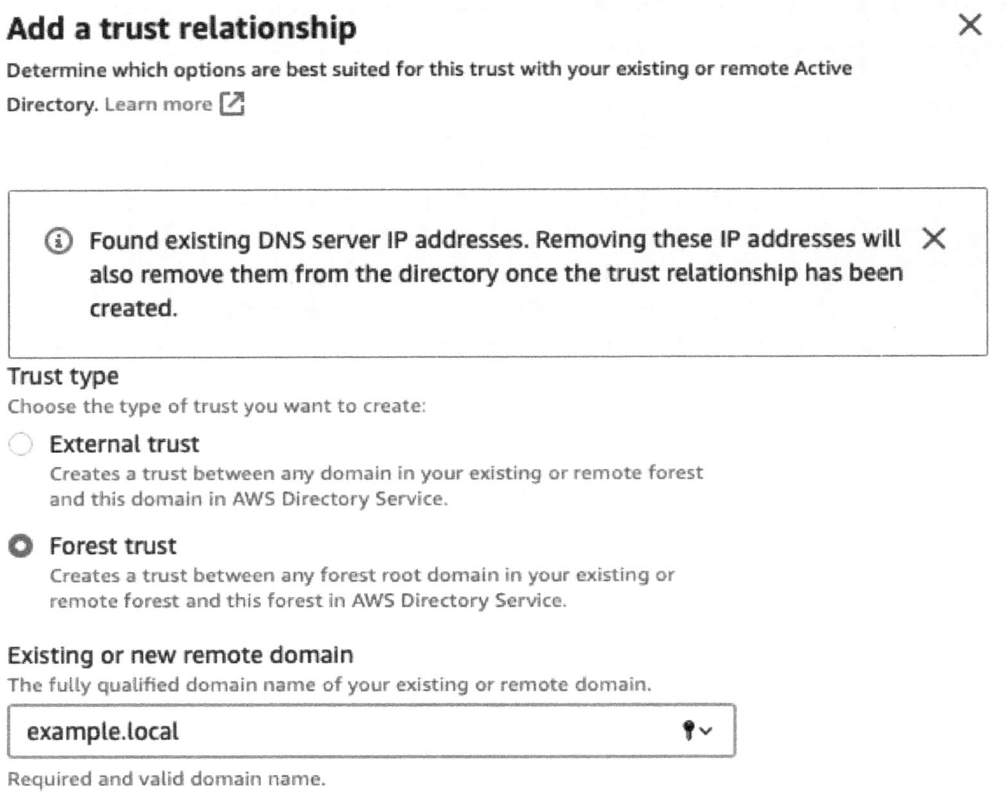

Add a trust relationship   ✕

Determine which options are best suited for this trust with your existing or remote Active Directory. Learn more 🗗

> ⓘ **Found existing DNS server IP addresses. Removing these IP addresses will** ✕
> **also remove them from the directory once the trust relationship has been**
> **created.**

Trust type
Choose the type of trust you want to create:

○ **External trust**
    Creates a trust between any domain in your existing or remote forest
    and this domain in AWS Directory Service.

◉ **Forest trust**
    Creates a trust between any forest root domain in your existing or
    remote forest and this forest in AWS Directory Service.

Existing or new remote domain
The fully qualified domain name of your existing or remote domain.

| example.local   ⚷ ⌄ |

Required and valid domain name.

Figure 11.41 – Setting up the trust type and domain target

4. Next, we will enter the trust password that we defined when we built the first half of the trust from example.local to corp.example.com. By providing the password, the on-premises domain will know the request to form a trust is legitimate. Additionally, we will set the trust directionality to two-way to match how we configured the trust that we set up earlier. Finally, we can see that our predefined conditional forwarder has been found and is already populated. We are ready to click on the **Add** button:

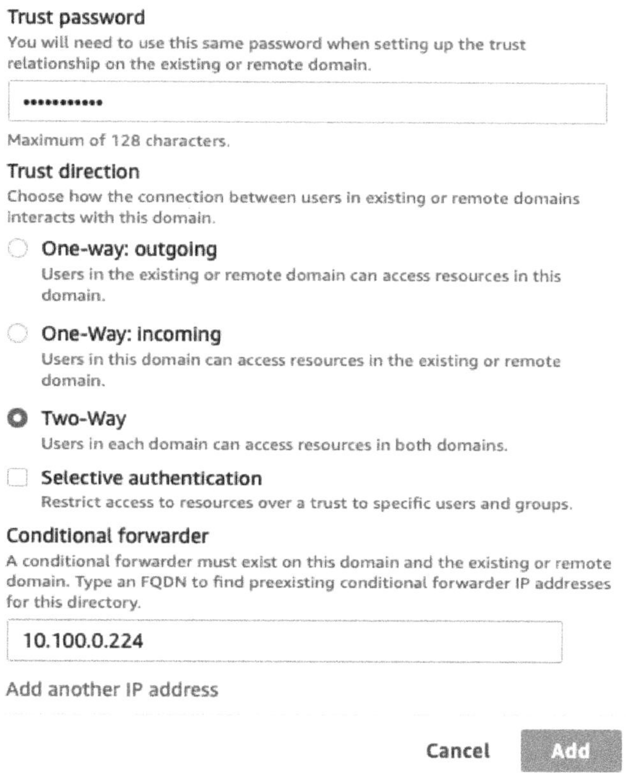

Trust password

You will need to use this same password when setting up the trust relationship on the existing or remote domain.

••••••••••

Maximum of 128 characters.

Trust direction

Choose how the connection between users in existing or remote domains interacts with this domain.

○ **One-way: outgoing**

Users in the existing or remote domain can access resources in this domain.

○ **One-Way: incoming**

Users in this domain can access resources in the existing or remote domain.

◉ **Two-Way**

Users in each domain can access resources in both domains.

☐ **Selective authentication**

Restrict access to resources over a trust to specific users and groups.

Conditional forwarder

A conditional forwarder must exist on this domain and the existing or remote domain. Type an FQDN to find preexisting conditional forwarder IP addresses for this directory.

10.100.0.224

Add another IP address

Cancel   Add

Figure 11.42 – Defining the trust directionality, shared secret, and conditional forwarder

After that, we are taken back to the **Directory details** screen. Here, we can now view the trust and its current status:

Networking details

VPC	Subnets	Status
AWS-DS-VPC01 | vpc-0d07113ba6d7d55cc ☑	AWS-DS-VPC01-Subnet01 | subnet-0910a11a56917cd93 ☑	⊘ Active
Availability zones	AWS-DS-VPC01-Subnet02 | subnet-0761b8490aab74b21 ☑	Last updated
us-east-1a, us-east-1b		Sunday, May 30, 2021
	DNS address	Launch time
	10.0.0.8, 10.0.1.138	Sunday, May 30, 2021

Trust relationships (1) Info

Trust relationships allow resources in this domain to be accessed by users located in an existing domain or vice versa.

Actions ▼

	Domain name ▲	Type ▽	Status ▽	Direction ▽	Selective auth ▽	Conditional forwarder IPs ▽	Last verified ▽
○	example.local	External	⊘ Creating	Two-Way	Disabled	10.100.0.228	Jun 5, 2021

Figure 11.43 – The trust being created

After it has been created, each domain will attempt to validate communications with each other and verify that the credentials and connection information provided are valid:

Figure 11.44 – The domain is verifying the trust

Finally, if all of the validations are successful and the domains are able to communicate, the trust is verified:

Figure 11.45 – The trust is verified and the resources and accounts can now be shared

Resources and accounts can now be shared across the domains. Applications that need access to non-administrative user information can query this domain to find the on-premises accounts.

Summary

In this chapter, we brought our user's identity information into AWS so that it could be consumed by applications hosted on AWS. First, we considered how administrative and non-administrative identity use cases differ. Then, we examined several different solution architectures to solve the challenge of bringing user identity information into AWS. Finally, we built a solution that enabled AWS-deployed applications to access user information through our organization's authoritative sources. We did this by using AWS Directory Services and building a trust between our on-premises Active Directory and a Managed AD domain created within our AWS account.

In the next chapter, we will discuss how to use AWS-native identity services, such as Amazon Cognito, to solve application identity use cases while still deferring to our external IDP as the authoritative source of user identity information.

Questions

1. What is the difference between an administrative account and a non-administrative account?

   a. The distinction varies based upon an organization's level of risk acceptance and compliance requirements. However, generally speaking, administrative accounts have access to privileged resources and are subject to heightened access and audit controls.

   b. There is no difference.

   c. Administrative accounts must be stored in Active Directory.

2. Why would a two-way trust allow AWS-hosted applications to access users and groups in an on-premises Active Directory?

Further reading

To learn more on the topic:

- *Tutorial: Setting up your base AWS Managed Microsoft AD test lab in AWS*: `https://docs.aws.amazon.com/directoryservice/latest/admin-guide/ms_ad_tutorial_test_lab_base.html`

- *Tutorial: Creating a trust from AWS Managed Microsoft AD to a self-managed Active Directory installation on Amazon EC2*: `https://docs.aws.amazon.com/directoryservice/latest/admin-guide/ms_ad_tutorial_test_lab_trust.html`

12
AWS-Hosted Application Single Sign-On Using an Existing Identity Provider

In the previous chapter, we looked at several solution architectures for non-administrative identity use cases. We defined our non-administrative use case as wanting to expose our organization's identity information to applications hosted on **Amazon Web Services (AWS)**, regardless of whether the account owner had access to the AWS backplane. Most organizations make a distinction between their administrative accounts and their standard user accounts, and often have distinct architectures for each of these use cases. Typically, standard application identity needs are satisfied through the use of standard user accounts. This chapter will focus on addressing the identity needs of AWS-hosted applications.

Whereas we can use native AWS services such as Amazon Cognito to solve application identity challenges on AWS, organizations often have policy or regulatory requirements that require them to demonstrate life cycle and access controls on the user accounts used with their applications. If an organization does not have an established pattern for correlating Amazon Cognito accounts to their organization's authoritative source of identity, applications and their users could sidestep important controls on access control.

Fortunately, Amazon Cognito user and identity pools allow app teams to create local account records and correlate them to a federated identity, meaning the existing user accounts that are federated from an existing **identity provider** (**IdP**). This model allows cloud-deployed applications flexibility to manage their application identity while using the centrally managed user identities that are governed by their organization. Let's explore how to configure an application with Amazon Cognito user pools against an external, authoritative federated provider using both **Security Assertion Markup Language 2** (**SAML2**) and **OpenID Connect** (**OIDC**) so that we can enjoy the convenience of Amazon Cognito for AWS-hosted applications without surrendering the governance and life-cycle controls we have in our authoritative identity systems. Then, we will look at our options for authorizing those users to access our AWS resources.

This chapter will address the following topics:

- Defining the use case and solution architecture
- Creating a user pool
- Connecting Amazon Cognito to an external IdP—SAML
- Connecting Amazon Cognito to an external IdP—OIDC
- Assuming roles with identity pools

Technical requirements

To get the most out of this chapter, you will need the following:

- An AWS account
- A SAML2- and **System for Cross-domain Identity Management** (**SCIM**)-compliant IdP such as Okta Identity Cloud, PingOne, or **Azure Active Directory** (**Azure AD**)
- A populated user directory to act as the user store for that IdP

Defining the use case and solution architecture

Before we begin connecting applications, user pools, and external IdPs, let's take a moment and visualize the solution we intend to build for the use case we want to solve. Once again, we have some familiar components in play for the Redbeard Identity organization, as shown in the following diagram:

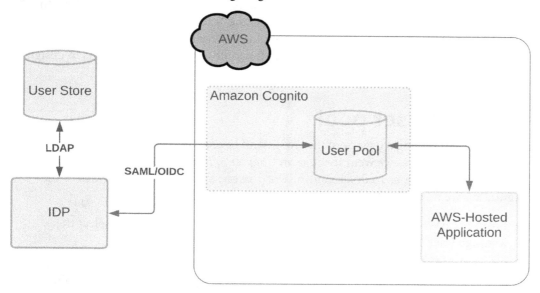

Figure 12.1 – Application references Cognito, which looks to the external IdP

In this design, the application will look to an Amazon Cognito user pool for its user information. The user pool will act as the application's user store, and detailed attributes will be provided at authentication time through the Amazon Cognito identity token. Since Amazon Cognito user pools provide a standards-compliant OIDC IdP, additional attributes can be accessed through the /userinfo endpoint as needed, if the application is sufficiently entitled and scoped to have that access. In order to ensure that the Redbeard Identity organization is still able to control access to its applications despite the AWS-hosted application looking to Amazon Cognito for user authentication and attributes, we can configure the user pool to be populated through Redbeard Identity's strategic IdP. This way, we will be able to populate the pool with the full complement of employee identity records, whether they have access to an AWS account or not, as well as ensure access to AWS-hosted applications is revoked when the user is terminated inside the external IdP's user store.

We will connect the external Redbeard Identity IdP to an Amazon Cognito user pool for our application identity use cases for applications hosted on AWS. We will only create a single user pool for our example, but other organizations will need to consider how they wish to proliferate Amazon Cognito user pools within their environments. Requiring each application team to connect to a singular user pool and auditing access on a per-client basis improves the security posture of the deployment; however, it ultimately undermines the speed and agility of the platform. As each AWS account can support 1,000 user pools, it is advised that administrators enforce user pools used with enterprise applications to ultimately federate to the enterprise's strategic IdP and leave the details of each user pool's creation and implementation to the business application teams. This approach balances centralized control of identity information with flexibility for application teams to design and implement their own per-application identity requirements.

Creating a user pool

We will begin by creating a user pool that we intend to use for all of our users. This will be a repeat of the process we went through in *Chapter 5*, *Introducing Amazon Cognito*, so we will not be as fastidious in documenting the process, aside from the specifics of the configuration we require to fulfill our use case. Proceed as follows:

1. From the **AWS Management Console**, go to **Amazon Cognito** and select the **Manage User Pools** option.

2. Select **Create a user pool**. This takes us through to the wizard. We name our pool and select the option to step through the settings, to make the changes we will need to configure this user pool instance as we want. The process is illustrated in the following screenshot:

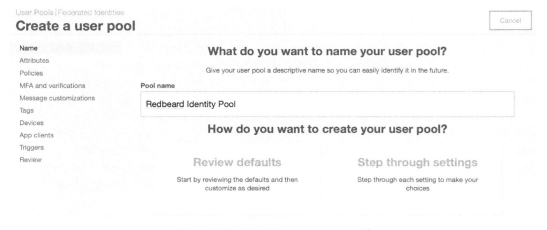

Figure 12.2 – Creating a new user pool

3. We will make several adjustments to the **Attributes** section. If we want our external IdP to be the authoritative source of user information for this user pool, we will need to ensure we include all of the attributes from the external IdP's user store that we will want to expose through the user pool. First, we want to make sure nobody can be configured with an **identifier (ID)** that is different from what gets sent from the external IdP, so we will set the user pool to use the email attribute as the ID. The process is illustrated in the following screenshot:

Email address or phone number - Users can use an email address or phone number as their "username" to sign up and sign in.

- ◉ Allow email addresses
- ○ Allow phone numbers
- ○ Allow both email addresses and phone numbers (users can choose one)

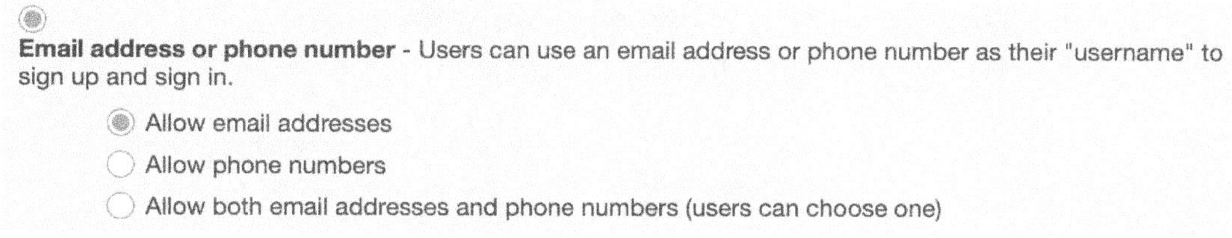

Figure 12.3 – Setting email attribute as the user pool ID

4. We then select the required attributes that we can send from our external IdP. As Amazon Cognito includes all of the standard attributes included as part of the OpenID specification, including many that may not be part of our organization's identity attribute schema, most of the available required attributes will not be required. The process is illustrated in the following screenshot:

Which standard attributes do you want to require?

All of the standard attributes can be used for user profiles, but the attributes you select will be required for sign up. You will not be able to change these requirements after the pool is created. If you select an attribute to be an alias, users will be able to sign-in using that value or their username. Learn more about attributes.

Required	Attribute	Required	Attribute
☑	address	☐	nickname
☐	birthdate	☐	phone number
☑	email	☐	picture
☑	family name	☐	preferred username
☐	gender	☐	profile
☑	given name	☐	zoneinfo
☑	locale	☐	updated at
☐	middle name	☐	website
☑	name		

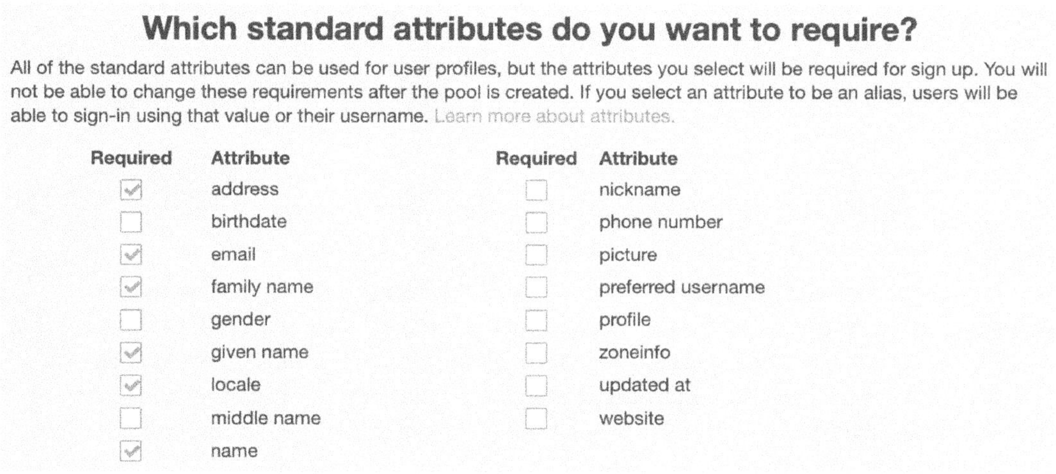

Figure 12.4 – Enabling required attributes that are available from the external IdP

5. We will be populating the user pool using **just-in-time** (**JIT**) provisioning from the external IdP. As is, the handful of required attributes we have selected will not provide rich identity data to the AWS-hosted apps that will be looking to this user pool for that information. As such, we will need to map additional custom attributes for each of the attributes that are available from the external IdP on the user pool account. The process is illustrated in the following screenshot:

Do you want to add custom attributes?

Enter the name and select the type and settings for custom attributes.

Type	Name	Min length	Max length	Mutable
string	title	1	256	✓

Type	Name	Min length	Max length	Mutable
string	userType	1	256	✓

Type	Name	Min length	Max length	Mutable
string	employeeNumber	1	256	✓

Figure 12.5 – Mapping additional attributes to enrich the accounts in the user pool

Once we have added all of the additional attributes, we can move on.

6. Under the **Policies** section, we want to disable the option for new users to sign themselves up for a user pool account. Since we want the external IdP to be the ultimate authentication authority for the applications that look to this pool, we do not want to introduce an uncontrolled class of users that only exists within the user pool. Every user within this user pool will have a corresponding record at the external IdP. We can proceed to the next step.

7. We can set remaining options to whichever values best suit our needs and wrap up the creation of the user pool.

8. Now, we can make cosmetic adjustments to our user pool's domain, including hosting it on our own subdomain. We will finish this before we move on to the next section of the configuration. The process is illustrated in the following screenshot:

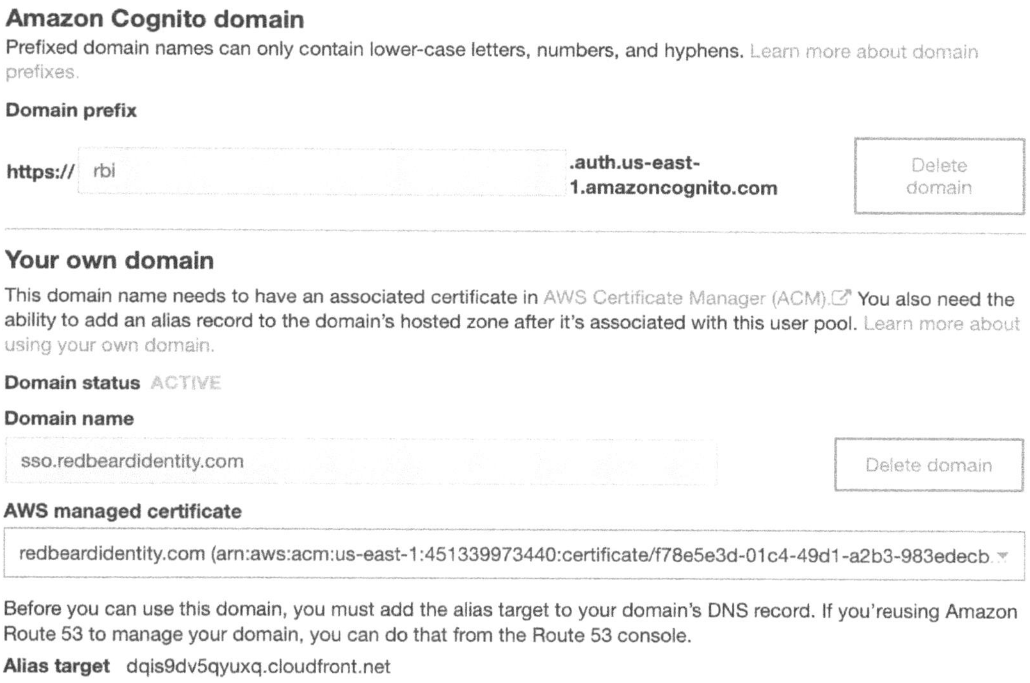

Amazon Cognito domain

Prefixed domain names can only contain lower-case letters, numbers, and hyphens. Learn more about domain prefixes.

Domain prefix

https:// rbi   .auth.us-east-   Delete
                                                  1.amazoncognito.com   domain

Your own domain

This domain name needs to have an associated certificate in AWS Certificate Manager (ACM). You also need the ability to add an alias record to the domain's hosted zone after it's associated with this user pool. Learn more about using your own domain.

Domain status ACTIVE

Domain name

sso.redbeardidentity.com   Delete domain

AWS managed certificate

redbeardidentity.com (arn:aws:acm:us-east-1:451339973440:certificate/f78e5e3d-01c4-49d1-a2b3-983edecb ▼

Before you can use this domain, you must add the alias target to your domain's DNS record. If you're using Amazon Route 53 to manage your domain, you can do that from the Route 53 console.

Alias target   dqis9dv5qyuxq.cloudfront.net

Figure 12.6 – Setting up the Amazon Cognito domain

We now have our new Redbeard Identity pool created and our domain configured. Next, we will build a connection between this user pool and our external IdP.

Connecting Amazon Cognito to an external IdP – SAML

Now that we have a user pool configured with attributes that match those found in our external IdP, we need to put some users inside it. We do not want users created directly inside the user pool as that would bypass the external IdP as the authoritative source of identity information for our users. To connect the external IdP with the user pool, we will need to configure our external IdP as an IdP for the user pool, as follows:

1. From the user pool, we can select the type of federated provider we want to add under the **Federation** menu. We will select the **SAML** option, as illustrated in the following screenshot:

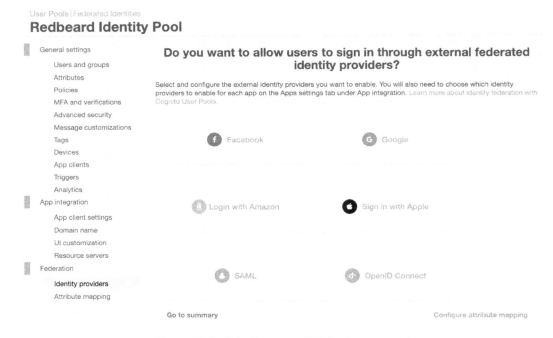

Figure 12.7 – Selecting a new IdP for the user pool

2. The configuration options are very sparse since it wants to import a metadata file. We will come back to the form shown in the following screenshot since we will need to build this connection on the external IdP side in order to create that metadata file first:

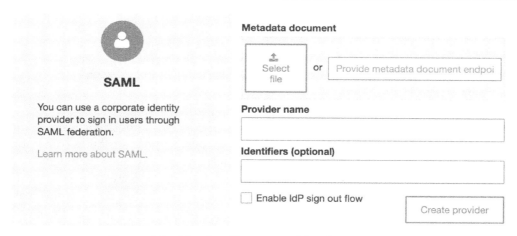

Figure 12.8 – Metadata file is required for IdP configuration

3. We can create a new SAML2 application inside the administrative control panel of the external IdP. We need to provide some specific information for the SAML flow to work—specifically, the **assertion consumer service (ACS) Uniform Resource Locator (URL)** and the entity ID. The user pool ACS URL will always be `https://<yourDomainPrefix>.auth.<region>.amazoncognito.com/saml2/IdPresponse`, or if we are using a custom subdomain, it will be `https://<subdomain>/saml2/IdPresponse`. The entity ID value will be `urn:amazon:cognito:sp:<yourUserPoolID>`. The process is illustrated in the following screenshot:

A   **SAML Settings**

General

Single sign on URL   ❷
> https://sso.redbeardidentity.com/saml2/idpresponse

☑   Use this for Recipient URL and Destination URL

☐   Allow this app to request other SSO URLs

Audience URI (SP Entity ID)   ❷
> urn:amazon:cognito:sp:us-east-1_rz2HyPFjt

Default RelayState   ❷
>

If no value is set, a blank RelayState is sent

Name ID format   ❷
> EmailAddress   ▾

Application username   ❷
> Okta username   ▾

Update application username on
> Create and update   ▾

Figure 12.9 – Creating a user pool application within the external IdP

> **Tip**
> We are using Okta Identity Cloud for the Redbeard Identity external IdP. The specific steps to configure these mappings, build a connection within the IdP, and export the metadata will be different depending upon the IdP in use at your organization.

4. As part of the application configuration, we will define attributes that get sent in the assertion, how those attributes will be named, and where those attributes come from in the IdP's user store. More importantly, we must match all of the required and custom attributes we defined in our user pool to attributes in our external IdP's user store, and include those values inside the assertion. It may be helpful to review that user pool's attributes list to ensure that every attribute is accounted for and mapped in this configuration. You can view this here:

ATTRIBUTE STATEMENTS

Name	Name Format	Value
email	Unspecified	user.email
name	Unspecified	user.displayName
given_name	Unspecified	user.firstName
family_name	Unspecified	user.lastName
address	Unspecified	user.postalAddress
custom:title	Unspecified	user.title
custom:userType	Unspecified	user.userType
custom:employeeNumber	Unspecified	user.employeeNumber
custom:costCenter	Unspecified	user.costCenter
custom:organization	Unspecified	user.organization
custom:division	Unspecified	user.division
custom:department	Unspecified	user.department
custom.managerid	Unspecified	user.managerId
locale	Unspecified	user.city

Figure 12.10 – Our attribute mappings

In the preceding screenshot, the attribute names on the left are how the attributes are named within the user pool, as well as in the SAML assertion from the IdP. The values on the right are where the IdP pulls the attribute values for each of those attributes in the user store.

5. As this user pool is designed to represent the full population of the Redbeard Identity organization, our IdP needs to allow everyone to access it instead of only provisioning access based upon group membership. We will assign this RBI user pool to *everyone* within the IdP to ensure that the entire organization will be accessible through this connection.

6. After defining the attribute mappings, providing the ACS URL, and entity ID, we can now generate the IdP metadata we need to load into our user pool. We return to the user pool Federation settings, select a SAML federated provider, load the file, and create a provider. The process is illustrated in the following screenshot:

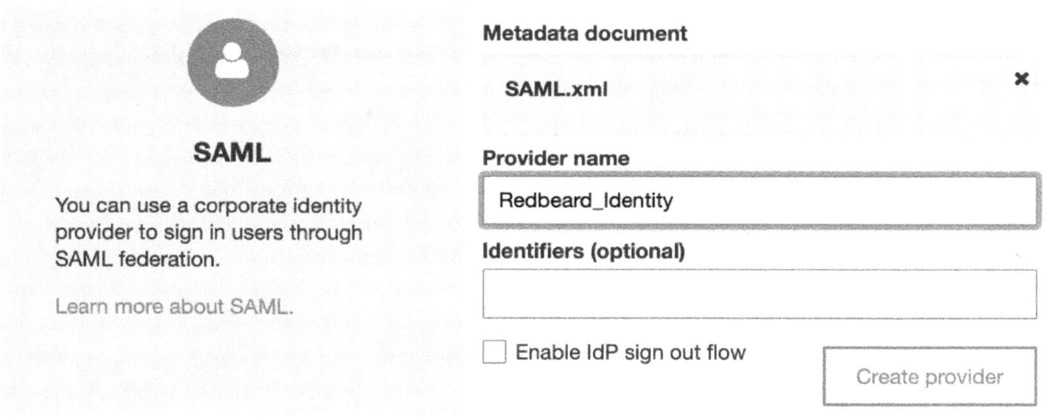

Figure 12.11 – Creating a federated provider in the user pool

7. Now that we have our IdP configured, we still need to tell our user pool which attributes from the IdP's assertion will correspond to each of the attributes within its own local user store. We do this by going to the **Attribute Mapping** section and selecting the federated provider we just created. This will be shown under the **SAML2** tab. There, we can enter the name of the attribute as it will appear in the assertion and correlate it to an attribute within the user pool schema. We must map all of the attributes. The process is illustrated in the following screenshot:

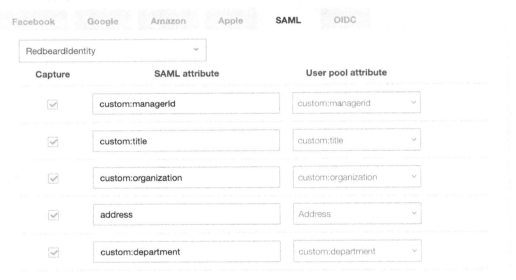

Figure 12.12 – Mapping assertion attributes to user pool attributes

With that, everything is now in place for the external IdP to authenticate and populate users for the user pool. However, there are still additional steps required when connecting applications to a user pool to ensure we constrain application access to user pool identities provided by the external IdP and not user pool native accounts.

Restricting application access to just the external IdP

The applications that will use this user pool will do so through an app client. App clients can be configured to support different **Open Authentication 2 (OAuth 2)** profiles, scopes, and callback **Uniform Resource IDs (URIs)**. In addition to those configuration settings, we have an additional setting now that we have enabled an external IdP, as illustrated in the following screenshot:

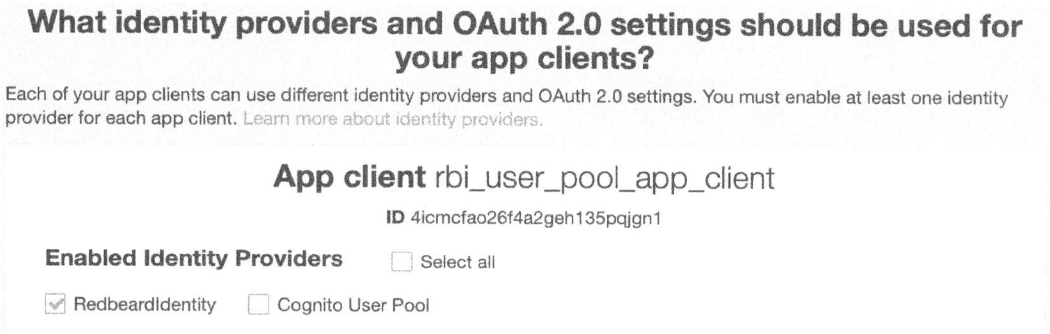

What identity providers and OAuth 2.0 settings should be used for your app clients?

Each of your app clients can use different identity providers and OAuth 2.0 settings. You must enable at least one identity provider for each app client. Learn more about identity providers.

App client rbi_user_pool_app_client

ID 4icmcfao26f4a2geh135pqjgn1

Enabled Identity Providers   ☐ Select all

☑ RedbeardIdentity   ☐ Cognito User Pool

Figure 12.13 – Option to enable different IdPs for the app client

Each app client can be configured to be used with the external IdP, the user pool, or both. We have our app client set to use the external IdP for reasons we have already repeatedly mentioned in this chapter. Specifically, our goal is to ensure all Redbeard Identity users are available for applications through this user pool, but we still want to retain the external IdP as the definitive source for user authentication, even though AWS-hosted applications will be looking at the user pool. By disabling the native user pool capability on the client and disabling self-registration into the user pool by users, we are able to enforce that control.

Populating the Amazon Cognito user pool through JIT provisioning

Finally, we have everything we need to expose our application users in AWS. We will use an OIDC test application to verify that a user has been created in the user pool.

> **Tip**
> This demonstration uses an application that walks users through the OIDC authorization code flow since the process is visible in a browser. Authorization code with **Proof Key for Code Exchange (PKCE)** is the best flow to use for production use cases.

8. Let's take a look at the user pool's user store at present. If we search it, there are currently no accounts available. If we felt so inclined, we could download the **comma-separated values (CSV)** template and bootstrap the user store with account information from the external IdP's user store. For now, we will be content with JIT provisioning. You can see the empty user store here:

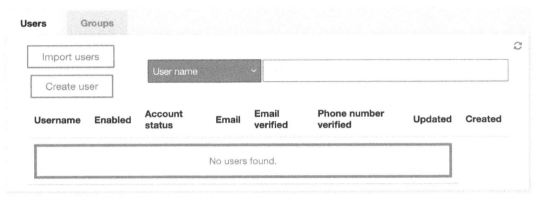

Figure 12.14 – User store is empty

9. We will populate the test application with the values we need to complete a code flow using our user pool and the app client that we defined to use the external IdP as its authentication source. We can bootstrap the configuration by using the **Discovery Document URL**. The format for that endpoint is shown in the following code snippet:

```
https://cognito-IdP.{region}.amazonaws.com/{userPoolId/.
well-known/openid-configuration
```

This means that we have the following value for our user pool's specific metadata endpoint:

```
https://cognito-IdP.us-east-1.amazonaws.com/us-east-1_
rz2HyPFjt/.well-known/openid-configuration.
```

Once we enter that value, we can click the **USE DISCOVERY DOCUMENT** button to populate the remaining values in the configuration, besides the client and scope information. You can see these values here:

OpenID Connect Configuration ⊗

Server Template	Custom ⇕
Discovery Document URL	https://cognito-idp.us-east-1.amazonaws.com/us-east-1_rz2HyPFjt/.w USE DISCOVERY DOCUMENT
	Use a discovery document to populate your server urls
Authorization Token Endpoint	https://sso.redbeardidentity.com/oauth2/authorize
Token Endpoint	https://sso.redbeardidentity.com/oauth2/token
Token Keys Endpoint	https://cognito-idp.us-east-1.amazonaws.com/us-east-1_rz2HyPFjt/.well-known/jwks.json

Remember to set https://openidconnect.net/callback as an allowed callback with your application!

OIDC Client ID	4icmcfao26f4a2geh135pqjgn1
OIDC Client Secret	1rnhs7lvh8sddr5vtnvqpnh8pg1oi44u6i236pnhhclu796i7tdi
Scope	openid profile email phone

Figure 12.15 – Configuring the OIDC test app

10. Once set up, we fire the code flow, as follows:

① Redirect to OpenID Connect Server

Request

```
https://sso.redbeardidentity.com/oauth2/authorize?
    client_id=4icmcfao26f4a2geh135pqjgn1
    &redirect_uri= https://openidconnect.net/callback
    &scope=openid profile email phone
    &response_type=code
    &state=464e8d669bc950c5e17ec834e2e97a46e4384775
```

START

Figure 12.16 – Requesting the authorization code

11. We get the authorization code and exchange it for an access token and a refresh token, as illustrated in the following screenshot:

Exchange Code from Token

Your Code is

b20c1cd2-1ef2-4554-bc42-a488ee8b9497

Now, we need to turn that access code into an access token, by having our server make a request to your token endpoint

Request

POST https://sso.redbeardidentity.com/oauth2/token
grant_type=authorization_code
&client_id=4icmcfao26f4a2geh135pqjgn1
&client_secret=1rnhs7lvh8sddr5vtnvqpnh8pg1oi44u6i236pnhhclu
&redirect_uri=https://openidconnect.net/callback
&code=b20c1cd2-1ef2-4554-bc42-a488ee8b9497

HTTP/1.1 200
Content-Type: application/json
{
    "id_token": "eyJraWQiOiJDUTVFUjFNSUg3Yk56bVowWkJuWktWQ001
    "access_token": "eyJraWQiOiJ3Q0loQzlSNHp0XC83Mm41OEt2c0hC
    "refresh_token": "eyJjdHkiOiJKV1QiLCJlbmMiOiJBMjU2R0NNIiⁱ
    "expires_in": 3600,
    "token_type": "Bearer"
}

NEXT

Figure 12.17 – ID, access, and refresh tokens

12. Next, we verify the signature on the ID token to ensure it came from the IdP, as illustrated in the following screenshot:

Verify User Token

Now, we need to verify that the ID Token sent was from the correct place by validating the JWT's signature

Your "id_token" is

```
eyJraWQiOiJDUTVFUjFNSUg3Yk56bVowWkJuWktWQ001QW42ZUtmblA0ek5MS
0DXvPrLmm_ph73K9lM4-
xFSVv9PN6OV9n_g4HHNxiBxP_7gfYT7CF1_WJnRLEvL7VamL28zzA0WhGmP1M
aSli3LQqhA0UxAP1fULA3I362-
_oGqtQ7MJsXWlOvVYYS1MUfvfWqf3V4v-GMZ-
uYqu6FoUN6_wlKbgLqUbDTq8voZjeeboLiSwaglAMkm-MhV_Wfg-
t9dBCjaEpOYtQarEkZfE4uL7_2XiXRP2FQBFoDQzE8LZgtKsgHzvmzbjisWDw
```

This token is cryptographically signed with the **RS256** algorithim. We'll use the public key of the OpenID Connect server to validate it. In order to do that, we'll fetch the public key from **https://cognito-idp.us-east-1.amazonaws.com/us-east-1_rz2HyPFjt/.well-known/jwks.json,** which is found in the discovery document or configuration menu options.

VERIFY

Figure 12.18 – Verifying the ID token

13. As the signature is valid, we can trust the decoded contents of the ID token. Here, we see all of the attributes we defined at the external IdP presented through the user pool:

Decoded Token Payload

```
{
  "at_hash": "SfhMkNiRUNSrHe_FnYFP6w",
  "sub": "8b0552a4-5003-4e88-ab31-33a6c3bc9616",
  "cognito:groups": [
    "us-east-1_rz2HyPFjt_RedbeardIdentity"
  ],
  "custom:department": "Identity Development",
  "iss": "https://cognito-idp.us-east-1.amazonaws.com/us-east-1_r
  "locale": "Richmond",
  "custom:userType": "Staff",
  "custom:employeeNumber": "S94577",
  "identities": [
    {
      "userId": "redbeardidentity+iamdev@gmail.com",
      "providerName": "RedbeardIdentity",
      "providerType": "SAML",
      "issuer": "http://www.okta.com/redbeardidentityuserpool",
      "primary": "true",
      "dateCreated": "1622314704121"
    }
  ],
  "auth_time": 1622321399,
  "exp": 1622324999,
  "iat": 1622321399,
  "email": "redbeardidentity+iamdev@gmail.com",
  "custom:title": "IAM Developer",
  "custom:organization": "Information Technology",
  "email_verified": false,
  "address": {
    "formatted": "901 E Byrd St Richmond VA 23219"
  },
  "custom:division": "Information Security",
  "cognito:username": "redbeardidentity_redbeardidentity+iamdev@g
  "given_name": "Iam",
  "nonce": "aNYrevez4nBGM7vmeCWzTdl1ZakPmdsk3A0_JTVOViADmVRhupab1
  "aud": "4icmcfao26f4a2geh135pqjgn1",
  "token_use": "id",
  "custom:costCenter": "30002",
  "name": "Iam Dev",
  "family_name": "Dev"
}
```

Figure 12.19 – The decoded ID token with all the user attributes we defined

14. Now that the user has authenticated through the user pool, their record has been populated within the Amazon Cognito user pool's user store, as we can see here:

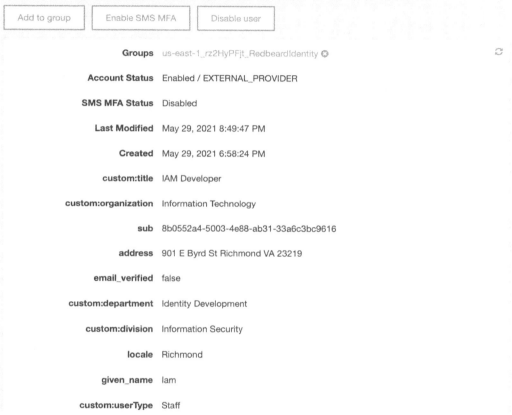

Figure 12.20 – The new user record in the user store

The user record will be updated from the authoritative IdP each time the user authenticates to an application that uses this user pool as its IdP.

Now that we have finished setting up the inbound federation with our authoritative IdP using SAML, we will repeat the process using OIDC.

Connecting Amazon Cognito to an external IdP – OIDC

Amazon Cognito user pools support the use of multiple external IdPs. It would be unusual, though not necessarily ill-advised, to connect the same external IdP to an Amazon Cognito user pool using both SAML and OIDC. We will connect our external IdP to OIDC in the interest of demonstrating how both protocols operate when used with an external IdP with a user pool. We'll proceed as follows:

1. From the user pool, we can select the type of federated provider we want to add under the **Federations** menu. We will select the **OpenID Connect** option. We can see a marker on the **SAML** option indicating an existing connection, as illustrated in the following screenshot:

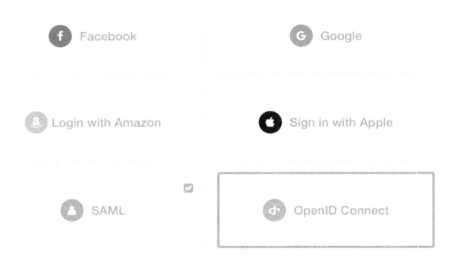

Figure 12.21 – Selecting a new IdP for the user pool

2. In the following screenshot, we see the required fields for configuring the new OIDC IdP. As we do not have all of these values yet, this means that we will need to create a client that the Amazon Cognito user pool will use with our external IdP in our external IdP as a first step:

Provider name

Client ID

Client secret (optional)

Attributes request method

GET

Authorize scope

Issuer

Run discovery

Identifiers (optional)

Create provider

Figure 12.22 – Required fields for a new OIDC IdP in Amazon Cognito

> **Tip**
> The steps shown for configuring the OIDC client and IdP attribute mappings for the client in this example were performed using Okta Identity Cloud. The experience may be different when using a different IdP platform, though the overarching objectives remain the same.

3. We sign in to our external IdP to create a client that Amazon Cognito will use with the IdP. First, we specify that this will be an OIDC application, as illustrated in the following screenshot:

Create a new app integration

Figure 12.23 – Defining the application type in the external IdP

4. Next, we specify the application type. The application type will determine the OIDC grant that this client will use. Amazon Cognito uses authorization code flow, which is enabled in our external IdP by selecting the **Web Application** option, as illustrated in the following screenshot:

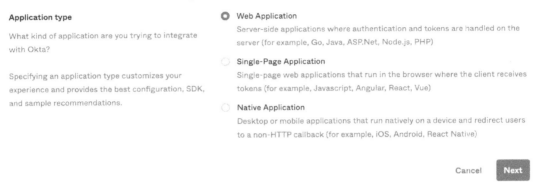

Application type

What kind of application are you trying to integrate with Okta?

Specifying an application type customizes your experience and provides the best configuration, SDK, and sample recommendations.

○ Web Application
Server-side applications where authentication and tokens are handled on the server (for example, Go, Java, ASP.Net, Node.js, PHP)

○ Single-Page Application
Single-page web applications that run in the browser where the client receives tokens (for example, Javascript, Angular, React, Vue)

○ Native Application
Desktop or mobile applications that run natively on a device and redirect users to a non-HTTP callback (for example, iOS, Android, React Native)

Cancel   Next

Figure 12.24 – Selecting the application type in the external IdP

5. Next, we give the application a user-friendly name, apply any branding, and add additional grants for our use case, as illustrated in the following screenshot. We are fine with just the defaults, so we can move on:

New Web App Integration

General Settings

App integration name

RBI User Pool

Logo (Optional) ❓

⚙️

Grant type

Learn More ↗

Client acting on behalf of itself
☐ Client Credentials

Client acting on behalf of a user
☑ Authorization Code
☐ Refresh Token
☐ Implicit (Hybrid)

Figure 12.25 – Client naming, branding, and options for additional flows

6. Next, we begin to configure some of the values that are unique to our Amazon Cognito user pool. We need to define at least one redirect URI where the external IdP will post the ID and access tokens upon successful user authentication. We need to enter our Amazon Cognito user pool domain or the domain alias that we configured for our user pool suffixed by /oauth2/IdPresponse, then we can finish the configuration. The process is illustrated in the following screenshot:

Sign-in redirect URIs

https://sso.redbeardidentity.com/oauth2/idpresponse   ×

Okta sends the authentication response and ID
token for the user's sign-in request to these URIs   + Add URI

Learn More ☐

Figure 12.26 – Defining the client's redirect URI

7. With the configuration completed, we are issued a client ID and secret that we can use for our Amazon Cognito user pool and our external IdP. We now have all of the values required to complete the OIDC IdP configuration in our user pool.

8. Returning to our Amazon Cognito user pool's **Federations** menu, let's click the **OpenID Connect** provider option once again, which will take us to the following screen. This time, we can insert all of the necessary values to complete the initial configuration:

Figure 12.27 – Defining the external OIDC IdP in the user pool

Let's take a moment and define each of the fields. **Provider name** is a user-friendly name for the OIDC provider within the Amazon Cognito user pool. **Client ID** and **Client secret** are the ID and shared secret used to authenticate the application at the external IdP and prevent data leakage. **Attributes request method** defines how the user pool should call the OIDC IdP's /userinfo endpoint, either through GET or POST. The scopes under **Authorize scope** are the scopes the client will use when calling the IdP. The openid scope is required for all OIDC authentication transactions, so it is a required field. The email, phone, and address scopes are also OIDC standards and provide claims for those attributes. Any additional scopes will depend upon the IdP used and the attributes mapped to each scope. For our external IdP, we will place all additional attributes behind the profile scope. **Issuer** is the ID for the IdP and usually appears as the base URL for the IdP. In our example here, the issuer is https://redbeardidentity.okta.com. Assuming our IdP is compliant, the Amazon Cognito user pool will be able to determine all of the endpoints necessary to execute the OIDC transactions by appending /.well-known/openid-config to the **Issuer** URI and querying the IdP's OIDC metadata. Finally, **Identifiers** are optional custom names that can be used instead of the IdP's name in the endpoint URLs, similar to the domain aliases we set up for our user pool.

Now that we have entered all of the values, we can save the OIDC IdP as a provider within our user pool. However, in order to ensure that the identity information that comes into the user pool is as complete as possible, we still need to adjust the attributes that the external IdP will send upon user authentication and map those attributes to the user attribute schema we defined for our user pool. Let's return to our IdP to begin enriching our claims. We'll proceed as follows:

9. From our external IdP, we open the profile editor to add additional attributes to our profile scope, as illustrated in the following screenshot:

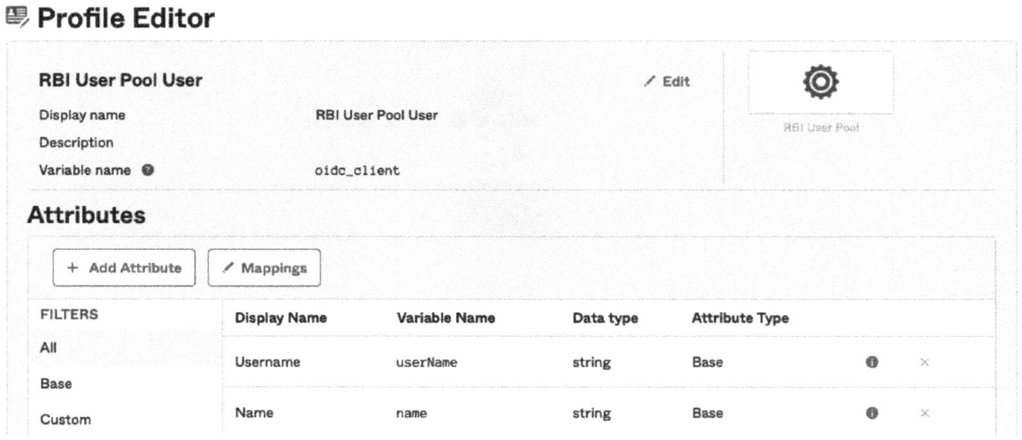

Figure 12.28 – Editing our profile scope's attributes

10. For each of the additional attributes that we need to populate in our user pool, we need to add a corresponding attribute to our profile mapping. We start by hitting the **Add Attribute** button and naming the attribute to be added, as illustrated in the following screenshot. We repeat this until we have all the attributes that are required to fully populate our Amazon Cognito user pool:

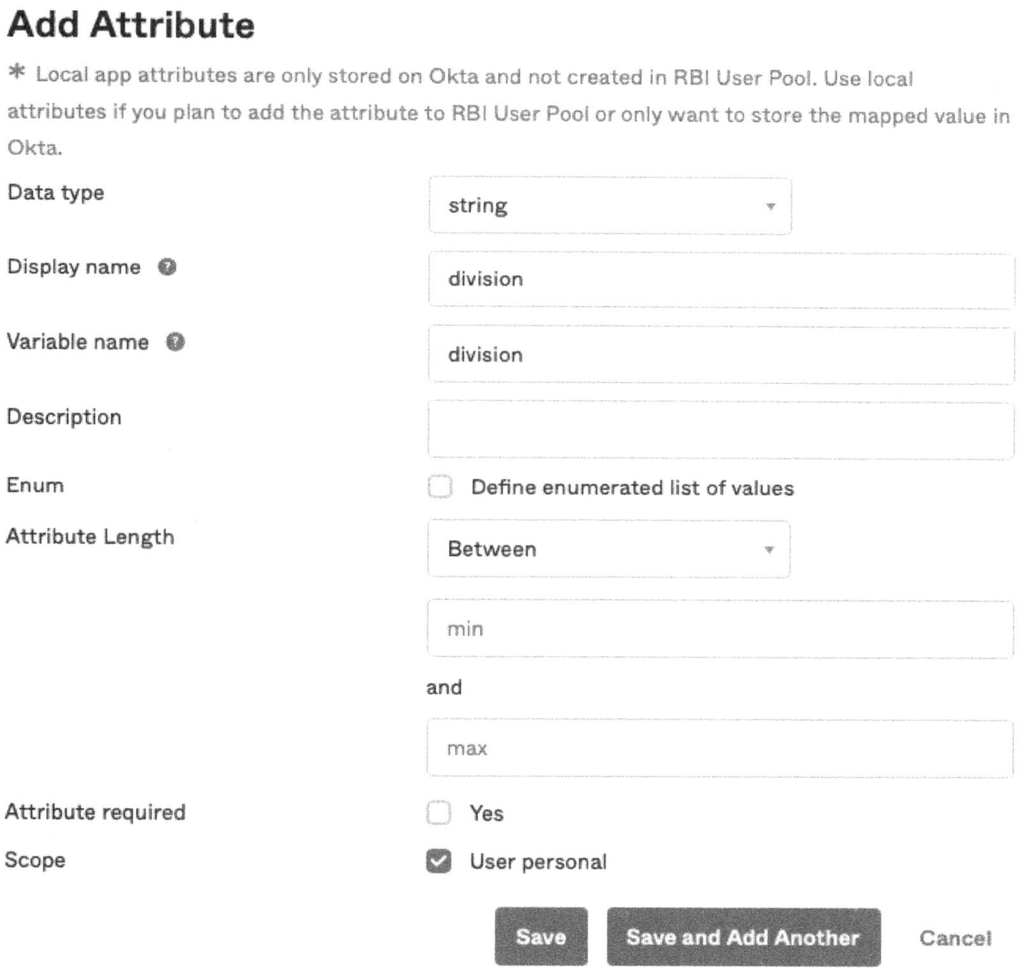

Add Attribute

✳ Local app attributes are only stored on Okta and not created in RBI User Pool. Use local attributes if you plan to add the attribute to RBI User Pool or only want to store the mapped value in Okta.

Data type	string ▾
Display name ❷	division
Variable name ❷	division
Description	
Enum	☐ Define enumerated list of values
Attribute Length	Between ▾
	min
	and
	max
Attribute required	☐ Yes
Scope	☑ User personal

Save   **Save and Add Another**   Cancel

Figure 12.29 – Adding attributes to our profile

11. Once we have all of our profile attributes listed, we then need to map them to attributes within our external IdP's user store. Once each attribute for the RBI user pool OIDC client's claim list has been mapped to an attribute in the external IdP's user store for fulfillment, as illustrated in the following screenshot, our work within the external IdP is complete:

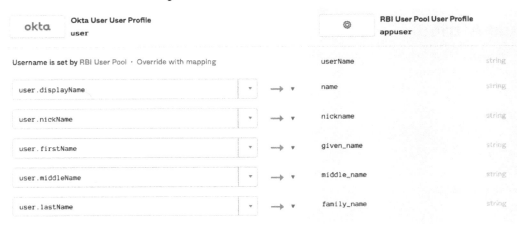

Figure 12.30 – Mapping attributes from the user store to fulfill the claims for the user store client

12. Now, we just need to configure the corresponding mapping within the Amazon Cognito user pool for our OIDC client. From the **Cognito** menu, under **Federation | Attribute mapping**, we will go to the **OIDC** tab. There, we will add and capture all of the attributes that will come in from the external IdP and map them to an attribute in our user pool, as illustrated in the following screenshot. We must name the attributes exactly as they will appear within the id_token or /userinfo endpoint, or they will not be found and added to the Amazon Cognito user pool directory:

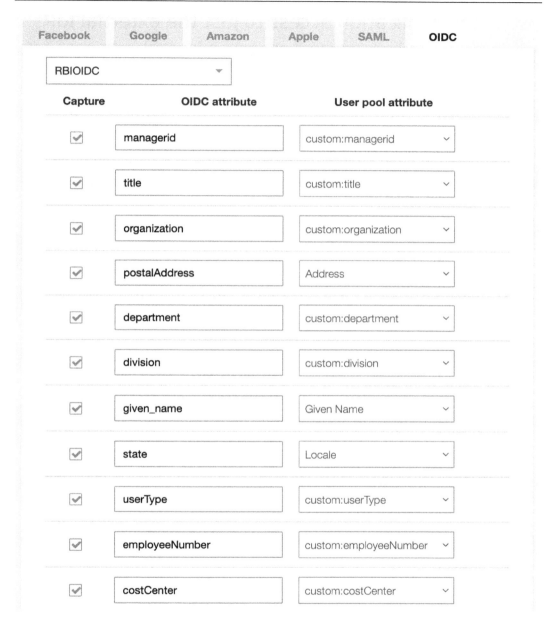

Figure 12.31 – Mapping external IdP claims to user pool attributes

That completes the configuration between the external IdP and the Amazon Cognito user pool. Next, we will define a new client in our user pool that will use this inbound federation for user authentication.

Restricting application access to just the external IdP

Inside our user pool, we already created one app client that defers to the external SAML IdP for user authentication. We could adjust that client's configuration to no longer respect the SAML IdP and, instead, look to the OIDC IdP and reuse that client. Alternatively, we could enable that client to use both the SAML IdP and OIDC IdP and allow users to select their IdP at authentication time. These patterns would be acceptable for applications where multiple organizations or sub-businesses with their own IdPs populate and authenticate into applications serviced by a single user pool. However, for the purposes of this example, we will assume that we want to enforce a single IdP as the authoritative source for each application. As such, we will create a new app client to use with our OIDC IdP configuration.

From our user pool, we can create new clients through the **App clients** menu. We simply click **Add another app client**, give it a name, and ensure the default client values work for our use case. The process is illustrated in the following screenshot:

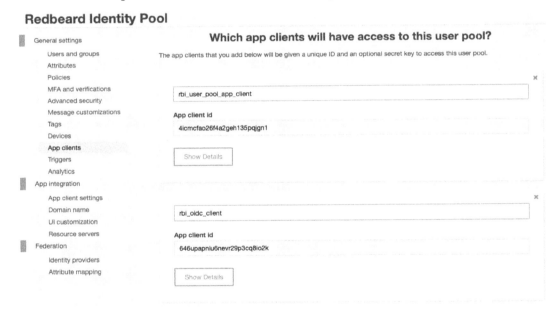

Figure 12.32 – Both clients for our user pool

We now have our previous `rbi_user_pool_app_client` client and the new `rbi_oidc_client` client available in our user pool. We can capture the new client's secret by hitting the **Show Details** button. We will record this in a secure location as we will be using it with our application momentarily.

Now that we have our `rbi_oidc_client` client, we need to make some adjustments to its configuration to ensure it only respects inbound federation through our external OIDC IdP. This is done by going to the **App client settings** menu, which will take you to the following screen:

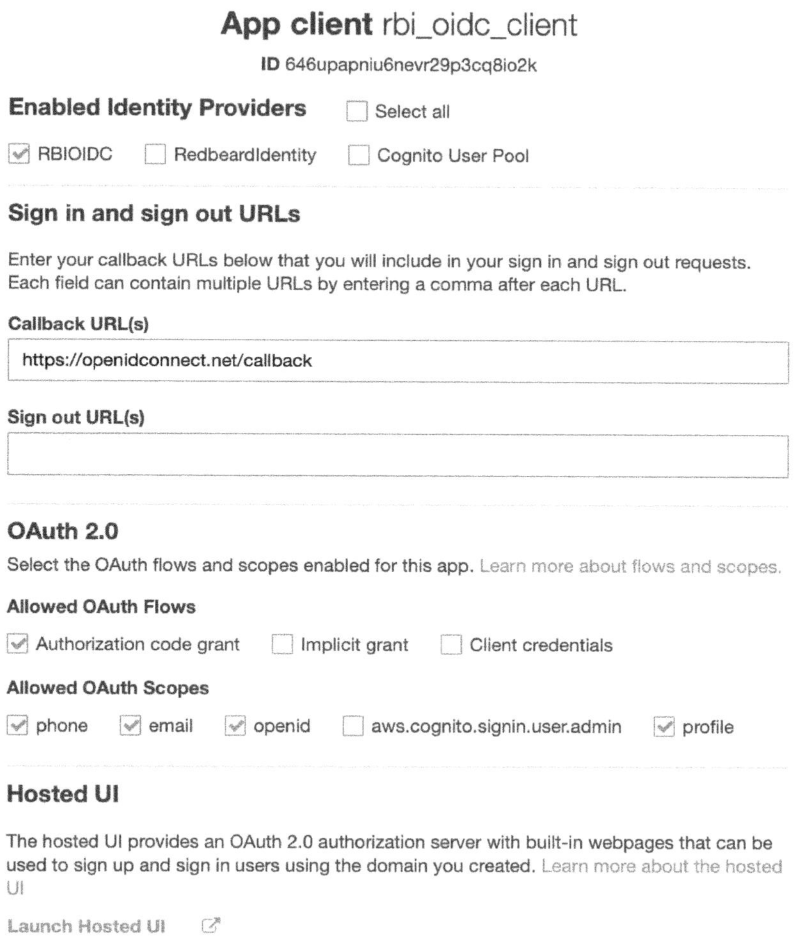

App client rbi_oidc_client

ID 646upapniu6nevr29p3cq8io2k

Enabled Identity Providers   ☐ Select all

☑ RBIOIDC   ☐ RedbeardIdentity   ☐ Cognito User Pool

Sign in and sign out URLs

Enter your callback URLs below that you will include in your sign in and sign out requests. Each field can contain multiple URLs by entering a comma after each URL.

Callback URL(s)

https://openidconnect.net/callback

Sign out URL(s)

OAuth 2.0

Select the OAuth flows and scopes enabled for this app. Learn more about flows and scopes.

Allowed OAuth Flows

☑ Authorization code grant   ☐ Implicit grant   ☐ Client credentials

Allowed OAuth Scopes

☑ phone   ☑ email   ☑ openid   ☐ aws.cognito.signin.user.admin   ☑ profile

Hosted UI

The hosted UI provides an OAuth 2.0 authorization server with built-in webpages that can be used to sign up and sign in users using the domain you created. Learn more about the hosted UI

Launch Hosted UI   ☐

Figure 12.33 – App client settings for rbi_oidc_client

Once there, we will scroll down until we find the correct client. Under **Enabled Identity Providers,** we need to leave all options deselected except for **RBIOIDC**, as this will ensure that the app client delegates all user authentication to the authoritative OIDC IdP instead of referring to the Amazon Cognito user pool directly. We also need to define the callback URLs for the application that will use this client, and we must make sure that all the correct scopes and OAuth flows are enabled. Once configured, as per the preceding screenshot, we are ready to test user creation and app authentication using this client.

Populating the Amazon Cognito user pool through JIT provisioning

Finally, we have everything we need to expose our application users to AWS-hosted applications using Amazon Cognito and our authoritative OIDC IdP. We will use an OIDC test application to verify that a user has been created in the user pool.

> **Tip**
> This demonstration uses an application that walks users through the OIDC authorization code flow since it is visible through a browser. Authorization code with PKCE is the best flow to use for production use cases.

Let's set up our test application, as follows:

1. From the test application's configuration page, we will enter values specific to our user pool and the app client we configured to use our external OIDC provider through the user pool. We should remember to use the **Discovery Document URL** field for our user pool to facilitate this setup. You can see the configuration in the following screenshot:

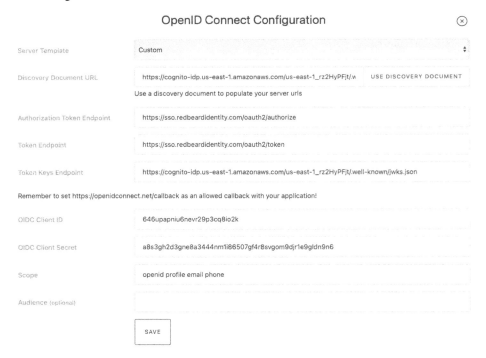

Figure 12.34 – Configuring the application to use the OIDC app client

2. Once set up, we fire the code flow, as follows:

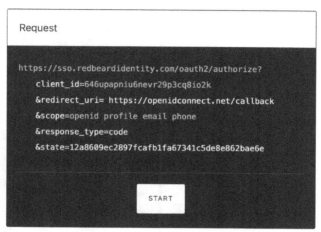

Figure 12.35 – Requesting the authorization code

3. We are redirected to the external OIDC IdP to authenticate the user. We will sign in as `Iam Dev`, as illustrated in the following screenshot, and continue:

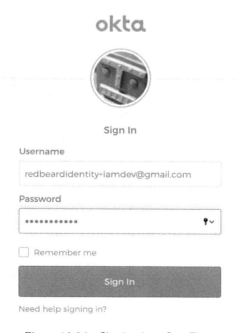

Figure 12.36 – Signing in as Iam Dev

4. We get the authorization code from our successful sign-in, as illustrated in the following screenshot. We can exchange it for an access token and a refresh token:

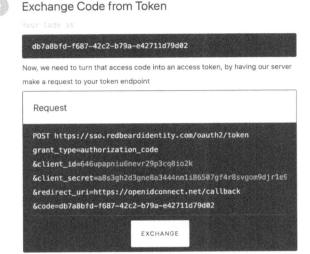

Figure 12.37 – Receiving authorization code

5. In exchange for the authorization code, we get an `id_token`, `access_token`, and `refresh_token` from the user pool, as illustrated in the following screenshot:

```
Request

POST https://sso.redbeardidentity.com/oauth2/token
grant_type=authorization_code
&client_id=646upapniu6nevr29p3cq8io2k
&client_secret=a8s3gh2d3gne8a3444nm1i86507gf4r8svgom9djr1e9
&redirect_uri=https://openidconnect.net/callback
&code=db7a8bfd-f687-42c2-b79a-e42711d79d02

HTTP/1.1 200
Content-Type: application/json
{
    "id_token": "eyJraWQiOiJDUTVFUjFNSUg3Yk56bVowWkJuWktWQ001
    "access_token": "eyJraWQiOiJ3Q01oQzlSNHpOXC83Mm41OEt2c0hC
    "refresh_token": "eyJjdHki0iJKV1QiLCJlbmMiOiJBMjU2R0NNIiⱳ
    "expires_in": 3600,
    "token_type": "Bearer"
}

                        NEXT
```

Figure 12.38 – ID, access, and refresh tokens received from the user pool

6. Next, we verify the `id_token` signature to ensure its cryptographic signature is valid, as illustrated in the following screenshot, thus proving the authenticity of the claims and the user's identity:

 ## Verify User Token

Now, we need to verify that the ID Token sent was from the correct place by validating the JWT's signature

Your "id_token" is

```
eyJraWQiOiJDUTVFUjFFNSUg3Yk56bVowWkJuWktWQ001QW42ZUtmblA0ek5MS
M4IWS1JuQKVVeN5CFQEjSZGZEJSBjzKqgIqLc0B8uxgKfbjZMmoIvVcQLNF2H
Ch3fkN8DjSbBRwdHemjgxyunShiPW_2jjlWmn3-
D7Y9sNl4lqYTNTRHmWcFmo5mJOXuEXlBo9SsowmrcgtDykwVokV_VoIhl3twl
lKv0jOk199TFijf2yLVLHZBzfnVi0ivbPN2CFXIlMfm4r1naM96pCM9zQ
```

This token is cryptographically signed with the **RS256** algorithim. We'll use the public key of the OpenID Connect server to validate it. In order to do that, we'll fetch the public key from **https://cognito-idp.us-east-1.amazonaws.com/us-east-1_rz2HyPFjt/.well-known/jwks.json,** which is found in the discovery document or configuration menu options.

VERIFY

Figure 12.39 – Verifying token signature

7. The signature is evaluated as valid, and we can decode the `id_token` to view the claims contained within it. This information is what will ultimately be passed to the application from the Amazon Cognito user pool after the user pool itself receives the claims and creates a user record, after validating claims from the external OIDC IdP. The process is illustrated in the following screenshot:

Decoded Token Payload

```
{
  "custom:managerid": "redbeardidentity+ceo@gmail.com",
  "at_hash": "PTsJOBaz5wUYALUXBLrO9w",
  "sub": "7e433c73-b564-4d89-8084-b6d2fd6bcfd6",
  "cognito:groups": [
    "us-east-1_rz2HyPFjt_RBIOIDC"
  ],
  "custom:department": "Identity Development",
  "iss": "https://cognito-idp.us-east-1.amazonaws.com/us-east-1_
  "locale": "VA",
  "custom:userType": "Staff",
  "custom:employeeNumber": "S94577",
  "identities": [
    {
      "userId": "00un7ree7x913DHwR5d6",
      "providerName": "RBIOIDC",
      "providerType": "OIDC",
      "issuer": null,
      "primary": "true",
      "dateCreated": "1623290576588"
    }
  ],
  "auth_time": 1623608823,
  "exp": 1623612423,
  "iat": 1623608823,
  "jti": "e3c7dac9-35aa-48e2-885b-940c9396e2d2",
  "email": "redbeardidentity+iamdev@gmail.com",
  "custom:title": "IAM Developer",
  "email_verified": false,
  "address": {
    "formatted": "901 E Byrd St Richmond VA 23219"
  },
  "custom:division": "Information Security",
  "cognito:username": "rbioidc_00un7ree7x913dhwr5d6",
  "given_name": "Iam",
  "nonce": "i7lfZLMwMzFfrnw_4MvffLJktyRDomyrlC97syxyvCs2yUG9ch3d.
  "origin_jti": "5fc1b7a3-4d19-4b91-93e2-f39aacd1c64c",
  "aud": "646upapniu6nevr29p3cq8io2k",
  "token_use": "id",
  "custom:costCenter": "30002",
  "name": "Iam Dev",
  "family_name": "Dev"
}
```

Figure 12.40 – Decoded claims from the user pool

8. We can now see this record in our user pool's directory, as follows:

Figure 12.41 – The user record created from the OIDC IdP in the user pool

If we look closely at this account, especially when compared to the one created by the external SAML IdP for the same user, we can already see some interesting formatting issues when using this method. Many of the standard IDs are alike, but the Amazon Cognito IDs are formatted far differently. Whereas we may be able to tune the formatting and data quality issues to align attributes and accounts across both the SAML and OIDC IdPs, it is much simpler to restrict each Amazon Cognito user pool to a single, authoritative external IdP instead of bifurcating the JIT provisioning flow from our external IdP. The downstream applications that look to the Amazon Cognito user pools as their source of identity are free to reference any attributes as their IDs—such as `email`—but we would avoid lots of account reconciliation issues by limiting our user pools to a single IdP if we were to deploy this model in a real organization.

Now that we have seen two methods to bring our organization's users into AWS while retaining control through our existing authoritative systems, let's look at our options for access control.

Assuming roles with identity pools

We have addressed our need for AWS-hosted apps to have baseline user authentication services available using Amazon Cognito user pools. This model allows us to continue to use our existing identity systems as the ultimate authoritative source for the users in those applications, even when those applications take advantage of services such as Amazon Cognito for their identity use cases. For applications with architectures that have deep integration into AWS services, Amazon Cognito identity pools can provide authorization to AWS resources such as Amazon **Simple Storage Service (S3)** buckets and Amazon **Relational Database Service (RDS)** databases. This allows the application users to indirectly interact with these services when using the application that is built to leverage them.

Let's consider a use case where the Redbeard Identity Sales team manages its sales reports through an application that is hosted on AWS. The reports are published to all other members of the organization through that application. Members of the Sales organization have full access to create, read, update, and delete reports, whereas all other Redbeard Identity workers only have read access to those reports. The application uses an Amazon S3 bucket as the report repository. The process is illustrated in the following diagram:

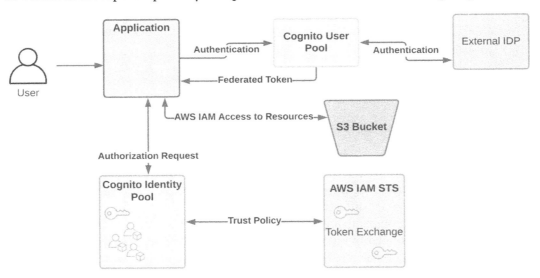

Figure 12.42 – Solutions architecture for application

We will use an Amazon Cognito identity pool and AWS **Identity and Access Management (IAM)** roles to generate temporary credentials that the Amazon Cognito users will use to interact with the reports in the Amazon S3 bucket for this use case.

> **Tip**
>
> We explored the features available through Amazon Cognito identity pools in *Chapter 5, Introducing Amazon Cognito*. Therefore, we will not review everything available in an Amazon Cognito identity pool in this chapter.

The first thing we will need to do is create an identity pool, as follows:

1. From the **AWS Management Console**, we go to **Cognito Service**, select **Manage Identity Pools**, and then **Create new Identity Pool**. This opens a wizard with which we will initially define the identity pool we will use with this application.

2. We will start by naming the identity pool. We will call it `RBI_External_IdP`.

3. Moving down to **Unauthenticated identities**, we will leave this unenabled to access resources, as illustrated in the following screenshot. We only want authenticated Redbeard Identity workers to access these resources:

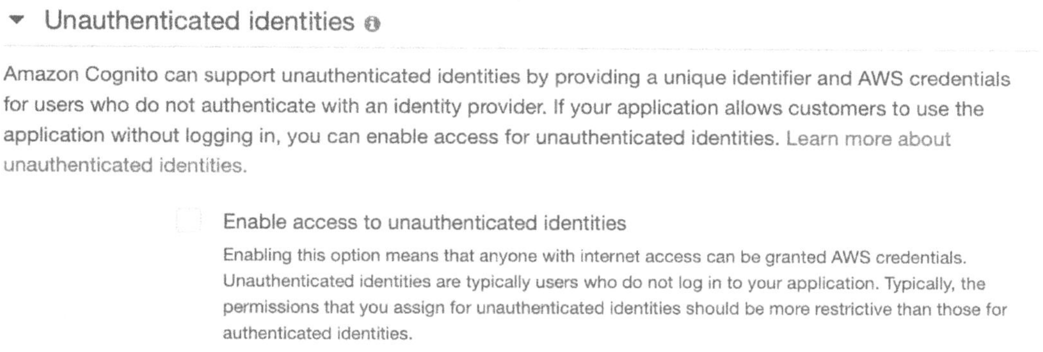

▼ Unauthenticated identities ❶

Amazon Cognito can support unauthenticated identities by providing a unique identifier and AWS credentials for users who do not authenticate with an identity provider. If your application allows customers to use the application without logging in, you can enable access for unauthenticated identities. Learn more about unauthenticated identities.

☐ Enable access to unauthenticated identities

Enabling this option means that anyone with internet access can be granted AWS credentials. Unauthenticated identities are typically users who do not log in to your application. Typically, the permissions that you assign for unauthenticated identities should be more restrictive than those for authenticated identities.

Figure 12.43 – Do not allow unauthenticated identities access

4. Next, under **Authentication providers**, we will enter the user pool ID and the app client ID we will use within that user pool for authenticating users, as illustrated in the following screenshot. We will use an app client ID that exclusively looks to the external IdP for authentication and provides information on the user's organization. We can now move to the next step of the wizard:

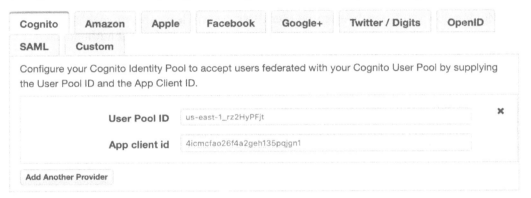

Figure 12.44 – Setting our user pool and app client ID as an authentication provider for the identity pool

5. The identity pool wizard asks for a role to assign to authenticated users and unauthenticated users. If we wish to use an existing AWS IAM role, we can, or we can let the wizard create new roles for our identity pools. We will let the wizard create new roles for us. We will be editing these momentarily within AWS IAM. It will create two roles called `Cognito_RBI_External_IdPAuth_Role` and `Cognito_RBI_External_IdPUnauth_Role`. The first is assigned as the default role for authenticated users, and the second is the default role for unauthenticated users. We can now complete the wizard and create our identity pool.

6. We now have our identity pool, but we still need to do some work so that it will fulfill our use case. Let's open it from the **Cognito Identity Pool** dashboard, and then click on the **Edit Identity pool** link in the upper right to look at our options for authorization.

7. From this **Edit Identity Pool** screen, let's expand the **Authentication providers** section. Under **Authenticated role selection**, there is a dropdown that allows us to pick how roles will be assigned to authenticated users. By default, it is set to **Use default role**. However, we can configure this identity pool to assign a distinct authenticated role based upon the value of the claim from the user pool at authentication time. Let's set the dropdown to **Choose role with rules**, as illustrated in the following screenshot:

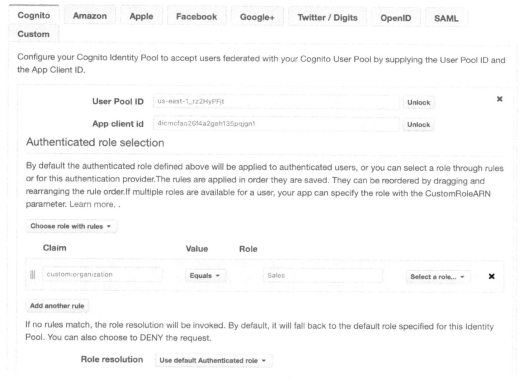

Figure 12.45 – Defining distinct authentication roles based on claims

We can fulfill our use case by building a special role for authenticated users with the `custom:organization` claim value of **Sales**. Any other authenticated user will default back to the default role for authenticated users.

8. We will need to adjust the default roles created for the identity pool and create a new role for the users with the **Sales** claim. Let's head to the **Roles** dashboard inside AWS IAM and review the policy on these roles, starting with `Cognito_RBI_External_IdPAuth_Role`.

We need to make sure that any authenticated user role for this identity pool has a trust policy that limits who can access that role. We can see the role's trust policy in the **Trust relationship** tab and by clicking the **Show policy document** link. The trust policy for `Cognito_RBI_External_IdPAuth_Role` is provided here:

```
{
    "Version": "2012-10-17",
    "Statement": [
        {
            "Effect": "Allow",
            "Principal": {
                "Federated": "cognito-identity.amazonaws.com"
            },
            "Action": "sts:AssumeRoleWithWebIdentity",
            "Condition": {
                "StringEquals": {
                    "cognito-identity.amazonaws.com:aud": "us-east-1:9fbe790a-13c5-4201-8368-eedaf5083caf"
                },
                "ForAnyValue:StringLike": {
                    "cognito-identity.amazonaws.com:amr": "authenticated"
                }
            }
        }
    ]
}
```

The important section is the `Condition` section, which constrains this role to only be assumed by authenticated members of the Amazon Cognito identity pool we created in *Steps 1-5* of this section.

9. Having validated the trust policy, let's now adjust the inline policy. Let's move to the **Permissions** tab and expand the policy document. We want to edit this document to limit access to just reading the items inside a specific Amazon S3 bucket. This policy statement already has some basic actions that came with the default role creation, so we will add an additional action to handle our Amazon S3 use case, as illustrated in the following code snippet:

```
{
    "Version": "2012-10-17",
    "Statement": [
        {
            "Effect": "Allow",
            "Action": [
                "mobileanalytics:PutEvents",
                "cognito-sync:*",
                "cognito-identity:*"
            ],
            "Resource": [
                "*"
            ]
        },
        {
            "Effect": "Allow",
            "Action": [
        "s3:GetObject"
            ],
        "Resource":
["arn:aws:s3:::sharedsalesreportbucket/reports/*"]
        }
    ]
}
```

10. Now, let's create a new role for those users who have the **Sales** claim. In the new role wizard, we will select **Web identity** for the **trusted entity** type. We will then select our Amazon Cognito user pool as the **identity provider** and define a condition that only allows authenticated users to assume this role. The process is illustrated in the following screenshot:

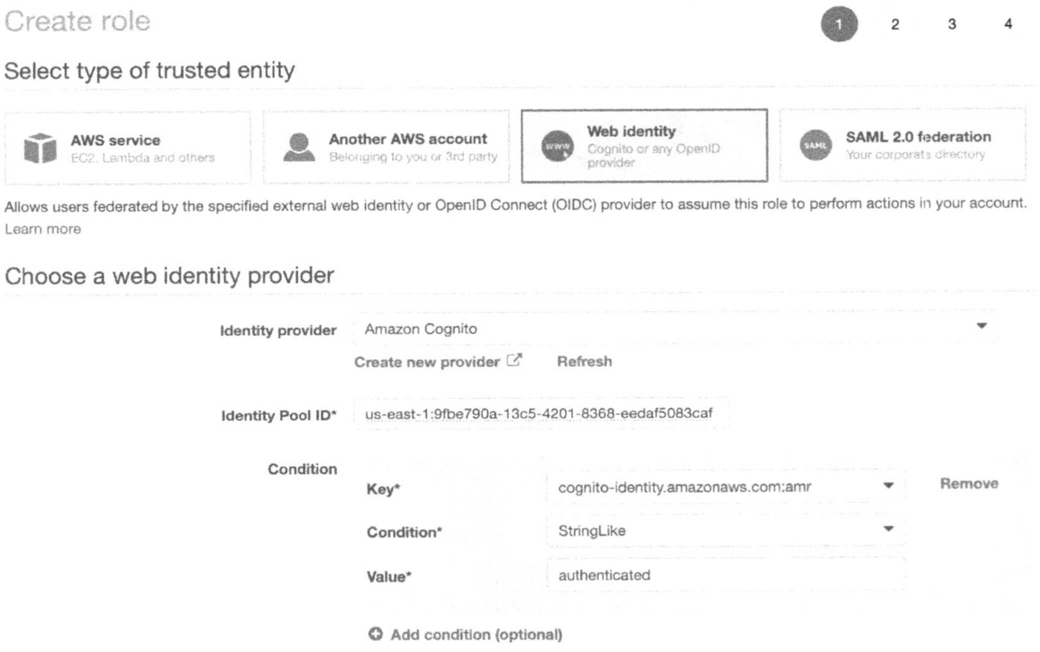

Create role   1   2   3   4

Select type of trusted entity

AWS service	Another AWS account	Web identity	SAML 2.0 federation
EC2, Lambda and others	Belonging to you or 3rd party	Cognito or any OpenID provider	Your corporate directory

Allows users federated by the specified external web identity or OpenID Connect (OIDC) provider to assume this role to perform actions in your account. Learn more

Choose a web identity provider

Identity provider	Amazon Cognito
	Create new provider ☐ Refresh
Identity Pool ID*	us-east-1:9fbe790a-13c5-4201-8368-eedaf5083caf
Condition	
Key*	cognito-identity.amazonaws.com:amr Remove
Condition*	StringLike
Value*	authenticated

⊕ Add condition (optional)

Figure 12.46 – Setting the trust policy for the new role

11. As we go through the wizard, we will need to create a new policy for our role. This will pop open a separate wizard with which we can create our new user-managed policy. We will define our new policy with the following statement, and save it as `SalesClaimsPolicy`:

```
{
    "Version": "2012-10-17",
    "Statement": [
        {
            "Effect": "Allow",
            "Action": [
                "mobileanalytics:PutEvents",
                "cognito-sync:*",
                "cognito-identity:*"
```

```
                    ],
                    "Resource": [
                        "*"
                    ]
                },
                {
                    "Action": ["s3:ListBucket"],
                    "Effect": "Allow",
                    "Resource":
    ["arn:aws:s3:::sharedsalesreportbucket"]
                },
                {
                    "Action": [
                        "s3:GetObject",
                        "s3:PutObject"
                    ],
                    "Effect": "Allow",
                    "Resource":
    ["arn:aws:s3:::sharedsalesreportbucket/reports/*"]
                }
            ]
        }
```

12. With our new policy saved, we can return to the role wizard, refresh to find `SalesClaimsPolicy`, and add it to our new role, as illustrated in the following screenshot:

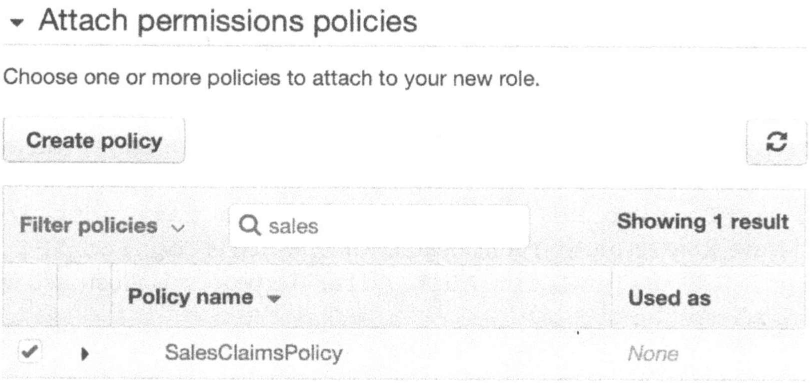

▾ Attach permissions policies

Choose one or more policies to attach to your new role.

Create policy			⟳

Filter policies ⌄	Q sales		Showing 1 result
	Policy name ▾		**Used as**
✔ ▸	SalesClaimsPolicy		None

Figure 12.47 – Adding the new policy to the new role

13. We will name the new role `Cognito_RBI_External_IdP_SalesClaim_Auth_Role` and exit the wizard.

14. Upon opening the new role, we can validate that the trust policy is what we want. The policy matches the trust policy of our other roles for the identity pool, as we can see here:

Edit Trust Relationship

You can customize trust relationships by editing the following access control policy document.

Policy Document

```
1 - {
2     "Version": "2012-10-17",
3 -   "Statement": [
4 -     {
5         "Effect": "Allow",
6 -       "Principal": {
7           "Federated": "cognito-identity.amazonaws.com"
8         },
9         "Action": "sts:AssumeRoleWithWebIdentity",
10 -      "Condition": {
11 -        "StringEquals": {
12            "cognito-identity.amazonaws.com:aud": "us-east-1:9fbe790a-13c5-4201-8368
                -eedaf5083caf"
13          },
14 -        "StringLike": {
15            "cognito-identity.amazonaws.com:amr": "authenticated"
16          }
17        }
18      }
19    ]
```

Figure 12.48 – Verifying the trust policy on the new role

15. Let's return to our identity pool's configuration settings and complete the setup. After opening up our **Settings** page and returning to the **Authentication providers** section for our identity pool, we will change the dropdown beneath **Authenticated role selection** from **Use default role** to **Choose role with rules**.

16. We will define one set of claims. The claim value must match what comes from the user pool, so if we are looking at the `organization` attribute for Redbeard Identity users, we will need to enter `custom:organization` as the claim name, as that is how that claim is mapped into the user pool's directory. We will leave the operator set to **Equals** and set the **Value** field to **Sales**. Under the **Role** dropdown, we will select `Cognito_RBI_External_IdP_SalesClaim_Auth_Role`. We will leave **Role resolution** at its default value, which is `Cognito_RBI_External_IdP_SalesClaim_Auth_Role`. The process is illustrated in the following screenshot:

Figure 12.49 – Configuring the role rules

17. Hit **Save changes**, and we are done.

The application now has its report functionality governed through Amazon Cognito identity pools. There are three potential outcomes when someone attempts to access the reports store, as we can see in the following diagram:

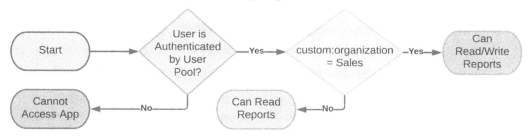

Figure 12.50 – Amazon Cognito-governed user flows

Any user who is not authenticated through Amazon Cognito will be unable to access the application entirely. A user who is authenticated by the user pool but does not have the **Sales** claim will assume the role that grants them temporary credentials to access the Amazon S3 bucket where reports are stored with read-only permissions. Principals with the claim are granted temporary credentials with an enhanced level of access. All AWS IAM roles that can be assumed through the identity pool are restricted to principals that were authenticated through the Amazon Cognito user pool we associated with the identity pool, and that limitation is enforced through the roles' trust policy.

Summary

In this chapter, we explored the authentication and authorization options available to applications hosted in AWS. We were able to provide identity information to those applications leveraging AWS identity services, particularly Amazon Cognito, while continuing to respect our organization's existing IAM infrastructure as the authoritative source for access control. We showed how to delegate authentication to an external provider using both SAML and OIDC when using an Amazon Cognito identity pool, and then explored how we could apply authorization controls to an AWS-hosted application by assigning distinct AWS IAM roles to Amazon Cognito identities based upon claims from that external IdP.

And with that, we have reached the end of the book. Congratulations on making it through! You now have a solid foundation of AWS identity knowledge that will make you better prepared to address your cloud identity challenges moving forward.

Questions

1. Why would an organization choose to federate their managed identities into an Amazon Cognito user pool for application identity?

    a. Allows the app team to use native AWS services for identity.

    b. Allows the organization to continue to enforce their compliance controls centrally, even though applications may not look directly to their identity systems for user information.

    c. They shouldn't; they should only connect apps directly to their organization's official IdP.

    d. A and B.

2. Why would we apply a trust policy that validates a principal was authenticated by the identity pool that is requesting temporary credentials for an Amazon Cognito user?

    Otherwise, non-authenticated users could be granted access to AWS resources within the account.

Further reading

Here are some resources for making applications SAML2- and OIDC-compliant through relying parties:

- OpenID Foundation resources page: `https://openid.net/developers/libraries/`

- Shibboleth **Service Provider** (**SP**) wiki: `https://wiki.shibboleth.net/confluence/display/SP3/Home`

- SimpleSAMLphp: `https://simplesamlphp.org`

- OIDC playground: `https://openidconnect.net`

Packt.com

Subscribe to our online digital library for full access to over 7,000 books and videos, as well as industry leading tools to help you plan your personal development and advance your career. For more information, please visit our website.

Why subscribe?

- Spend less time learning and more time coding with practical eBooks and Videos from over 4,000 industry professionals

- Improve your learning with Skill Plans built especially for you

- Get a free eBook or video every month

- Fully searchable for easy access to vital information

- Copy and paste, print, and bookmark content

Did you know that Packt offers eBook versions of every book published, with PDF and ePub files available? You can upgrade to the eBook version at packt.com and as a print book customer, you are entitled to a discount on the eBook copy. Get in touch with us at customercare@packtpub.com for more details.

At www.packt.com, you can also read a collection of free technical articles, sign up for a range of free newsletters, and receive exclusive discounts and offers on Packt books and eBooks.

Other Books You May Enjoy

If you enjoyed this book, you may be interested in these other books by Packt:

Keycloak - Identity and Access Management for Modern Applications

Stian Thorgersen, Pedro Igor Silva

ISBN: 978-1-80056-249-3

- Understand how to install, configure, and manage Keycloak

- Secure your new and existing applications with Keycloak

- Gain a basic understanding of OAuth 2.0 and OpenID Connect

- Understand how to configure Keycloak to make it ready for production use

- Discover how to leverage additional features and how to customize Keycloak to fit your needs

- Get to grips with securing Keycloak servers and protecting applications

Mastering Identity and Access Management with Microsoft Azure

Jochen Nickel

ISBN: 978-1-78913-230-4

- Apply technical descriptions to your business needs and deployments
- Manage cloud-only, simple, and complex hybrid environments
- Apply correct and efficient monitoring and identity protection strategies
- Design and deploy custom Identity and access management solutions
- Build a complete identity and access management life cycle
- Understand authentication and application publishing mechanisms
- Use and understand the most crucial identity synchronization scenarios
- Implement a suitable information protection strategy

Packt is searching for authors like you

If you're interested in becoming an author for Packt, please visit `authors.packtpub.com` and apply today. We have worked with thousands of developers and tech professionals, just like you, to help them share their insight with the global tech community. You can make a general application, apply for a specific hot topic that we are recruiting an author for, or submit your own idea.

Share Your Thoughts

Now you've finished *Implementing Identity Management on AWS*, we'd love to hear your thoughts! Scan the QR code below to go straight to the Amazon review page for this book and share your feedback or leave a review on the site that you purchased it from.

https://packt.link/r/1800562284

Your review is important to us and the tech community and will help us make sure we're delivering excellent quality content.

Index

W

Y

www.ingramcontent.com/pod-product-compliance
Lightning Source LLC
Chambersburg PA
CBHW081454050326
40690CB00015B/2790